RIGHT SIZING THE PEOPLE'S LIBERATION ARMY: EXPLORING THE CONTOURS OF CHINA'S MILITARY

Edited by

Roy Kamphausen
Andrew Scobell

September 2007

This publication is a work of the U.S. Government as defined in Title 17, United States Code, Section 101. As such, it is in the public domain, and under the provisions of Title 17, United States Code, Section 105, it may not be copyrighted.

*****

The views expressed in this report are those of the authors and do not necessarily reflect the official policy or position of the Department of the Army, the Department of Defense, or the U.S. Government. This report is cleared for public release; distribution is unlimited.

*****

Comments pertaining to this report are invited and should be forwarded to: Director, Strategic Studies Institute, U.S. Army War College, 122 Forbes Ave, Carlisle, PA 17013-5244.

*****

All Strategic Studies Institute (SSI) publications are available on the SSI homepage for electronic dissemination. Hard copies of this report also may be ordered from our homepage. SSI's homepage address is: *www.StrategicStudiesInstitute.army.mil*.

*****

The Strategic Studies Institute publishes a monthly e-mail newsletter to update the national security community on the research of our analysts, recent and forthcoming publications, and upcoming conferences sponsored by the Institute. Each newsletter also provides a strategic commentary by one of our research analysts. If you are interested in receiving this newsletter, please subscribe on our homepage at *www.StrategicStudiesInstitute.army.mil/newsletter/*.

ISBN 1-58487-302-7

TABLE OF CONTENTS

Foreword v
  General John M. Shalikashvili

Part I: Framing the Issue............................1

1. Introduction......................................3
   Roy Kamphausen and Andrew Scobell

2. Framing the Problem: China's Threat
   Environment and International Obligations.........19
   Michael R. Chambers

3. China's National Military Strategy:
   An Overview of the "Military Strategic Guidelines" ..69
   David M. Finkelstein

Part II: PLA Strategic Systems 141

4. "Minding the Gap": Assessing the Trajectory
   of the PLA's Second Artillery
   Evan S. Medeiros 143

5. PLA Command, Control, and Targeting
   Architectures: Theory, Doctrine,
   and Warfighting Applications 191
   Larry M. Wortzel

Part III: The PLA Ground Force 235

6. "Preserving the State": Modernizing
   and Task-Organizing a "Hybrid" Ground Force.....237
   Cortez A. Cooper III

7. PLA Ground Force Modernization
   and Mission Diversification:
   Underway in All Military Regions................281
   Dennis J. Blasko

Part IV: The PLA Air Force.........................375

8. Future Force Structure of the Chinese
   Air Force.. 377
   Phillip C. Saunders and Erik Quam

9. Right-Sizing the PLA Air Force:
   New Operational Concepts Define a Smaller,
   More Capable Force 437
   Kevin M. Lanzit and Kenneth Allen

Part V: The PLA Navy..............................479

10. The Strategic and Operational Context
    Driving PLA Navy Building......................481
    Michael McDevitt

11. Right-Sizing the Navy: How Much
    Naval Force Will Beijing Deploy? 523
    Bernard D. Cole

Part VI: The Wrap Up.............................557

12. The "Right Size" for China's Military:
    To What Ends? 559
    Ellis Joffe

About the Contributors 573

FOREWORD

The U.S. Army War College and the National Bureau of Asian Research (NBR) are two organizations with which I have a strong connection. I was in the class of 1976 at Carlisle and I currently serve on the Board of NBR where I am closely aligned with the Strategic Asia Program.

As such, I was quite pleased when the NBR joined the continuing efforts of the U.S. Army War College's Strategic Studies Institute to study developments in China's People's Liberation Army (PLA) as cosponsors of the 19th PLA Carlisle Conference from October 6-8, 2006.

Right-Sizing the People's Liberation Army: Exploring the Contours of China's Military is the ninth volume in this series published by the Strategic Studies Institute and represents the collective scholarly efforts of those who contributed to the 2006 conference. The book addresses how the leadership of China and the PLA view what size of PLA best meets China's requirements. Among other things, this analytical process makes important new contributions on the question of PLA transparency, long an issue among PLA watchers.

Throughout my professional career, both during and since my service in the military, a great deal of emphasis has been put on understanding not only how, but also why a military modernizes itself. Some of the determining factors are national policies and strategy, doctrine, organizational structure, missions, and service cultures. While this list is not exhaustive, it does begin to paint a picture of just how broad and deep military interests run.

I had a number of meetings with the Chinese military leadership during my time as Chairman of the Joint Chiefs. They were very interested in learning from our experiences in Operation DESERT STORM, and specifically in missile defense. It is my belief to this day that they were trying to learn how they might engage Taiwan should the military option be called into play.

Bearing this in mind, it is important when we look at the structure and strategy for growth within the Chinese military that we not restrict ourselves to the lens of our Western focus. Rather, we need to see the world as China sees it. We need to see a world in which the "Taiwan issue" as well as that of North Korea and others are not viewed as short-term concerns, but fit into how China sees itself in a long-term leadership role in the region and in the world.

This latest volume makes an important contribution to this effort. The authors—and the 65 conference participants from academia, think tanks, the U.S. Government, and overseas whose observations were of great help—have each done a terrific job and are to be thanked for their efforts. Only through sustained, systematic efforts such as this can we begin to understand how China's military modernization might affect the Asia-Pacific security situation in the years to come.

JOHN M. SHALIKASHVILI
General, USA Retired
Former Chairman, Joint Chiefs of Staff

PART I:

FRAMING THE ISSUE

CHAPTER 1

INTRODUCTION

Roy Kamphausen and Andrew Scobell

China is the emerging power having "the greatest potential to compete militarily with the United States and field disruptive military technologies that could over time offset traditional U.S. military advantages absent U.S. counter strategies," according to the 2006 *Quadrennial Defense Review* from the U.S. Department of Defense.[1] With military spending and modernization that have persisted with little or no abatement or recantation for well over a decade, China has the entire international community wondering to what ends such growth will be put.

HOW BIG, HOW CAPABLE, AND WHY?

To answer this question, the National Bureau of Asian Research (NBR) and the U.S. Army War College's Strategic Studies Institute (SSI) assembled scholars and People's Liberation Army (PLA) analysts for the 2006 PLA Conference at Carlisle Barracks, Pennsylvania.[2] Only a year earlier, Secretary of State Condoleezza Rice had observed in an interview that China's "military buildup looks outsized for its regional concerns."[3] The question before the PLA Conference raises another important question: What would a "right sized" PLA look like? In other words, in terms of China's national security strategy, regional and global requirements and expectations, and domestic drivers, what might an armed force consistent with Beijing's legitimate self-defense requirements be composed of and how large would it be? The PLA Conference, by exploring the right size for PLA missions, functions, and organization, provided insight into future Chinese defense planning, strategic intentions, and potential PLA missions.

This volume—harvesting the fruit of research and discussion from the 2006 PLA Conference—considers the force structure of the PLA and China's latest training, doctrinal, and procurement efforts across the arms and services of its military forces. Organized on a service-by-service basis, this assessment provides new insights into the drivers behind the size, posture, and arming of the Chinese military. Though China's military intentions have long been shrouded in a veil of secrecy, the chapters herein draw vital information from a diverse assortment of Chinese and American sources to illuminate these hidden contours, offering perspectives and conclusions with far-reaching implications for policymakers and defense leaders in the United States and worldwide.

One key theme emerging from this volume is that, as far as modernization is concerned, the PLA is by no means monolithic. A service-by-service analysis reveals that while doctrines may be aligned under the rubric of a broad national military strategy, some service programs have larger handicaps—that is, are further behind national requirements—than others. In these instances, "surprise" modernization programs may be likely to emerge.

A second critical theme, and one that cuts across all service programs, is the growing importance of the human dimensions of the PLA. As modernization continues and systems become more complex, the human elements—education, training, personnel management, etc.—will be increasingly critical to the development of the armed forces. The might of a military, after all, is only as strong as the people comprising it and the strategies they undertake.

CHINA'S THREAT ENVIRONMENT AND NATIONAL MILITARY STRATEGY

Chapter 2, contributed by Michael Chambers, explores China's threat environment, analyzing the extent to which exogenous factors are driving China's military growth. China's immediate external threat environment appears relatively benign—in spite of the constant latent tension with Taiwan—while domestic problems appear more likely to pose a threat to the legitimacy of the Chinese Communist Party's (CCP) rule, and could necessitate the use of military force. These problems include, *inter alia*, separatist threats from Taiwan, Xinjiang, and Tibet, and popular protests over inequality and poor regional governance.[4] Even so, such threats would probably not in their own right justify force modernization and expansion.

Against this backdrop of unique peacetime circumstances—a domestic threat environment that does not immediately require force modernization, a favorable external threat environment, and relatively few international commitments—what factors propel China's determination upon military growth? Chambers argues that, in the short term, China's primary concern is continued unease over conflict with Taiwan. Despite improvements in cross-Strait relations since 2006, China must nevertheless remain primed for conflict and continue to pursue a strategy of proactive deterrence. Over the longer term, military growth could be linked to a broader "great power" strategy to secure global economic interests and trade routes, and to build "muscle" to reinforce its diplomatic efforts—

particularly vis-à-vis the United States. The rise of China as a global economic and trading power makes the protection of critical sea lines of communication, energy assets, and other maritime interests imperative for the country's future.

The question then follows: Does China possess a systematic plan for reaching these modernization goals? In Chapter 3, "China's National Military Strategy," David Finkelstein's ground-breaking primary source research clearly demonstrates that China does have a plan: the current *National Military Strategic Guidelines* promulgated in January 1993 by Jiang Zemin. These so-called "Military Strategic Guidelines for the New Period," Finkelstein argues, direct all PLA modernization: "Every modernization program, every reform initiative, and every significant change that the PLA has undergone... for over a decade, are the results of some of the fundamental decisions made when the new guidelines were promulgated."

In an effort to catalogue the PLA's bold military buildup and get to the source of what China is seeking to achieve, Finkelstein revisits the vast yet cryptic body of literature on PLA reform from the past 13 years—from PRC *Defense White Papers* to military speeches to U.S. Government reports. He begins by identifying three "pillars" of PLA reform and modernization: the acquisition of new weapons and combat capabilities; institutional and systemic reform; and the development of new warfighting doctrines. Taken together, changes in these three areas have helped the PLA become a more capable and professional force.

China's National Military Strategy, derived primarily from the military strategic guidelines issued to the PLA by the CCP Central Committee's Military Commission, addresses a number of key points—six of which Finkelstein elaborates upon more explicitly: (1) the strategic assessment; (2) adjusting the content of the strategic concept of active defense; (3) articulating the strategic missions and strategic objectives; (4) guidance for military combat preparations; (5) identifying the main strategic decision; and (6) determining the focus for "Army building." Assessing these six components for the "New Period," Finkelstein finds that China's new guidelines can be reduced to five major tasks: (1) defending national territory and sovereignty, (2) securing the nation's maritime rights and interests, (3) maintaining the unity of the motherland, (4) ensuring internal stability, and (5) maintaining a secure and stable external environment—especially along China's periphery.

Finkelstein reminds us, however, that such guidelines and tasks have *not* been arrived at overnight. Rather, they are the result of a series of carefully calculated decisions on funding, prioritization, and compromise. These pragmatic steps, moreover, are based on the types of studied assessments that any professional military establishment would be expected to undertake—there is nothing, as Finkelstein writes, "foreign, strange, exotic, or exceptional" about them. Ultimately, the PLA's overarching military strategy is to develop the operational capabilities that will enable China to deter conflict and, if deterrence fails, to prosecute conflicts successfully—an objective that will allow the country to achieve its larger national goals.

The PLA's Strategic Forces.

With the understanding that the "Military Strategic Guidelines for the New Period" provides basic guidance for each of the PLA's services and tactical areas, a service-by-service evaluation lends deeper insight into the modernization process. The first component of China's warfighting architecture is comprised of the PLA's strategic systems. Among these systems are China's strategic missile forces, discussed in Chapter 4, "'Minding the Gap': Assessing the Trajectory of the PLA's Second Artillery," written by Evan Medeiros. He analyzes the current trajectory of the PLA branch known as the Second Artillery, examining the doctrine-capabilities relationship between nuclear and conventional missile forces and asking what this comparison reveals about the Second Artillery's future evolution. Using Chinese military writings drawn from military books and journals from 1996 to the present, Medeiros finds that there is broad consistency between doctrine and capabilities within the PLA's nuclear and conventional forces. However, while both are rapidly evolving, they are still, to some extent, playing "catch-up" to meet the dynamic, ever-changing doctrinal requirements.

China's nuclear doctrine, argues Medeiros, has not radically changed despite shifts in the global nuclear landscape. Instead, it has responded by developing new missile systems focused on retaliatory capabilities, deterrence, and survival. To date, there are few indications that the growth of China's nuclear missile force will extend beyond prevailing doctrines. China's policy assertions, most prominent of which is China's "no first use" (NFU) pledge, are an important factor as well. Medeiros argues that the pledge contains a degree of conditionality such that observers ought to spend less time attempting to figure out whether or when Beijing might lift the pledge altogether, and more time in understanding those situations in which Beijing might justify a first strike within its NFU pledge.

China's Army will build forces to support China's overall deterrence posture and, if deterrence fails, to assume warfighting capabilities. China's recent assignment of domestic land troops to UN peacekeeping operations in Lebanon, Sudan, and Liberia is evidence of Beijing's growing confidence in executing overseas operations, and of its goal to augment the prestige of the PLA via domestic, regional, and complex "real world" missions.

Blasko also reminds us that, over time, the human element will be as important as the technical one. In coming years, party loyalty—institutionalized through the political commissar and the party committee systems—may be tested as a more sophisticated personnel force begins to question certain ideological aspects of training to a greater degree than their predecessors. With the current trends in personnel downsizing, the increasing incidence of corruption, and the consequent low morale among troops, human resource development becomes even more critical. Though some of these problems will be mitigated as the PLA shrinks and hones its technical expertise, the success of the Army in combat operations will directly correlate with the abilities of the PLA officers and NCO corps to plan and execute a new joint doctrine.

China's Air Force.

Chapter 8 by Phillip Saunders and Erik Quam on the future force structure of the PLA Air Force (PLAAF) utilizes a scenario-based approach to interpret impending developments. They review PLAAF missions, cataloguing past and present capabilities and limitations, and then examine the newest systems and the Air Force's future aspirations. Finally, they analyze how decisions and trade-offs are

including disaster relief efforts, local civil develo
projects, and back-up support for local s
operations. More mission-specialized than ever
the PLA will be strengthened by a well-traine
commissioned officer (NCO) corps, carry ou
integrated joint and combined arms operatio1
place greater responsibility on personnel witl
tech expertise. It will benefit as well from a nu1
key technological developments: space-based
and aerial surveillance platforms that can pas
real time data to a variety of PLA systems; aut
"logistic interaction platforms" that facilita
projects; and a newer, more advanced helicopt
that will project presence more easily and rapi

According to Cooper, the Chinese lea
appears convinced that hard military pow
accompany and undergird a "peaceful rise."
the battlefield is in the East China Sea or the
peninsula, the Chinese intend to have availabl
organized, technologically superior armed f
can be expeditiously deployed. While Beijing
touts the PLA as possessing the right size for d
missions, its continental force is becoming inc1
capable of conducting offensive operatio
within and beyond its borders. The United S
other potential foreign competitors must no
complacent when considering the security
that could face them as a result of a more off
oriented PLA.

In the bigger picture, Army force modern
analyzed by Cooper appears to correlate ne
trends identified by the 2006 PRC *Defense W*
In Chapter 7, "PLA Ground Force Moderniz
Mission Diversification," Dennis Blasko outl
of these developments. Over the next few

made within the Air Force, e.g., between foreign versus domestic production, and high-technology versus low-cost systems. They conclude that perceptions of the international threat environment and budget concerns will have a significant influence on the size of the PLAAF and the speed of its modernization.

Since assessing these possibilities is hardly an exact science, Saunders and Quam sketch three alternative scenarios to illustrate a range of potential outcomes: (1) an expedited effort to maximize capability; (2) a high-technology Air Force; and (3) a domestically-produced Air Force. A common denominator in all of these scenarios, however, is that budget realities, technological limitations, and regional relations — especially with the United States and Taiwan — will act as constraints on the PLAAF's modernization efforts. These factors already seem to point to a Chinese Air Force that is smaller, yet more technologically capable, in the foreseeable future.

In Chapter 9, Kevin Lanzit and Kenneth Allen delve further into the specifics of this smaller yet more capable Air Force, evaluating how institutional and doctrinal developments — underway since the 1990s — are being operationalized by the PLAAF. Such reforms include significant changes in leadership, force structure, organizational and enlisted personnel structure, education, and training. New mission requirements, force structure modernization, and the addition of advanced information and weapons technology are inducing the PLAAF to rethink old concepts of air doctrine, restructure command elements, and revamp its education and training programs.

Many still predict that a comprehensive force structure modernization will take at least 10-15 years to complete. Weighed down by lingering hardware deficiencies, obsolete aircraft, and delays in fielding command and control and air surveillance aircraft, this process cannot be expected to occur overnight, or even over the course of a few years. Introducing and operationalizing reforms in air doctrine to accommodate hardware advances will require at least a decade of sustained effort. Furthermore, the human element—training a new generation of military professionals who are proficient with these new systems—will be among the greatest continuing challenges facing the PLAAF.

China's Navy.

Whither the PLA Navy (PLAN)? What factors have been driving developments in force posture, size, and capabilities of the PLAN over the past 15 years? In Chapter 10, Michael McDevitt argues that the recent expansion of China's maritime operations is indicative of a belief among Chinese leaders that the strategic interests of the state can be secured only with a robust naval force. While this approach represents a departure from historical Chinese naval thinking, McDevitt proposes five factors animating such change: (1) a shift in the major strategic direction of the PLA from coastal to offshore defense; (2) a maritime strategy aligned with the continental strategic tradition of China; (3) the need to deter Taiwan's bid for independence and, if necessary, combat a U.S. relief effort in the Strait; (4) the unique set of circumstances in which international seaborne trade drives China's economic growth; and (5) China's increasing dependency on oil and resources transported by sea.

According to McDevitt, the rough blueprint for the PLAN's growth seems to derive not from the Western-style blue water Navy—which, in addition to being expensive, would represent a departure from China's continentalist military tradition—but instead from a Soviet-style anti-access/sea denial model. Similar geographic circumstances, affordability, and access to Soviet-developed technology and concepts make this design particularly congruous with and amenable to Chinese interests. This approach to Navy building comports not only with the Chinese land-based military worldview but as well with Beijing's new political message of a peaceful and nonthreatening rise.

Pressure from the United States for China to become a "responsible stakeholder" and other such exhortatory signals from the West seem to have encouraged Chinese leaders to begin thinking about including peacetime operations among their missions set. Such sorties will require the PLAN to learn how to deploy and sustain surface combatants, amphibious ships, and support ships on distant stations for extended periods of time—and possibly to acquire an aircraft carrier. According to McDevitt, these signals could lay the groundwork for a "second iteration" of the Navy in which China furthers its own interests while demonstrating that it too can be a responsible partner and good neighbor in the world community's humanitarian undertakings.

What will this Navy look like in 10 years, and how will it be disposed? Bernard Cole answers these questions in Chapter 11, "Rightsizing the Navy: How Much Naval Force Will Beijing Deploy?" Cole suggests that 10 years from now, the character of naval forces will be linked closely with Beijing's perceptions of the interests, resources, and intentions of the international environment—particularly those countries in maritime Asia. Cole uses a scenario-building approach similar to that employed by Saunders and Quam to project three alternative "maritime futures": the first involves

Taiwan, the second the East China Sea, and the third the Straits of Malacca.

In the event of a conflict with Taiwan, the PLAN would employ maritime forces to execute any number of operations, ranging from the restriction of seaborne trade to a full-scale amphibious invasion. However, its principal role would be to isolate the battlefield by deploying submarines to prevent intervention by outside forces.

In the case of the East China Sea, where China has contested territorial interests over the Diaoyutai (Senkaku) islands, a number of analysts have concluded that a conflict in the near future is not unlikely. Both Beijing and Tokyo currently have a military presence in the area, and intensive patrolling by their ships and aircraft amplifies the possibility of an inadvertent rift. Military action by either side would likely involve surface combatants supported by long-range aircraft and submarines, while planning for such a conflict would entail upgrading joint and integrated operations doctrine, as well as the PLAN's submarine force.

In the third scenario, the invaluable sea lines of communication through the Straits of Malacca are compromised, stanching the flow of key resources and posing a direct threat to the economic welfare of China. Here, the Navy would have to increase the number of state-of-the-art warships and sea replenishment ships to support those surface ships permanently stationed in the Strait. In this scenario, as in the other two, the PLA Navy is large, capable, and prepared for joint operations. Thus, by 2016, the Navy might be expected to become dominant among East Asian navies, and a formidable competitor to the U.S. Navy in Asia's maritime theater.

To What Ends?

Cataloguing the advances in China's strategic systems and doctrines, plus its Army, Air Force, and Navy capabilities, enables us to see how the Chinese military's size and composition accord with its rhetoric, and to evaluate the parameters within which China determines the appropriate scale for the PLA. In accord with a military strategy based on deterrence and denial, and focused on enhancing China's broader national objectives, the rapid modernization undertaken by the PLA over the past 10-15 years has been calculated and gradual, yet comprehensive and broad.

In Chapter 12, "The 'Right Size' for China's Military: To What Ends?" Ellis Joffe takes one last look at the drivers of PLA modernization. While many analysts will agree that the primary impetus for China's military buildup over the past decade or so has been the need to deter Taiwan from pursuing independence, the specter of war along the Strait has diminished dramatically since 2006—and with it the immediate need for military muscle. Why then does China continue to build and update its armed forces?

Joffe reiterates that the development of a more robust military force is part and parcel of China's unalterable, long-range objective of achieving "great power" standing among the international community. Because this goal is longitudinal and generational, efforts to build a force commensurate with the country's international prominence are seen as a gradual process, composed of a series of short-term steps and activities. The first of these objectives, writes Joffe, might be the attainment of a preeminent position in the East Asian region. Driving this aspiration are the same forces—physical presence, nationalism, and economic power—that underlie the country's global motivations. Although China's present-day regional strategy has converged on Taiwan—preventing the province from seceding and interdicting any U.S. naval intervention—PLA forces have so far failed to demonstrate an outright intention to compete with U.S. dominance in the Western Pacific. Instead, China has been engaging in defensive maneuvers designed to protect the country's maritime security interests—and to hedge against increasing U.S. military might in the region.

Thus, while the speed and scope of China's military modernization may change in the future, its direction will probably not. Joffe reminds us that China's security interests vis-à-vis the United States will continue to drive PLA missions, functions, and organization for years to come.

ENDNOTES - CHAPTER 1

1. U.S. Department of Defense, *Quadrennial Defense Review Report*, February 6, 2006, p. 29, available at *www.defenselink.mil/pubs/pdfs/QDR20060203.pdf*.

2. Former Secretary of Defense Donald Rumsfeld raised similar questions during a speech he made in Singapore in June 2005. He asked: "Since no nation threatens China, one must wonder: 'Why this growing investment [in defense]? Why these continuing large and expanded arms purchases? Why these continued deployments?'"

3. Secretary Condoleezza Rice, "Interview With the CBS News Editorial Board," September 12, 2005, New York City, available at *www.state.gov/secretary/rm/2005/53033.htm*.

4. For an interesting report on popular dissent in China by a recent first-hand observer, see Guy Sorman, "The Truth About China," *The Wall Street Journal*, April 20, 2007, p. A15.

CHAPTER 2

FRAMING THE PROBLEM: CHINA'S THREAT ENVIRONMENT AND INTERNATIONAL OBLIGATIONS

Michael R. Chambers

What is the proper size and structure of the People's Liberation Army (PLA)? The answer to this question will be based at least in part on the nature and source of the threats to the security of the People's Republic of China (PRC). As government officials and military officers plan for the size, structure, and equipping of a military force to defend their country, many factors will be taken into consideration. But the first will be the requirement to defend the nation against immediate and potential threats to its security.

This being the case, a first step toward understanding the "proper" size and structure of the PLA will be to understand the threat environment that China faces. This chapter argues that there are few direct, immediate external threats to the PRC today against which it needs to prepare. However, there are several potential threats worrying the Chinese civilian and military leadership. The principal external threat that the Chinese perceive is from the United States. Over the long term, the United States is seen as a potential constraint on China's rise to great power status. In the short term, the United States poses a potential threat to China's territorial integrity via its support for Taiwan's ability to defend itself prior to a peaceful resolution to the cross-Strait situation. U.S. involvement in a Taiwan conflict scenario also interacts with other external security threats to the PRC, such as China's sense of threat to its energy security specifically and the more general security of its maritime lines of communication. Disruption of Chinese access to the sea lanes, and in particular to China's energy imports, by the United States during a conflict over Taiwan could seriously harm the continued growth of the PRC economy, thereby undermining one of the key elements of continued legitimacy for the Chinese Communist Party (CCP). Prudent leadership in Beijing will take the steps necessary to enhance the PLA's capabilities to cope with these threats. Other external security challenges confronting Beijing include border, island, and offshore mineral rights disputes that threaten the territorial integrity of the PRC, nontraditional security threats such as the international narcotics trade, and regional instability that could negatively impact the regional environment for China's continued strategic economic development — or even lead to refugee flows into the PRC.

However, the most pressing threats to China's security are to be found not in the international arena, but at the domestic level. These are the threats of separatism to China's territorial integrity — specifically Taiwan,[1] Tibet, and the Uyghurs and other Muslim groups of Xinjiang — and the threats to the legitimacy of the CCP's rule from popular protests over corruption, illegal taxes, and illegal land-grabs. Reactions to the separatist threats, particularly the Taiwan issue, could require the use of military force, and thus will require planning for adequate numbers of troops and appropriate weapons systems. However, the political threats are more of a police issue than a military issue, and it will be hard to justify increasing the size or enhancing the weaponry of the PLA to meet these threats.

A second factor that might affect the size and structure of the Chinese military will be the PRC's international commitments. The U.S. military has been shaped to meet its commitments to its allies in Europe and Asia; to what extent do China's alliance commitments affect the shaping of its military forces? Or its participation in regional security organizations, such as the Shanghai Cooperation Organization (SCO) or the Association of Southeast Asian Nations (ASEAN) Regional Forum (ARF)? Might Chinese involvement in United Nations (UN) peacekeeping operations also help to determine the size and structure of the PLA? As will be argued below, China's international commitments do not have a significant impact on determining the size and structure of the PLA. Nor do they directly address the more significant external security challenges facing the PRC, although some of these obligations and commitments address these threats indirectly. Several of these commitments, such as the SCO, do help to address domestic-level threats, such as separatism.

If the PRC's international commitments have little effect on the sizing of the PLA, and its most pressing security threats are more domestic and political in nature than external and military, then how are we to understand the motivations shaping Chinese military modernization? In the near term, clearly it is Beijing's desire to win any conflict over Taiwan—and this involves coping with a U.S. military intervention. Over the longer term, it would appear that Chinese aspirations and ambitions to become a global great power might be the driving force. To play such a role, the PRC would want to develop the capabilities to secure its global economic interests and trade routes, and to have military capabilities to provide muscle behind its diplomacy. In addition, it would need military capabilities to defend against or deter those of other great powers—including the United States.

The next section of this chapter will briefly consider what it is that China seeks to "secure" with its military force, and point to a potential expansion of its definition of its national interests. Then the chapter will turn to analyzing the threat environment, beginning with external threats to Chinese security and following with the internal security environment. The PRC's international obligations and commitments will then be examined before turning to the conclusion.

SECURING WHAT? CHINA'S SECURITY INTERESTS

Before examining the potential and real threats to China's security, we should consider precisely what these security interests are that might be threatened. According to the white paper, "China's National Defense in 2006," the PRC's national defense policy is defined as:

- Upholding national security and unity, and enhancing the interests of national development. (This includes security against external and internal threats.)
- "Achieving the all-round, coordinated, and sustainable development of China's national defense and armed forces."
- "Enhancing the performance of the armed forces with informationization as the major measuring criterion."

- "Implementing the military strategy of active defense" such that the PLA is prepared to win "local wars under conditions of informationization" and to enhance "national sovereignty, security, and interests of development."
- "Pursuing a self-defensive nuclear strategy" to deter other countries from using or threatening to use nuclear weapons against China.
- Fostering an international security environment conducive to China's peaceful development.[2]

Similarly, Chinese officials have summarized the goals of China's security policy as intended to safeguard the PRC's sovereignty, unity, and territorial integrity, as well as maintain the conditions for China's economic development—including a stable and peaceful international and regional environment.[3]

What is clear from these and other similar statements is that Beijing is not concerned simply with the survival of the nation. Comments in the 2004 defense white paper (and in the 2000 and 2002 versions of these white papers), as well as in major speeches (such as Jiang Zemin's report to the 16th Party Congress in 2002) that "peace and development remain the dominating themes of the times," suggest that the general security situation remains favorable for the PRC.[4] In his report, former President and Party Secretary Jiang asserted, "It is realistic to bring about a fairly long period of peace in the world and a favorable climate in areas around China."[5] This confidence is even more pronounced in the 2006 defense white paper, which states: "China's overall security environment remains sound. . . . Its overall national strength has considerably increased, as has its international standing and influence. . . . Balancing developments in both domestic and international situations, it is well prepared to respond to complexities in the international security environment."[6]

China is not in any danger of being overrun by a military adversary at the moment. However, Beijing is concerned about the loss of territory through separatist actions and/or foreign aggression, and seeks to maintain the unity of its national territory. Taiwan and Tibet are the two most prominent examples of territory that the PRC wants to keep integrated with the national territory, but there are also disputes concerning islands and waters in the East China Sea (the Diaoyu Islands) and in the South China Sea (the Spratly Islands), as well as border territories that remain under dispute (such as along the Sino-Indian border). Beijing is also concerned about "unity" in the sense of maintaining social stability under the Party's political leadership. In October 2006, the CCP Central Committee adopted a resolution on "Major Issues Regarding the Building of a Harmonious Socialist Society"; according to a *Xinhua* commentary in December 2006, this was the first major Party document to address the issue of "mass incidents" (riots and protests) and to make this an important task for the Communist Party. The same commentary also labeled these mass incidents as "the most outstanding problem that seriously disturbs social stability" and stated that such disturbances represented a "major threat" to the Party's ability to govern the country.[7] Not surprisingly, China's communist rulers want at all costs to maintain the existing regime (or political system). Regime security could be threatened by domestic turmoil and opposition, or by the efforts of foreign powers to undermine the legitimacy of the Communist Party's rule (short of military invasion and occupation, foreigners would have to work with domestic forces to effect regime change).

Besides these military and political interests that need to be secured, the Chinese are also seeking to safeguard their economic interests. This should not be surprising either, since most governments seek to do the same, and economic performance is often linked to the success and legitimacy of a government or ruling party. However, what is interesting here is that China's economic interests seem to be expanding beyond its own territory. China is now the third largest trading power in the world (since 2004) and the fourth largest economy (since 2005). Its booming economy is highly dependent on trade for its success, and the PRC has become heavily dependent on the sea lines of communication.

As noted in the 2004 defense white paper, the PLA is tasked with protecting China's "maritime rights and interests," a task for which it is not yet fully prepared. In a December 2006 meeting of senior Party members in the PLA Navy (PLAN), President Hu emphasized that China is a maritime power, and asserted that the PRC "should endeavor to build a powerful people's navy that can adapt to its historical mission during a new century and a new period." Hu further claimed that the PLAN has an "important" and "glorious" task of protecting China's "authority and security and maintaining our maritime rights."[8]

Similarly, an editorial in *Jiefangjun Bao* to commemorate the paper's 50th anniversary noted that conceptions of national interests had already extended from national territory, seas, and airspace to include further out into the deep seas, outer space, and the electromagnetic sphere, arguing that the PRC needed to develop the capabilities to secure these interests. The *Jiefangjun Bao* editorial then nailed down its point: "In order to accelerate national development and safeguard national security, China has great strategic interests in the deep seas and in outer space."[9] In addition, many Chinese analysts are now also calling for the PRC to play a more active role in maintaining international and regional peace and stability, as these are prerequisites for China's continued economic growth, itself a critical source of legitimacy for the CCP.[10]

Thus, China's security today is predicated not only on safeguarding its national territory from attacks by major powers, but also on maintaining territorial integrity against separatist threats and on preserving domestic social stability. And it requires a peaceful and stable regional and international environment as well as secure access to economic resources and to the sea lines of communication. The growth of China's economic interests as it becomes a global economic power is prompting a concomitant expansion of China's security interests as it emerges as a great power. As we will see below, these new challenges may well help shape the size and structure of the PLA even though they do not represent direct, immediate external threats to China's security.

THE EXTERNAL THREAT ENVIRONMENT

As of mid 2007, China faces few direct, immediate external threats to its security. Consequently, "China's overall security environment remains sound," and the PRC is able to point to "peace and development" as the "principal themes" in the international security environment, as well as to the stability of the overall security situation in the Asia-Pacific region.[11] Nevertheless, there are a number of external security challenges that the Chinese face and for which Beijing feels it must plan, including the acquisition of new weapons systems. Such challenges include territorial disputes that continue to threaten the territorial integrity of the PRC; the security of the maritime lines of communication, upon which China is increasingly dependent as a major global trading power and one of the largest global energy importers; nontraditional security threats such as the international narcotics trade; and potential regional instability (for example, on the Korean peninsula) that could threaten the continued vibrancy of the Chinese economy either by undermining the sense of stability and security in the East Asian region that incubates China's growth, or by more directly harming it with an influx of refugees that disrupt the local Chinese economy.

While each of these security challenges is felt to require attention from the leadership in Beijing, the principal external security challenge is the one posed by the United States. In the long term, the United States poses a potential threat to strategically contain or encircle China; in the short term, American support for Taiwan represents a potentially powerful obstacle to Beijing's efforts to reunify the island with the PRC. Moreover, the U.S. factor (particularly American intervention in a conflict with Taiwan) intermingles with several of the other external challenges, aggravating China's sense of potential threat.

The United States as Potential Threat.

Beijing is clearly worried about a hegemonic and unilateralist United States, as was made evident in the 2004 and 2006 defense white papers.[12] This is because the United States, as the "lone superpower" in the world today, is the sole country with the military and economic wherewithal to thwart China's rise to great power status not only within the Asia-Pacific region but also globally. Based on American political, economic, and military influence, it is feared that Washington might attempt to contain the PRC's rise, particularly through strategically encircling it. The increased U.S. military presence in Central Asia and Pakistan as a result of the war on terrorism and the invasion of Afghanistan aggravated such fears, as has the intensification of U.S.-Japanese security cooperation in the last few years.[13] China's pursuit of a "good neighbor" policy since 1997 and its efforts to promote East Asian regionalism, particularly through the ASEAN Plus Three Process (the regional dialogue involving the 10 members of ASEAN plus China, Japan, and South Korea), are intended to create a ring of friendly states surrounding the PRC that could serve as a strategic buffer against pressures that the United States might exert in a containment ploy.[14] Astute Chinese diplomacy has thus proved beneficial in trying to blunt the potential threat from the United States.

Likewise, Chinese trade policy has intentionally sought to create interdependencies between the PRC and its neighbors as a means to further bind their interests to those of China in ways that would preclude their siding with American containment efforts. However, as noted by Rear Admiral Yang Yi, director of the Institute of Strategic Studies at the National Defense University, "compared with the political, diplomatic, and cultural means of safeguarding China's interests, China's military force lags far behind. As a responsible big power, China needs to build a military force worthy of its international status."[15] The PLA thus sees itself as requiring adequate weapons systems to deter or blunt possible U.S. containment efforts through military pressures. But such considerations also suggest the role of China's ambitions, not just current security challenges, in shaping the PLA.

While strategic encirclement or containment by the United States is a potential problem, the more pressing security challenge is the possibility of U.S. intervention to protect Taiwan in the event of a cross-Strait conflict. The United States has promised to assist Taiwan should the PRC launch unprovoked aggression against the island, with President George W. Bush promising in April 2001 that Washington would do "whatever it takes" to help Taiwan defend itself in such an event.[16] With the PLA pledged to protect the territorial integrity of the nation, including preventing the formal separation of Taiwan from China, Chinese civilian and military leaders feel the need to plan for the possibility of confronting American military forces in the event of war with Taiwan. Such an eventuality requires Beijing to acquire modern weapons systems that could defeat the U.S. Seventh Fleet and other forces that would be used against the PLA in a Taiwan conflict, as well as the weapons systems necessary to project adequate power to subdue the island. And as the PLAN has changed its strategy to implement an area-denial maritime strategy in China's littoral areas to thwart a U.S. intervention, new naval assets need to be acquired. As Thomas Christensen has noted, it is not necessary for the PLA to directly match American military capabilities in the Western Pacific in order to pose problems for U.S. forces; instead, the Chinese could use asymmetrical capabilities to meet these challenges.[17] This course is likely a key rationale behind the PRC's testing of an anti-satellite weapon in January 2007—to demonstrate to the United States that it already possesses such capabilities. Whether conventional or asymmetric, the PLA will develop and deploy new capabilities to meet the potential threat from the United States.

The U.S. challenge in a Taiwan conflict scenario also blends into some of the other external security challenges facing the PRC. In particular, it is possible that the United States would try to disrupt Chinese shipping through the critical sea lanes of the South China Sea and the Strait of Malacca, hampering Chinese trade but especially Chinese oil imports. As discussed below, this is one of the critical aspects of China's maritime and energy security. Should such a disruption occur, it would cause serious harm to the Chinese economy, thereby undermining the legitimacy of the Communist Party's rule (which has come to rest heavily on economic performance as well as nationalism as the twin pillars of its continued legitimacy in the reform era). To cope with this potential threat, China will need to acquire power projection capabilities that it currently lacks. Recent improvements in the PLAN's submarine forces—including purchases of *Kilo*-class submarines from Russia and indigenously produced *Yuan* (Type 094) and *Shang* (Type 093) class submarines—could help in this regard, but aircraft carriers might also be necessary.

Despite these apprehensions, the United States remains only a potential threat to China: although the two countries are not truly friends, neither are they truly adversaries. The United States and the PRC have a mix of common and conflicting interests, which lead to broad areas for potential cooperation even while they lead to areas of tension and dispute. Yet because of such security challenges, Beijing, as a matter of prudence, perceives the need to develop and modernize the PLA to address the potential threats China faces from the United States. However, the actual level of (potential) threat may not be as significant due to the successful diplomacy that Beijing has engaged in over the past decade. China's "good neighbor" diplomacy as well as its trade diplomacy have created degrees of interests in the PRC's neighbors such that many of them—including U.S. allies South Korea and Australia as well as many Southeast Asian countries—would be hesitant to side with the United States against China in a conflict over Taiwan. They also desire to avoid being put into a position where they would be forced to choose. As a result, Beijing has succeeded in weakening American alliances in the region, and in complicating American access to military facilities that could prove critical in the event of U.S. intervention in a Taiwan conflict. These developments mitigate some of the potential threat that the United States might pose to China, although they do not eliminate it.

Territorial Disputes.

Since the end of the Cold War, the PRC has settled several of its border disputes with neighbors, including with Russia, Kazakhstan, and Vietnam (land and coastal maritime borders, although agreement on the latter is being implemented slowly). The resolution of these disputes has greatly reduced the potential threat to China's territorial integrity. However, there are still some border and territorial disputes that remain unresolved, and these are potentially quite troubling for the PRC.

The first of these is with India. The Sino-Indian border dispute dates back to the 1950s and disagreements over the legitimacy of the McMahon Line. By 1959, tensions along the border escalated into armed combat, and in 1962 China and India fought a brief border war in which the Chinese forces soundly defeated the Indian forces before halting and returning home.

Although there have been numerous discussions of the border dispute over the years, in 2003 Beijing and New Delhi agreed to negotiate a political framework for resolving the dispute, and in 2005 agreement was reached during Chinese Prime Minister Wen Jiabao's visit on a set of principles intended to guide further negotiations on the framework.[18] Despite a few small steps forward and continued calls for a swifter resolution, little real progress has been made. In fact, on the eve of President Hu Jintao's November 2006 visit to India, the PRC ambassador to India roiled the waters by claiming that India's Arunachal Pradesh state was Chinese territory.[19] President Hu and Prime Minister Manmohan Singh sought to reestablish calm, agreeing during the visit to accelerate work on negotiating a resolution of the border. This political desire for a resolution, coupled with the overall improvement in Sino-Indian relations over the last few years, would seem to remove or at least undercut the Sino-Indian border dispute as a rationale for development and modernization of the PLA.

A second dispute involves China's claims to the South China Sea and the Spratly Islands. In addition to their geostrategic position along critical sea lanes through the South China Sea, the PRC is interested in the Spratlys due to the rich fishing grounds near the islands and the possible energy deposits in the seabed under them. Beijing's claims to all of the Spratly Islands—and perhaps to the whole of the South China Sea—run afoul of claims by Vietnam to all of the Spratlys and by the Philippines, Malaysia, and Brunei to specific Spratly islands within their exclusive economic zones. (Taiwan has also asserted claims to the islands that mimic those of the PRC.) In addition, China's possible claims to the Sea itself seem to challenge Indonesia's claims to waters around its Natuna Island.[20]

The Spratly Islands have been a source of conflict in the past between China and its neighbors. The PRC took seven islands from the Vietnamese in a brief naval clash in 1988, and the Chinese seized Mischief Reef from the Philippines in 1995, aggravating regional fears at that time of a "China threat." Indeed, the PLAN has built up and fortified the "fishermen's shelter" on Mischief Reef and has built an airstrip on Woody Island in the Paracel Islands (an island group north of the Spratlys and disputed between the PRC and Vietnam).

Overall, however, the Chinese have taken steps in the last several years to dampen the potential for conflict in the South China Sea. In 2002, China agreed with the other claimants to the Declaration on Conduct of Parties in the South China Sea, an interim step while the various disputants continued to work out a formal code of conduct (which is still not completed). Equally significant, in September 2004 Chinese and Philippine oil companies agreed to a joint exploration project in an area which they dispute, and in March 2005 the Vietnamese joined the project.[21] These steps, combined with the success of China's good neighbor policy, have helped to ameliorate the sense of threat that China feels over these disputed territories. Still, the disputes have not been truly resolved but only set aside for the time being, and the PLA may need to continue thinking about developing the air and naval capabilities (e.g., long-range aircraft, aircraft carriers, advanced warships) required to safeguard the Chinese claims. Moreover, should oil be discovered in significant quantities under the Spratlys, the current cooperation could devolve into forcible competition to strengthen national claims to the islands and waters.

While China's border dispute with India and its territorial disputes with Southeast Asian countries over the Spratlys are largely in abeyance, the same cannot be said for the Sino-Japanese dispute over the Diaoyu Islands and the East China Sea. Although the contention over the Diaoyu (Senkaku in Japanese) Islands has existed for decades, the dispute heated up in the late 1960s with reports of potential energy deposits nearby. The jostling became fairly intense in the mid-to-late 1990s when the third UN Convention on the Law of the Sea (UNCLOS III) came into force (1994). Both China and Japan sought to establish their 200 nautical mile exclusive economic zones in the East China Sea and over the islands with less than 400 nautical miles separating the undisputed territory of the two countries. To bolster Japan's claims to the islands, a group of Japanese citizens constructed a lighthouse on the largest of the islands; Chinese responded with attempts to raise Chinese flags on the disputed territory. More recently, the scene of Sino-Japanese contention has been the Chunxiao gas field in the East China Sea, lying just on China's side of the boundary as recognized by Japan but with part of the deposit possibly under Japanese territory. The Chinese started to develop this energy source in 2003, and the China National Offshore Oil Company (CNOOC) announced in July 2006 that the project had become operational.[22] To reaffirm China's claims to the gas field in the face of Japanese protests against the gas project, in September 2005 the PLA sent five naval vessels to visit the area, and one of the warships trained its gun on a Japanese surveillance aircraft.[23]

What makes this territorial dispute with Japan particularly worrisome, unlike those with India or the ASEAN countries, is that political tensions with Japan have been increasing over the last couple of years rather than subsiding—although such tensions eased in late 2006 following the hastily arranged October 8 summit meeting that year between new Japanese Prime Minister Shinzo Abe and the Hu-Wen leadership team. Since late 2004, we have seen a *Han*-class submarine intrude into Japanese territorial waters, three weekends of anti-Japanese riots in major Chinese cities during the spring of 2005, and the temporary suspension of Japanese economic aid to the PRC in the spring of 2006 in response to Chinese actions and statements against Japan. Japan also is engaging in competition with China for influence in the region, especially in Southeast Asia, although it appears frequently to be one or two steps behind the Chinese. Finally, Japan is seeking to become a "normal" country, a country that is not severely constrained in the use of its military to work with friends and allies for common purposes.

As a result of these factors, many Chinese analysts are identifying Japan as a potential threat to Chinese security, arguing that increasing militarism and nationalism in Japan and closer cooperation with the United States—especially in the event of a Sino-American conflict over Taiwan—point to Japanese efforts (in conjunction with the United States) to contain the PRC's rise.[24] Concerns about the future of Sino-Japanese relations and of a possible threat to China's security from this neighbor were a common theme during discussions with several civilian foreign policy and security analysts in Beijing in August 2006 (these analysts specialized in East Asia generally, or in Northeast or Southeast Asia).[25]

Based on these mounting security concerns vis-à-vis Japan, we can expect to see the PRC develop military capabilities to deter Japanese involvement (even if only in a support role) in a conflict over Taiwan that involved the United States, and capabilities to safeguard

Chinese interests in the East China Sea and the Diaoyu Islands. Such capabilities could certainly include the new *Zubr*-class air-cushioned landing craft that China has recently agreed to purchase from Russia, which would enhance the PLA's amphibious capabilities.[26]

Maritime and Energy Security.

China's emergence as a major trading power has led it to develop a heavy dependence on sea lines of communication (SLOCs). The phenomenal growth of the Chinese economy (averaging approximately 9 percent annual growth since 1979) has been driven by Chinese exports, which in turn depend on the import of components and raw materials. Maintenance of the growth and development of the economy is a crucial aspect of the continued legitimacy of the Communist Party's rule, with disruption of Chinese trade not only affecting the economic security of the PRC, but also the secure rule of the current regime. Yet, since the PLA's naval capabilities are not adequate to protect Chinese shipping in the East Asian SLOCs (especially into the southern reaches of the South China Sea and the Southeast Asian straits), China must rely on the United States to provide the public good of the freedom of the seas and the security of the sea lanes. However, if there is a Sino-American conflict over Taiwan, there is the strong possibility that this public good could become an excludable good to China. Thus, prudent military planners in Beijing will seek to enhance Chinese military capabilities to protect the sea lanes that Chinese shipping uses.[27] The need to pursue SLOC security helps to explain the priority placed on building naval power as reflected in the 2004 PRC defense white paper, and in President Hu's December

2006 call to build a "powerful people's navy that meets the demands of carrying out our army's historic missions in the new century and new stage."[28] It also bolsters the 2006 defense white paper's emphasis on gradual extension of the PLAN's "strategic depth."[29]

This dependence is particularly salient with respect to China's oil imports. The PRC imports approximately 40 percent of its oil, and of that amount 80 percent comes through the Strait of Malacca. In December 2003, President Hu Jintao expressed "extreme concern" over this vulnerability to China's oil supplies because the country would face a "predicament" should some incident happen and/or foreign countries blockade the Strait.[30] Chinese security analysts remain concerned about the possibility of piracy or terrorism in the Strait, as well as the possibility of the Strait being blocked by another country's naval forces.[31] Hu called on the country to find ways to alleviate this "Malacca Dilemma," including energy conservation and diversification of sources.

Beijing has also pursued discussions with the three Strait countries (Indonesia, Malaysia, and Singapore) on cooperation to maintain security in the Strait; although Indonesia and Malaysia have been hesitant to allow a significant role for any outside power in the Strait, avenues for cooperation that do not impinge on the sovereignty of the three littoral states are being explored. In order to enhance the security of the Malacca Strait, the PLA will need to build up its power-projection capabilities to extend down into that area (e.g., aircraft carriers and long-range aircraft). It is also possibly seeking bases in friendly countries along the route from oil sources in the Persian Gulf; this potential "string of pearls" strategy has received a fair bit of attention in Western news media, although it is not clear that China actually has an explicit strategy along these lines.[32]

As the Chinese have sought to diversify their sources of oil and gas, they have turned increasingly to Russia and Central Asia as suppliers. In 2005, Russia supplied 10 percent of China's crude oil imports, and in March 2006 China and Russia signed an agreement to supply Russian gas to the PRC.[33] Beijing has also provided a gift of $400 million to Russia for a feasibility study on constructing a spur of the East Siberian-Pacific Ocean (ESPO) pipeline to China.[34] In Central Asia, the Chinese have helped to build an oil pipeline from Kazakhstan to the PRC which is now operational, purchased a major oil company in Kazakhstan (PetroKazakhstan) in October 2005, and struck a deal with Turkmenistan in April 2006 to purchase gas from that country beginning in 2009 and build a pipeline to deliver it.[35] Chinese oil companies are seeking additional deals as well.

As China becomes a larger player in the Central Asian energy game, it will develop important assets that it will desire to secure against terrorism or other types of threats. To do so, the PLA will need to further develop its land- and air-based power projection capabilities—thus pulling the military in a somewhat different direction from sea- and air-based power projection capabilities for dealing with threats to Chinese maritime interests. In addition, China might find itself increasingly in rivalry with Russia, which has sought to maintain its virtual monopoly on Central Asian energy exports as a way to continue its political influence in the region. At the moment, Sino-Russian relations are quite good, with growing economic interdependence as well as continuing Chinese purchases of Russian weaponry to modernize the PLA. However, if Sino-Russian rivalry in Central Asia becomes intense, this development could join with Russian dissatisfaction over trade asymmetries, Chinese immigration, or other factors to create a more assertive Russia, resulting in the perception of Russia as a threat to Beijing. Should that happen, the Russian contributions to China's military modernization would likely be terminated, and the PRC's whole security calculus would have to be revised.

Nontraditional Security Threats.

As is true of many countries, China is facing challenges from nontraditional security threats. Some of these, like the spread of infectious diseases, are not a military threat, and although the military could play a role in humanitarian efforts to cope with an outbreak, the size and structure of the PLA will not be affected by them. Others, however, do have military aspects. One of these is transnational crime such as international drug trafficking. China borders two of the most prolific drug productions centers in Asia: Afghanistan (heroin and opium) and Myanmar, formerly known as Burma (heroin and opium, and increasingly methamphetamines). Many of these drugs are making their way into China. Although combatting illegal drug trafficking is primarily a police issue, China has involved the PLA at times to strengthen control over its border in order to staunch the flow of drugs into the country, such as it did along its border with Myanmar in August-September 2003. Yet, such a threat will not drive the modernization of the PLA.

Like counternarcotics, combatting terrorism is another of the nontraditional security challenges facing the PRC that is more characteristically a matter of police work but that can also involve the use of military forces. For China, fears of terrorism are linked closely to the threat of separatism, particularly in the Xinjiang region, and Beijing is concerned about the links among Muslim separatists (particularly the East Turkestan Islamic Movement, or ETIM) and al-Qai'da. It is also concerned about separatists in Xinjiang receiving assistance from allies in neighboring countries, such as Kazakhstan, Kyrgyzstan, and Tajikistan. Because of the transnational nature of terrorism, China has been working with its partners in the Shanghai Cooperation Organization (SCO) to fight terrorism in the Central Asian region. The SCO held multilateral joint anti-terrorism exercises in 2003, and is planning another exercise for 2007. China also held bilateral anti-terrorism exercises with Kazakhstan in August 2006, which involved police and special operations forces, and with Tajikistan in September 2006.[36]

Combatting terrorism will likely have some small impact on the structure of the PLA and some of the hardware it seeks to acquire or develop, particularly in terms of the missions that special operations forces would train for and the surveillance equipment they would need. However, addressing this threat will be handled as much by the People's Armed Police as by the PLA, and is not likely to have a major impact on the size and structure of the PLA.

Regional Instability.

Instability in neighboring countries is the final external security challenge for the PRC. Such political, economic, and social troubles would affect the general regional environment for China's security, including China's need for strategic economic development: turbulence in neighboring countries could affect the general business climate in the region, with spill-over effects in China. As Beijing has said repeatedly, it desires peace and stability in its region of the globe in order for it to concentrate on economic development. More worrying for the Chinese leadership than this general environmental impact, however, is the possibility that instability in a neighbor bordering the PRC could have a direct impact on the Chinese economy. This could occur particularly through refugee flows into China; significant influxes of refugees could negatively impact the local economy for the affected Chinese border provinces. Because of the importance of continued strong economic growth for the legitimacy of Communist Party rule and for China's drive to become a great power, Beijing needs to guard against such borderland disruptions.

Central Asia is one area where China will not want to see turmoil because it could flow into already restive Xinjiang. Political disturbances in Central Asia could also harm Chinese economic interests and assets (e.g., in the energy and mineral sectors) in those countries, or they could link with the terrorist threat. Likewise, China's promotion of closer economic integration between its southern provinces and the countries of the Mekong subregion will increase Beijing's desire for stability in those Southeast Asian neighbors. But the Korean peninsula is the place of most concern to the PRC because of the current nuclear crisis there. Estimates over the past several years have suggested that there are already tens of thousands of North Koreans living in northeastern China; political turmoil in North Korea, economic collapse, or war with the United States and South Korea could send tens of thousands more refugees streaming across the border, straining a local economy that was previously one of the PRC's "rust belts."

Diplomacy has been and will continue to be the principal method for addressing these security challenges, as Beijing has shown with the North Korean situation. Economic tools—such as increasing trade with, aid to, and investment in neighbors—could also be used to promote stability in these countries. However, military forces might be involved as well. In August-September 2003, as tensions were rising over North Korea's nuclear program, Beijing replaced its border police with regular PLA troops, with reports suggesting that between 15,000 and 150,000 PLA troops had been stationed on the Sino-North Korean border.[37] In July 2006, following Pyongyang's test of several ballistic missiles, China reportedly beefed up its border with additional regular troops.[38] While the exact purpose of these military moves is not known for certain, one of the reasons commonly assumed is that China wanted to prevent massive refugee flows from crossing the border.

Such a rationale is completely defensive in nature. However, as China becomes more proactive in the region on achieving great power status, or as the Chinese leadership accedes to becoming a "responsible stakeholder" in the region with concomitant responsibilities for maintaining neighborhood peace and stability, then we could imagine that Beijing would use expeditionary forces to quell nearby turmoil before it crossed national borders. The reported stationing of 4,000 Chinese troops in southern Sudan to protect Chinese oil interests there may be a sign of things to come.[39]

As this change of mindset occurs, the PLA will need to develop adequate power projection capabilities, including not only weapon platforms but also troop transport capabilities. According to the Pentagon's 2006 report on the Chinese military, the PLA's expeditionary forces currently consist of three airborne divisions, two amphibious infantry divisions, two marine brigades, about seven special operations groups, and one regimental-size reconnaissance element in the Second Artillery missile force. The capabilities of these units are steadily improving, not least through the introduction of new equipment.[40]

To summarize, there are few direct, immediate threats to the national security of China that originate from beyond its borders. Nevertheless, there are several external security challenges that Beijing will likely address, and many of these—such as coping with the United States, protecting Chinese claims to islands in the East and South China Seas, and safeguarding the PRC's energy supplies and access to SLOCs—will affect the size and structure of the Chinese military. Challenges of this nature will require that the PLA acquire new weapon systems as it modernizes its forces.

THE INTERNAL SECURITY ENVIRONMENT

Domestic threats present more of a direct threat to the security of China, and to the leadership. As noted in a December 2006 *Xinhua* commentary, "The prevention and proper handling of mass incidents is a major test for the CPC's [Communist Party of China] governing ability."[41] Separatism, based on ethnic differences or the unresolved civil war with the Republic of China on Taiwan, threatens the territorial integrity of the country. Domestic turmoil and social unrest generated by corruption and official malfeasance threaten to undermine the legitimacy of the Communist Party's rule and of the political system as a whole. Because the international security environment is relatively peaceful and non-threatening for the PRC, some analysts of Chinese security policy see Beijing placing greater emphasis on these two domestic-level security threats than on external threats in their list of security priorities.[42] Whether these are indeed the top two security issues for the Chinese leadership or not, they are critical challenges to Beijing. However, only the issue of separatism truly involves the potential use of military forces, and thus has the potential to affect the size and structure of the PLA. Similarly, the PLA has traditionally played a role in addressing domestic natural disasters and humanitarian crises, and has even served to an extent as a social security net. While the PLA will likely continue to serve these functions, their impact on the size and structure of the military are not likely to be strong—especially as the PRC further develops the People's Armed Police and a social security system.

The Separatist Threats.

As enunciated in China's defense white papers and in agreements of the SCO, Beijing is concerned about the "three forces" of terrorism, separatism, and extremism.[43] Yet all three of these forces seem to boil down to the same broad threat—the placing in jeopardy of China's borders by groups willing to use political violence to pursue their goal of tearing away a piece of Chinese territory. The primary geographical areas of concern for the Chinese leadership are Taiwan, Tibet, and Xinjiang. All three of these areas have links to external actors (and many would argue that Taiwan is an external actor), and military force has been used in the past by Beijing in efforts to exert control over these regions.

The threat in Xinjiang is ethnic and religious separatism. Although the ethnically Han Chinese have become a narrow majority of the population there recently, the ethnically Turkic Uighurs—Muslims who retained their faith during the communist era—used to be the dominant group in the region. During the tumult of the 1930s-1940s, the people of what is now Xinjiang managed twice to establish short-lived autonomous East Turkestan republics. In 1950, the PLA reestablished Chinese control over this region. Since then, the people of Xinjiang have had an uneasy relationship with Beijing due to their desire for greater autonomy or even independence.

Tensions between the Uighurs and the Chinese authorities escalated during the 1990s as Muslim separatists seeking to re-create East Turkestan engaged in bombings of civilian targets such as buses and markets, as well as government institutions in Xinjiang; they also allegedly bombed a bus in Beijing in March 1997. During the 1990s, these separatists also became linked to al-Qai'da and the Taliban regime in Afghanistan, where some Uighur members of the ETIM (one of the most radical of the Muslim separatist groups) received terrorist training. With funding, training, and arms from abroad, the Muslim separatists in Xinjiang have forced Beijing to use military as well as police forces to suppress this threat.[44]

The Chinese have also engaged in security cooperation with their SCO partners to address the threat posed by these Muslim separatists, including periodic joint anti-terrorism exercises, such as those with Kazakhstan and Tajikistan in 2006. In addition, Chinese authorities have sought to increase the presence of ethnically Han Chinese in Xinjiang in order to shift the local balance of power away from the Uighurs, and have also encouraged a "Go West" program of economic development to try to bring more prosperity to this northwestern region of the country.

The situation in Tibet today is not as ominous as the one in Xinjiang, although in the past Tibet has presented a more serious threat. Like Xinjiang, Tibet enjoyed de facto autonomy from central Chinese authority in the early-to-middle 1900s, and was brought under Beijing's control in 1950 by the PLA. Tensions rose during the 1950s between the Tibetans and the Communist Party leadership, leading to a revolt in 1959, the flight of the Dalai Lama to India, and a crackdown by the PLA. Tibetan resistance continued through the 1960s, supported by the United States and the Tibetan government-in-exile in India. Tensions rose again as Tibetans clamored for independence in the late 1980s, leading to suppression of protest and the declaration of martial law in Lhasa in 1989. Since then, Beijing has sought to infiltrate ethnic Han into Tibet, encouraged economic policies to bring enhanced prosperity to the region, and carefully monitored events for signs of protest. It has also occasionally engaged in quiet discussions with representatives of the Dalai Lama, who claims only to want autonomy for Tibet rather than independence. But the Chinese leadership, not trusting the Dalai Lama, remains wary of pro-independence sentiments in Tibet.

In both Xinjiang and Tibet, Beijing has used the PLA in the past to quell disturbances, but the development of the capabilities and training of police forces means that uprisings in either area today will first be a police issue, and will involve the military only if the turmoil escalates. The Chinese leadership is also employing economic and population policies to address the situation in these two areas. And since the PLA already possesses the capabilities to deal with these contingencies when military force is required, it is unlikely that separatism in Tibet or Xinjiang will have a major impact on the future size and shape of the PLA.

This is far from the case with regard to Taiwan. Because the Taiwan issue originates in the unresolved civil war from 1946 to 1949, military force is very much relevant. The PRC's largest concentration of forces today is arrayed against Taiwan, with the threat of military action should the government in Taipei declare independence and thus formally separate from China. Beijing's fears that Taiwan might seek independence contributed to the 1954-55 and the 1958 Taiwan Strait crises, and the PRC engaged in military coercion (including live-fire missile tests) during 1995-96 in a futile attempt to thwart the reelection of Taiwanese President Lee Teng-hui when it feared he was becoming pro-independence. Concerns about the desire of current President Chen Shui-bian to push for independence led the Chinese government to pass an anti-secession law in March 2005. The threatened use of military force to deter Taiwanese independence is a critical part of Chinese policy toward the island. Because of Taiwan's security relationship with the United States, the PRC must also prepare for an American intervention in a conflict over Taiwan, and because of tightening U.S.-Japan relations and Sino-Japanese tensions, there is also a possibility of Japanese involvement (through logistical support) in such a U.S. intervention.

However, threatened military force is not the only policy of Beijing toward Taiwan. The PRC is also actively using diplomacy to prevent Taiwan independence and to isolate the regime internationally. The most important element of this part of the strategy is to play on the American desire not to allow either side of the Strait to change the status quo unilaterally — Washington would rather avoid war between China and Taiwan. Thus, as President Chen appeared in late 2003 to be moving toward a position more strongly in favor of independence, Beijing worked on Washington to warn Chen against any such moves. President George W. Bush did so while standing next to PRC Premier Wen Jiabao at the White House. In addition to diplomacy, China is using trade with and investment from Taiwan to create economic interdependence as a way to bind the Taiwanese economy so tightly to the mainland economy that declaring independence would have enormous economic costs to the island. While the diplomatic and economic strategies seem to be working at the moment, military strategists in Beijing feel the need to plan for the possibility of their failure. Thus, the PLA will continue to increase and modernize the Chinese military forces in preparation to use force to halt Taiwanese independence.

The Threat of Domestic Turmoil.

Social unrest and mass protests are growing in China, presenting a serious threat to the legitimacy of the Party's rule by an increasingly assertive public. According to official government statistics, the number of "mass incidents" in 2004 — including protests, riots, and mass petitioning — reached 74,000, compared to 58,000 in 2003 and only 10,000 in 1994. In 2005 there were 87,000 "disturbances of public order," up 6.6 percent from 2004.[45] Many of these protests have turned violent in the last few years, due either to the actions of the protestors or to overly zealous police forces, and news of these violent incidents has frequently leaked out to the foreign media. The causes of such protests and disturbances are many, but common ones include official corruption; illegal land seizures or extremely low compensation for land confiscated for industrial or commercial development; illegal taxes on farmers by local officials; serious environmental pollution by factories and inadequate responses from the companies or government officials; and laid-off workers demanding welfare payments.

Because the Communist Party is supposed to represent the interests of the people, and is seen as doing anything but representing the people in these cases of unrest, the legitimacy of Party rule is potentially at risk from the increasing assertiveness of the people in voicing complaints against the government and Party. This social unrest could also negatively impact local economic development, which is a key goal of the leadership.

As noted earlier, the Party leadership is clearly aware of this security threat, and makes noises about addressing it. In summarizing the lessons of his 13 years of leadership at the 16th Party Congress in 2002, out-going Party Secretary Jiang Zemin's sixth lesson was: "Ensure stability as a principle of overriding importance and balance reform, development, and stability. Stability is a prerequisite for reform and development."[46] In order to achieve social stability, corruption has to be tackled. Jiang warned his colleagues: "If we do not crack down on corruption, the flesh-and-blood ties between the Party and the people will suffer a lot and the Party will be in danger of losing its ruling position, or possibly heading for self-destruction."[47] Premier Wen Jiabao noted in his March 2006 report on the work of the government that officials still needed to clean up corrupt practices, provide better social security services, better protect the environment, reform the rural tax system, and take other steps in order to better protect the "vital interests" of the people so that China can achieve stability and a "harmonious society."[48] Likewise, in a speech commemorating the 85th anniversary of the founding of the CCP, President and Party Secretary Hu Jintao reminded his colleagues that corruption threatened to undermine the Party's hold on power.[49] The October 2006 Party resolution on "Major Issues Regarding the Building of a Harmonious Socialist Society" and the December 2006 *Xinhua* commentary on the need to correctly handle "mass incidents" point to the seriousness with which the leadership is taking this issue.[50]

Beijing has also been seeking to address the issue of income disparities between the coastal and inland regions of western China through its "Develop the West" program. In early July 2006, the Chinese government announced plans to spend an additional $21 billion on major industrial, infrastructural, and social projects in western China, adding to the nearly $125 billion spent since 2000.[51]

Dealing with social stability is an issue for the police rather than the PLA. It is also an issue to be dealt with through political and legal reforms as well as economic policies. Addressing the issue of domestic turmoil is not going to be a critical factor affecting the PLA. Nevertheless, it is a major security challenge to the Chinese leadership.

CHINA'S INTERNATIONAL COMMITMENTS AND OBLIGATIONS

Efforts to address the security threats that China faces will shape the size and structure of the Chinese military. Addressing the challenge of the United States—either strategic encirclement or American intervention in the Taiwan Strait—is a major driver of PLA modernization and weapons acquisitions. Defending Chinese claims to islands in the East and South China Seas, or preparing for an invasion of Taiwan in the event that Taipei declares independence, will require enhanced amphibious capabilities. These tasks plus the need to protect Chinese shipping in the SLOCs will require air and naval power-projection capabilities. Securing China's energy interests in Central Asia, or playing the role of regional stabilizer there or on the Korean peninsula, will require air- and land-based power-projection capabilities.

The same cannot be said for China's international obligations and commitments, with the exception of peacekeeping. As the United States and other countries have discovered, peacekeeping requires skills different from those of warfighting, and a different mind-set. China has increased its participation in UN peacekeeping operations over the last several years, and if this trend continues the PLA may create more specialized units for peacekeeping operations (it has already created some engineering units for de-mining operations and has special peacekeeping training facilities). But the PRC's other international commitments, whether they be to allies or international organizations, are not of a nature or extent such that they will help to shape the size and structure of the PLA.

For the most part, China's international obligations and commitments do not directly address the important external security challenges facing the PRC. To some extent, such commitments may address these threats indirectly (e.g., the role that Pakistan may play in providing port and communication facilities from which the PLAN can monitor China-bound shipping from the Persian Gulf), and they may address domestic security threats such as the threat of separatism. Although China's alliances with North Korea and Pakistan were created to deal with specific external security threats (from the United States and India, respectively), the commitments in these two alliances are not terribly strong, and the threats themselves are no longer as salient as they once were. In addition, the international organizations to which the PRC belongs (such as the SCO and the ASEAN Regional Forum) require little of their members beyond talking.

China's Alliance Commitments.

China today has only two remaining alliances: the formal alliance with North Korea (Democratic People's Republic of Korea, or DPRK), and the informal alliance (or entente) with Pakistan.[52] I am defining "alliance" here to mean a relationship of security cooperation between two (or more) countries directed against a common adversary and which entails a level of defense assistance in the event that one of the allies is attacked by the common adversary. China has a patron-client type of relationship with Myanmar as well as security cooperation whereby Beijing provides arms to Myanmar and receives access to certain military facilities. However, China has apparently not promised any degree of defense assistance to Myanmar in case it is attacked by a common enemy. Sino-Myanmar relations are thus not truly relevant to the discussion here.

The 1961 Treaty of Friendship, Cooperation, and Mutual Assistance between China and North Korea formalized the informal alliance that had existed between these two neighbors since 1950. If either party was attacked (presumably by the United States or its

South Korean ally, the Republic of Korea [ROK]), the other was committed to "immediately render military and other assistance by all means at its disposal."[53] So far, this provision has never been activated. Moreover, it is unclear how China would respond today if the United States attacked North Korea. With its pursuit of a nuclear weapons program and its indifference to Chinese interests, the DPRK has become a rogue ally, distinguished by saber-rattling and truculence in its dealings with pereceived enemies. This is especially true since Pyongyang's July 2006 missile tests and its October 2006 nuclear test, but it was evident even by 2003 when China moved troops to their common border. Despite the existence of the formal alliance, as early as the mid-1990s and the first North Korean nuclear crisis Beijing was warning its neighbor that the Chinese would not bail them out if they got themselves into hot water.[54] During the current nuclear crisis, China has attempted to restrain the more outlandish behavior of its ally: the PRC has pressured the North Koreans to attend talks, has refused to provide weapons that Pyongyang felt it was entitled to by virtue of the alliance, and allegedly requested the termination of the mutual assistance clause of the alliance.[55] It also participated in UN Security Council-imposed sanctions following Pyongyang's nuclear test. Rather than maintain or uphold some military commitment in this case, China seems to be trying to minimize its obligations to the DPRK.[56]

The Sino-Pakistani entente has no codified treaty to formalize the relationship, but an understanding was reached between the two countries around the turn of the year 1962 that China would help Pakistan if it was attacked by India. China made credible noises to this effect during the 1965 India-Pakistan war, but rattled its saber at India too late in the 1971 war. Realizing that it would be difficult to provide extended deterrence to protect Pakistan, or to punish India for aggression against Pakistan, China decided to provide Pakistan with the wherewithal to defend itself and maintain a balance of power against India. In the 1980s, China conveyed to Pakistan the blueprints and possibly some materials to make a nuclear bomb; in the 1990s, Pakistan acquired technology and parts to enhance its missile capabilities from the PRC. In the 2000s, China is helping Pakistan to build a second major port city at Gwadar, further from the border with India and providing more strategic depth to its friend and ally.[57] Yet even as the PRC is providing Pakistan the defensive capabilities to deter or defend against Indian aggression, it is trying to restrain its ally from provoking conflict in the subcontinent. These Chinese efforts at restraining its ally were evident during the 1999 Kargil crisis, as well as the crisis of 2001-02.

As of mid 2007, it is totally improbable that the United States would invade China via North Korea. It is also most unlikely that India will attack Pakistan, and the trend in Sino-Indian relations suggests that tensions between these two countries are on the wane. Thus, the original rationales for the alliances are perhaps no longer valid. Still, they have not completely dissipated. China still wants North Korea as a buffer against the U.S. presence in Northeast Asia, and it wants Pakistan to continue to serve as a strategic counterweight to India. Moreover, Pakistan can serve as a means for the PRC to address one of its external security challenges, albeit indirectly: the new port at Gwadar might be one of the "string of pearls," serving as a basing facility for Chinese combat vessels in the PLAN's efforts to safeguard the SLOCs from the Persian Gulf.

Chinese Participation in UN Peacekeeping.

Because sending troops on UN peacekeeping missions is purely voluntary, China has no formal obligation or commitment to provide them. However, as one of the five Permanent Members of the UN Security Council, and as a rising great power, the PRC feels a sense of obligation to contribute to such endeavors as a way to demonstrate that it is a responsible member of the international community. It also demonstrates China's commitment to the UN as the proper venue for addressing international security issues, rather than allowing "certain countries" to take up that role unilaterally. China first contributed military observers to UN peacekeeping operations in 1990, and has contributed a total of approximately 6,800 personnel to 21 UN peacekeeping operations since then.[58] Chinese commitments jumped from 358 personnel as of December 31, 2003, to 896 as of June 30, 2004. As of December 31, 2006, China had 1,666 personnel involved in peacekeeping missions. Moreover, in mid-September 2006 the Chinese government pledged to increase its commitment to the UN peacekeeping mission in Lebanon (UNIFIL) by over 800 personnel to a total of 1,000.[59] Coming in the wake of Israel's incursion into southern Lebanon to fight Hezbollah, this Chinese offer was seen to demonstrate the PRC's desires to be a responsible international power; it also scored points for Beijing with Arab and Muslim countries. Despite this pledge, however, Chinese participation in UNIFIL stood at 343 troops as of February 28, 2007, far short of its September 2006 pledge.

If China fulfills its pledge to increase its contribution to UNIFIL (and does not reduce its contributions to other missions), just over 2,400 PLA personnel (troops, police, and military observers) will be involved in UN peacekeeping operations. While this is a new high for the PRC, it is still a small number of personnel compared to the total PLA strength of approximately 2.3 million troops. As earlier noted, since peacekeeping requires somewhat different skills than fighting wars, Beijing decided to create specialized units for peacekeeping operations (e.g., engineering units for de-mining operations), and it joined the UN Standby Arrangement System in 2002 with pledges of engineering, medical, and transport teams.[60] Beijing has also established two peacekeeping training facilities to enhance its participation in such operations: one at the PLA's International Relations Academy in Nanjing (which is related to China's defense intelligence and defense attaché programs) and the China Peacekeeping Police Training Center (which trains civilian police in Langfang, Hebei Province).[61] With dedicated units and related training facilities, the PRC's voluntary commitments to UN peacekeeping are already having some effect on the structure of the PLA, and will continue to do so as China's participation increases. Nevertheless, this effect will likely be quite minimal in the grand scheme of things.

Involvement in International Organizations and Strategic Partnerships.

China's involvement in international security organizations such as the SCO and the ARF, and in strategic partnerships (with countries such as Russia and Indonesia), does not and will not have much of an impact on the size and shape of the PLA. These relationships involve security cooperation, but as of yet do not entail military commitments that would require specific weapon systems, skills, or structures of the PLA (as we might find in an alliance). Rather, China's participation in these arrangements entails *political* commitments. Promises to cooperate on security issues for the most part mean holding dialogues and discussions to improve general relations, which include security relations. Security and strategic dialogues, joint military exercises, and even working together on the production or purchase of weapons might be part of the broad avenue of cooperation. Yet, in the case of some of China's strategic partners, they are not sure what exactly is involved in the relationship: for example, many Indonesians are still waiting to see specific content to the special relationship declared in the spring of 2005.[62]

As for the SCO and ARF, there is certainly content to security cooperation in these organizations. The ARF was inaugurated in 1994 as a means to promote dialogue and discussion on security issues in the Asia-Pacific region and to foster the development of confidence-building measures (CBMs) among the members of the dialogue with the hope of eventually achieving preventive diplomacy. Unofficially—from the perspective of the other members—it was created as a mechanism to acculturate a rising and potentially threatening China into the norms of international society in the region. Although the Chinese were initially hesitant about participation in the ARF, they have come to be quite active participants. As such, the Chinese have made commitments to the other members of the ARF, but these are commitments to engage in dialogue and to strive to build confidence in dealings with each other, in part through attendance at and hosting of the various intercessional workshops and seminars on topics ranging among CBMs, peacekeeping, search-

and-rescue operations at sea, training for disaster relief, and the law of armed conflict.

Similarly, the SCO involves commitments to cooperate in discussing regional security issues with an eye toward dampening potential problems. Originating in 1996 as the "Shanghai Five" and involving efforts by China, Russia, Kazakhstan, Kyrgyzstan, and Tajikistan to build confidence and reduce troop levels along their mutual frontiers, the SCO so-named was launched in June 2001 with the addition of Uzbekistan and with the additional goals of combating international terrorism and international drug trafficking, among other transnational issues. SCO members have drafted and signed the Shanghai Convention on Combating Terrorism, Separatism, and Extremism, and they have created the Regional Anti-Terrorism Structure (RATS) to assist their cooperative efforts to combat terrorism. The SCO has also hosted multilateral and bilateral antiterrorism exercises. Yet there is no formal commitment of troops under the SCO for specific common tasks, only a commitment to cooperate with other SCO members on the common security threats. Thus, China's participation in the SCO—like its participation in the ARF—should not have significant impact on the size, weapon systems, or structure of the PLA.

CONCLUSION

In examining China's security environment, it seems that threats to the PRC's security will have a stronger impact on the size, shape, and weaponry of the PLA as it modernizes than China's international security commitments and obligations will. To a large extent, this is because these commitments are essentially political in nature rather than military, and even when there is Chinese military participation, relatively small numbers of troops are involved. This is true even of Chinese participation in UN peacekeeping operations, where the nature of these activities has led to the creation of special training facilities as well as dedicated peacekeeping units, but with the numbers of personnel involved remaining small compared to the total number of troops in the PLA. Nor will the PRC's alliance commitments have a large effect: Beijing has done more to try to restrain than to support Pakistan and North Korea over the last few years, and China's current relatively benign relationships with the targets of both alliances (India and the United States) reduce the salience of the alliances.

While threats to China's security will have more impact on the shape and size of the PLA than will the PRC's international security obligations, there are few direct and immediate external threats confronting the PRC today. Rather, any such "threats" are more in the nature of potential security challenges. Nevertheless, these external issues interact with the more salient domestic threats—such as separatism—to become matters of concern to the Chinese civilian and military leaderships. In fact, the primary external security challenge (the United States) combines with the primary separatist threat (Taiwan) as the most powerful forces shaping in Beijing's thinking on how to modernize the PLA over the near term. The challenge from the United States also interacts with other external threats, such as maritime security and energy security, to compound the relevance of those challenges.

More than external security challenges, domestic security threats are the most pressing for Beijing, in particular the threats of separatism (especially Taiwan) and domestic turmoil. However, the threat of domestic instability is really a political issue better suited for the police (and the government's economic policies and legal system) than for the PLA—unless the Chinese people were to rise up in mass rebellion against the CCP. Even the separatist threat, in the cases of Xinjiang and Tibet, is more of a police issue than a military one, and the PLA already possesses the capabilities to address such challenges should it become involved. Thus, the separatist threat in these regions will not have a significant impact on the size and shape of the PLA. Taiwan, on the other hand, is affecting the shape of the modernization of the PLA and will continue to do so. Not only does Taiwan have a capable military of its own, but the separatist threat it poses links with the external challenges of the United States, Japan, and maritime vulnerability to drive the near-term modernization of the PLA. Because Beijing has not ruled out the use of force against Taiwan, it must be prepared to match or otherwise address the military capabilities of these hi-tech, "informationalized" potential adversaries.

While countering such external security challenges and the Taiwan threat requires military planning, Beijing has used astute diplomacy and economic/trade policies to mitigate the severity of these threats. Economic interdependence and common interests on a host of international issues have encouraged the PRC and the United States to develop a cooperative working relationship despite the potential for conflict over Taiwan. Beijing's conscious effort to reverse the escalating tensions with Japan during late 2006 is another case in point. And even with regard to Taiwan, Beijing seems more assured in mid 2007 than it has been in recent years, as evidenced by the more confident and less strident tone of the 2006 defense white paper compared to the previous two editions. Security challenges, China has discovered, can be managed through political and economic relationships and policies, not just military capabilities.

Finally, the PRC's conception of its threat environment is in a period of gradual expansion to embrace considerations beyond border and territorial defense. The rise of China as a global economic and trading power is prompting the Chinese leadership to perceive interests that are more far-flung than previously conceived. Protection of Chinese energy and other ocean-going cargo along the Southeast Asian SLOCs, even from ports as distant as the Persian Gulf, and protection of Chinese energy industry assets in Central Asia or farther afield (such as the Sudan)—are emerging economic and maritime imperatives shaping the longer-term vision for modernizing the PLA. Likewise, China's aspirations to be a great power may influence its perceived need to provide stability throughout the Asia-Pacific region. Such interests will require China to develop power-projection capabilities well beyond those the PLA currently possesses. Without any serious external threats driving the modernization of the PLA, these aspirations—along with the desire to develop the capabilities to address external security challenges and the Taiwan threat—will be key drivers for the continued modernization of the PLA.

The author thanks Luke Amerding and Hsieh Pei-Shiue for their research assistance.

ENDNOTES - CHAPTER 2

1. Admittedly, including Taiwan as a domestic security issue rather than as an international issue will be controversial. However, this chapter is focusing on Beijing's views of its threat environment, and since it officially sees the Taiwan issue as an internal issue, I will treat a PRC-Taiwan conflict as a domestic-level conflict.

2. Information Office of the State Council of the PRC, "China's National Defense in 2006," Beijing, December 2006, Chapter II.

3. For example, see Liu Xiang and Zhang Bihong, "Chinese Official Zhang Zhijun Expounds on China's Security Policy at Munich Conference on Security Policy on 5 February," *Xinhua*, February 5, 2006, via World News Connection (WNC).

4. Information Office of the State Council, "China's National Defense in 2004," Chapter I; *idem*, "China's National Defense in 2000," Beijing, October 2000, Chapter I; *idem*, "China's National Defense in 2002," Beijing, December 2002, Chapter I; Jiang Zemin, "Build a Well-Off Society in an All-Round Way and Create a New Situation in Building Socialism with Chinese Characteristics," Report to the 16th Congress of the Chinese Communist Party, November 11, 2002, available at *english.people.com.cn/200211/18/eng20021118_106983.shtml*.

5. Jiang, "Build a Well-Off Society," Part IX.

6. "China's National Defense in 2006," Chapter I.

7. "Commentary: China Strives to Prevent and Handle Mass Incidents," *Xinhua*, December 9, 2006, via LexisNexis.

8. David Lague, "China Airs Ambitions to Beef Up Naval Power," *International Herald Tribune*, December 28, 2006; Cao Zhi and Chen Wanjun, "Hu Jintao Emphasizes . . . a Powerful People's Navy that Meets the Demands of Our Army's Historic Mission," *Xinhua* Domestic Service, December 27, 2006, via WNC.

9. Editorial Department, "On Our Military's Historic Missions in the New Century, New Stage—Written on the 50th Anniversary of the Founding of '*Jiefangjun Bao*'," *Jiefangjun Bao* (Internet version), February 17, 2006, via WNC.

10. For example, see the interview with Rear Admiral Yang Yi in Tao Shelan, "Military Expert: In China's Peaceful Development It Is Necessary to Uphold the Dialectical Strategic Thinking of Making the Country Rich and Building Up Its Military Strength," *Zhongguo Xinwen She*, May 16, 2006, via WNC. Also, author interviews with civilian foreign policy and security analysts, Beijing, August 2006.

11. Information Office of the State Council, "China's National Defense in 2006," Chapter I.

12. The 2004 white paper notes that "tendencies of hegemonism and unilateralism have gained new ground, as struggles for

strategic points, strategic resources, and strategic dominance crop up from time to time." The 2006 white paper has similar language and also notes specifically that the United States is striving to "enhance its military capability in the Asia-Pacific region." Information Office of the State Council, "China's National Defense in 2004," Chapter I; and *idem*, "China's National Defense in 2006," Chapter I.

13. On the U.S.-Japan relationship serving to help contain the PRC, see, for example, Xu Feng, "US Factor in Japan Becoming a Military Power," *Liaowang*, August 13, 2005, via WNC; and Editorial Department, "Sino-Japanese Relations Are Facing New Test," *Ta Kung Pao*, August 16, 2005, via WNC. On the concerns with the U.S. military presence in Central Asia, see, for example, Zhou Liang, "Demand for Troop Pullout from Central Asia Makes United States Tense with a Great Sense of Urgency," *Liaowang*, August 12, 2005, via WNC.

14. Author interviews with foreign policy analysts, Beijing, April 2004. See also Robert Sutter, "Asia in the Balance: America and China's 'Peaceful Rise'," *Current History*, September 2004, p. 287.

15. See Tao Shelan, "Military Expert: In China's Peaceful Development, It Is Necessary to Uphold the Dialectical Strategic Thinking of Making the Country Rich and Building Up Its Military Strength," *Zhongguo Xinwen She*, May 16, 2006, via WNC.

16. David E. Sanger, "U.S. Would Defend Taiwan, Bush Says," *New York Times*, April 26, 2001, p. A1.

17. Thomas J. Christensen, "Posing Problems Without Catching Up," *International Security*, Vol. 25, No. 4, Spring 2001. See also Jason Bruzdzinski, "Demystifying *Shashoujian*: China's 'Assassin's Mace' Concept," in Andrew Scobell and Larry Wortzel, eds., *Civil-Military Change in China: Elites, Institutes, and Ideas after the 16th Party Congress*, Carlisle, PA: Strategic Studies Institute, U.S. Army War College, 2004.

18. See Pallavi Aiyar, "China, India: No Ground Given in Border Talks," *Asia Times On-line*, June 27, 2006.

19. "Boundary Issue Stops India-China Ties Attaining Full Potential," *Press Trust of India*, December 17, 2006.

20. Beijing issued a map in the early 1990s that appeared to include these waters off Natuna, but as the Chinese claim was

made with a dashed line, the precise location of the claim was unclear. Indonesian requests for clarification have never been answered.

21. T. J. Burgonio, "Accord on Spratlys Launches GMA's China Visit," *Philippine Daily Inquirer*, September 2, 2004; and "Philippines, China, Vietnam to Conduct Joint Marine Seismic Research in South China Sea," *Xinhua News Agency*, March 14, 2005.

22. "Production Begins at Chunxiao Gas Field," *China Daily*, August 8, 2006.

23. Mark J. Valencia, "The East China Sea Dispute; Prognosis and Ways Forward," PacNet No. 47A, Pacific Forum CSIS, September 15, 2006, p. 1.

24. For example, see Xu Feng, "US Factor in Japan Becoming a Military Power"; Editorial Department, "Sino-Japanese Relations Are Facing New Test"; "Japan's War Capability Exceeds Defense Need," *People's Daily Online*, August 5, 2005. See also Liu Tianchun, "Luelun Riben Zhengduo Nengyuan zhi Zhan" ("A Brief Study of Japan's Fight for Energy"), *Zhongguo Shehui Kexue Yuan Yanjiushengyuan Xuebao* (*Journal of the Graduate School of the Chinese Academy of Social Sciences*), 2004, No. 6, pp. 44-52.

25. Author interviews, Beijing, August 2006.

26. Wendell Minnick, "China to Buy Armed Hovercraft," *Defense News* (online), posted September 11, 2006.

27. Tao Shelan, "Military Expert: In China's Peaceful Development..."; "Military Expert Proposes Readjusting China's Security Strategy," *Zhongguo Tongxuan She*, December 22, 2005, via WNC; and Wu Pin, "PLA General Deems Maritime Security as an Outstanding Issue," *Ta Kung Pao* (Internet edition), September 15, 2006, via WNC. For calls to beef up the PLAN, see also Guo Xinning, "Lun Zhongguo Goujia Anquan Zhanlue Fangzhen," ("On the Guiding Principle of China's National Security Stategy"), *Waijiao Pinglun* (*Foreign Affairs Review*), No. 88, April 2006, pp. 31-36; and Fang Kun, "Zhanlue Diyuan yu Zhongguo Haijun Jianshe" ("Stratgeic Geography and the Development of China's Navy"), *Shijie Jingji yu Zhengzhi* (*World Economics and Politics*), 2004, No. 8, pp. 43-48.

28. "China's National Defense in 2004"; and Cao Zhi and Chen Wanjun, "Hu Jintao Emphasizes . . .," *Xinhua* Domestic Service, December 27, 2006, via WNC.

29. "China's National Defense in 2006," Chapter II.

30. Wen Han, "Hu Jintao Urges Breakthrough in 'Malacca Dilemma'," *Wen Wei Po* (Internet version), January 14, 2004.

31. Author interviews, Beijing, August 2006. It should be noted that these Chinese analysts were reluctant to express worry about the United States attempting to block the Strait in the event of a Sino-American conflict, arguing that this would happen only in a worst-case scenario.

32. On the "string of pearls," see Booz Allen Hamilton, *Energy Futures in Asia: Final Report*, prepared for the Director of Net Assessment, Office of the Secretary of Defense, Washington, DC, November 2004, pp. 51-53. For discussion of this report and the strategy, see Bill Gertz, "China Builds Up Strategic Sea Lanes," *Washington Times*, January 18, 2005; and David Zweig and Bi Jianhai, "China's Global Hunt for Energy," *Foreign Affairs*, Vol. 84, No. 5, September-October 2005, pp. 34-35.

33. David E. Sanger, "China's Oil Needs Are High on U.S. Agenda," *New York Times*, April 19, 2006, p. A1.

34. "CNPC to Issue $400 Mln Grant to Build ESPO Pipeline Branch to China," *Interfax*, March 22, 2006.

35. "Kazakhstan-China Oil Pipeline Opens to Commercial Operation," *Xinhua*, July 11, 2006; Stephen Blank, "Turkmenistan Completes China's Triple Play in Energy," *China Brief*, Vol. VI, issue 10, May 10, 2006, pp. 6-8.

36. On the Sino-Kazakh joint anti-terrorism exercise, see Xu Jingyue and Wei Wu, "China Border Official Hails Standard of Sino-Kazakh Antiterrorism Drill," *Xinhua Domestic Service*, August 28, 2006, via WNC. On the Sino-Tajik exercise, see Tan Jie and Li Fengming, "China-Tajikistan Joint Anti-Terrorism Military Exercise Concludes Satisfactorily," *Jiefangjun Bao* (Internet version), September 26, 2006, via WNC.

37. "China Puts Soldiers on North Korea Border," Associated Press, via *New York Times* online updates, September 15, 2003.

38. Author interviews, Beijing, August 2006; and "China Allegedly Deploys More Troops along North Korean Border—HK Paper," BBC Monitoring, July 27, 2006, via LexisNexis.

39. Michael Elliott, "The Chinese Century," *Time*, January 22, 2007, p. 37.

40. Office of the Secretary of Defense, "Annual Report to Congress: Military Power of the People's Republic of China, 2006," Washington, DC: Department of Defense, July 2006, pp. 29-30.

41. "Commentary: China Strives to Prevent and Handle Mass Incidents," *Xinhua*, December 9, 2006.

42. For example, see Thomas J. Christensen, "China," in Richard J. Ellings and Aaron L. Friedberg, eds., *Strategic Asia 2001-02: Power and Purpose*, Seattle: National Bureau of Asian Research, 2001, pp. 27-28 and 30-32.

43. See, for example, Information Office of the State Council, "China's National Defense in 2004," Chap. I; and the SCO's Shanghai Convention on Combating Terrorism, Separatism, and Extremism, available at *www.setsco.org*.

44. For example, in January 2007, Chinese police raided an alleged terrorist training camp in western Xinjiang run by ETIM. See Maureen Fan, "Raid by Chinese Kills 18 at Alleged Terror Camp," *Washington Post*, January 9, 2007, p. A12.

45. Irene Wang, "Incidents of Social Unrest Hit 87,000," *South China Morning Post*, January 20, 2006, p. 4.

46. Jiang, "Build a Well-Off Society."

47. *Ibid*.

48. Wen Jiabao, "Full Text: Report of the Work of the Government," *Xinhua*, March 14, 2006, via WNC.

49. Joseph Kahn, "China's Leader Pushes Doctrine While Warning of Corruption," *New York Times*, July 1, 2006, p. A4.

50. "Commentary: China Strives to Prevent and Handle Mass Incidents."

51. David Lague, "China Plans to Spend $21 Billion on Its West," *International Herald Tribune*, July 4, 2006, p. 12.

52. For a discussion of ententes, see Michael R. Chambers, "Dealing with a Truculent Ally: A Comparative Perspective on China's Handling of North Korea," *Journal of East Asian Studies*, Vol. 5, No. 1, pp. 35-75.

53. Sino-Korean Treaty of Friendship, Cooperation, and Mutual Assistance, published in *Peking Review*, Vol. 4, No. 28, July 14, 1961, p. 5.

54. Andrew Scobell, *China and North Korea: From Comrades-in-Arms to Allies at Arm's Length*, Carlisle, PA: Strategic Studies Institute, U.S. Army War College, March 2004, p. 19.

55. Chambers, "Dealing with a Truculent Ally," pp. 50-51.

56. Despite the PRC assertion in September 2006 that it had no plans to revise the alliance treaty with North Korea, China's compliance with the UN sanctions suggests true Chinese feelings toward the DPRK. See "China Says No Plan to Revise Pact Committing Military Aid to N. Korea," *Japan Economic Newswire*, September 14, 2006, via LexisNexis.

57. For a discussion of the Gwadar project, see John Garver, "The Future of the Sino-Pakistani *Entente Cordiale*," in Michael R. Chambers, ed., *South Asia in 2020: Future Strategic Balances and Alliances*, Carlisle, PA: Strategic Studies Institute, U.S. Army War College, November 2002, pp. 414-18.

58. "China's National Defense in 2006," Appendix V.

59. Details on Chinese contributions to peacekeeping missions are available at the UN Department of Peacekeeping Operations website, *www.un.org/Depts/dpko/dpko/*. For the Chinese commitment to boost its contributions to UNIFIL, see David Lague, "Chinese Military on the Global Stage: Beijing Finds Benefits in Peacekeeping Role," *International Herald Tribune*, September 20, 2006, p. 1.

60. Pang Zhongying, "China's Changing Attitude to UN Peacekeeping," *International Peacekeeping*, Vol. 12, No. 1, Spring 2005, pp. 91-92.

61. See Bates Gill and James Reilly, "Sovereignty, Intervention, and Peacekeeping: The View from Beijing," *Survival*, Vol. 42, No. 3, Autumn 2000, pp. 51-52; and "Annan Tours China's New Peacekeeping Training Centre," UN News Services, October 12, 2004, available at *www.un.org*.

62. Author interviews with Indonesian foreign policy analysts, Jakarta, August 2006.

CHAPTER 3

CHINA'S NATIONAL MILITARY STRATEGY: AN OVERVIEW OF THE "MILITARY STRATEGIC GUIDELINES"

David M. Finkelstein

INTRODUCTION AND BACKGROUND

A Decade of Impressive Change.

In the immediate wake of Tiananmen in 1989 and for at least 2 years thereafter, the Chinese People's Liberation Army (PLA) was deeply engaged, for obvious reasons, in an intense period of political work. The seemingly singular focus of the Chinese armed forces on strengthening Party-Army ties and on ideology was cause for foreign students of Chinese military affairs to wonder whether the first decade of the 1990s, like the 10 years during the Great Proletarian Cultural Revolution, would turn out to be another "lost decade" from the standpoint of military modernization and enhanced professionalism. Clearly, this turned out *not* to be the case. If anything, the decade of the 1990s should be viewed as a period during which the PLA made tremendous strides as a professional military force.

At mid 2007, the PLA is over 3½ years beyond a seminal decade of focused and sustained efforts to modernize. For more than a decade, the armed forces of China have been undergoing transformative adjustments of such a profound nature relative to their past that one group of Chinese military authors considers this ongoing period of reform to constitute the PLA's "Third Modernization."[1]

Since the end of Beijing's ninth Five Year Plan in 2000, many of the outputs and "deliverables" of this remarkable period of change have become evident to foreign observers of Chinese military affairs. Ardent analysts, scholars, and other observers of the PLA are familiar with the long list of changes that have unfolded, and it is not the purpose of this chapter to provide a comprehensive accounting of them. At the same time, for those less familiar with what has transpired, a brief overview may usefully provide some appreciation of the scope and scale of the changes undergone or underway.[2]

For the sake of brevity, almost all the reforms or modernization efforts the PLA has engaged in over the past 13 years can be treated under at least one of what I refer to as "The Three Pillars" of PLA reform and modernization. They are:

Pillar 1: The development, procurement, acquisition, and fielding of new weapons systems, technologies, and combat capabilities. Under this pillar, one would cite:

- End item purchases from Russia such as SU-27 and SU-30 aircraft, *Kilo* Class submarines, *Sovremenny* destroyers, and precision-guided munitions (PGMs);
- Indigenously produced conventional weapons systems such as Chinese-made submarines and surface vessels, armor, and communications equipment;
- Production of conventional missiles and upgrading the quality and survivability of China's nuclear arsenal; and

- Basic research and development in which the PLA is engaged domestically to produce information-age military technologies, to include the creation of a fourth general department (the General Equipment Department, 1998) in yet another rectification of the military research and development (R&D) establishment.

Pillar 2: The vast array of institutional and systemic reforms. These include critical changes to the PLA's corporate culture that are focused on raising the levels of professionalism of the officer corps and enlisted force (especially NCOs) and making them more adept at employing and maintaining new battlefield technologies. This pillar also encompasses the myriad organizational changes aimed at optimizing the force, many of which came into effect in the mid-to-late 1990s. Under this pillar, one could list:

- Major changes to the officer professional military education system;
- The creation—for the first time—of a corps of professional NCOs;
- More stringent requirements for officer commissioning, the diversification of the sources of commissioning, and the standardization of criteria for promotion; and
- Force structure adjustments that include a significant new emphasis on the Navy, Air Force, and strategic rocket forces, the downsizing of staffs, the consolidation of ground force units at the division and brigade levels, and new battlefield logistics paradigms.

Pillar 3: The development of new warfighting doctrines for the employment of these new capabilities.[3] In 1999 the PLA revised its operational-level doctrine from its previous emphasis on ground force-centric *combined arms operations* to one emphasizing *joint operations* in the aerospace, maritime, and electromagnetic battle space dimensions. This new operational doctrine is aimed at shifting the PLA:

- From a focus on operational planning to prosecute protracted wars on the mainland to short-duration high-intensity joint campaigns off China's littoral;
- From focusing on an enemy's weakest forces to attacking and destroying the enemy's most vital assets;
- From the concept of mass to the concept of concentration of firepower; and
- From static defenses to mobile offenses.

In short, the attention of the PLA is now doctrinally fixed on being able to prosecute short campaigns inflicting shock and paralysis (vice long wars of attrition) to level the technological playing field at the inception of hostilities by concentrating PLA's best capabilities against the enemy's most important assets.

Taken in their totality, the programs instituted by the PLA to date constitute a set of significant strides in modernization and reform—efforts that will enable the PLA to become over time a more capable force in an operational sense and a more professional one in an institutional sense. None of this happened overnight. It is the result of a series of carefully made decisions, sustained focus, increased levels of funding, prioritization, and incrementally implemented changes and adjustments over time.

Many of the most important changes under Pillars 2 and 3, and a good number of the major deliverables under Pillar 1 (though clearly not all), are well-known and in the public domain. These programs are universally acknowledged and understood among those who regularly follow the Chinese armed forces. There is, however, much less certainty, less agreement, and much room for discussion as to the larger rationale for what is unfolding year by year.

What is the PLA trying to achieve and, more importantly, why is it trying to achieve it? What calculations, assumptions, and assessments are driving Beijing to enact these changes in its military forces? What objectives does the leadership of the PLA seek to achieve? It is not just the cadre of international scholars and specialists of the PLA at universities and research institutes who are discussing these questions. Such questions continue to be asked in various quarters of the U.S. Government, especially within the Department of Defense (DoD). The answers being reached, and especially the assumptions underlying those answers, are laden with potentially profound foreign policy and force structure implications for the United States — witness the attention given China in the Pentagon's most recent *Quadrennial Defense Review*.

These fundamental questions about the rationale and objectives of PLA modernization were included in the important policy front piece to the Pentagon's *Annual Report to Congress: Military Power of the People's Republic of China* in 2006, as quoted here:

> China's leaders have yet to adequately explain the purposes or desired end-states of their military expansion . . . this lack of transparency prompts others to ask, as Secretary of Defense Rumsfeld did in June 2005: Why this growing investment? Why these continuing large and expanding

arms purchases? Why these continuing robust deployments?[4]

However, while the 2006 DoD report (and its predecessors) provides significant details about the manifestations of PLA modernization, it leaves the larger questions open-ended, putting the burden on China to answer them. At the 2005 and 2006 Shangri-La (International Institute for Strategic Studies [IISS]) conferences in Singapore, then U.S. Secretary of Defense Donald Rumsfeld posed these larger-order questions about the aims of PLA modernization in his addresses, as alluded to in the quotation above.

In essence, these larger-order questions are inquiring whether Beijing has a particular national military strategy that is guiding the modernization and transformation of the PLA; and if it does have such a strategy, what is it?

This chapter, not surprisingly, will argue that China does in fact have the equivalent of a national military strategy, the elements of which are contained in the "Military Strategic Guidelines" issued to the PLA.

Revisiting China's National Military Strategy.

The first efforts. Attempting to understand the elements of China's national military strategy is a challenge with which this author has grappled in the past. In 1998, for a conference jointly hosted by the RAND Corporation and the Council of Advanced Policy Studies, I prepared a paper titled "China's National Military Strategy."[5] A little more than a year later (2000), based on new developments in PLA programs, I published a revised and updated version of that paper as a CNA Corporation monograph.[6]

Almost 9 years have passed since that first paper was written. To some degree, given what was understood at the time, the first effort has stood up reasonably well. A strategic level context was provided for what was beginning to unfold in PLA modernization. The "Two Transformations" program was identified and explained (possibly for the first time in Western writings about the PLA). Differentiations were made between the PLA's strategic objectives and its implementing programs, and a linkage was established between nondefense PRC national objectives and national-level military objectives. Moreover, the paper and its sequel adjudged that significant institutional changes and adjustments could be expected in the future, so that a modicum of predictive power inhabited those papers that has stood the test of time.

Yet, in retrospect, there are some issues in those first papers that were either incorrectly drawn or incompletely understood. Among these shortcomings was the discussion of the active defense (*jiji fangyu*) in terms of an "operational concept" or a "doctrine" (more on this to follow). In the category of omissions, a very significant term that came up in some of the data reviewed at the time was glossed over, its significance simply not being fully appreciated—to wit, the phrase "military strategic guideline."

Finally, the approach used to explain China's national military strategy in those previous papers can be seen today as inadequate, although at the time it made sense. In those first papers, a Western analytic framework was employed as a way to lend structure and rigor to my speculative efforts to visualize what China's actual national military strategy *might* look like were it published in the public domain. The approach, here quoted from the original paper, was as follows:

> This paper . . . offers a notional, annotated national military strategy for China. It is notional because the PRC has not published a detailed national military strategy. It will ask and answer the question: "If China, like the United States, published a national military strategy, what would it look like and what would it say?"
>
> The U.S. Army War College (AWC) model of military strategy serves as the superstructure of the following analysis of the PLA. While the PLA would certainly *not* use an American construct to articulate its national military strategy, this model is nevertheless a useful tool for the descriptive and analytic purposes of this paper. Useful frames of reference from the planners on the Joint Staff who produce the Pentagon's national military strategy as well as several universal military concepts such as "center of gravity" and other terms are also used. Into these "frames" we shall place Chinese "lenses" to articulate a vision of the bigger picture.

In other words, in order to derive the PRC's national military strategy in the absence of an officially published document, the outputs and programs associated with PLA modernization that were then knowable were identified and analyzed. They were then placed into a non-Chinese framework to explain the larger whole. What was presented, therefore, was a national military strategy for the PRC that was inferential and deduced. By and large, this is the methodology that is still used by DoD in its annual report to Congress.

New possibilities. In the intervening years since those first papers were written, China has *not* all of a sudden published an official "National Military Strategy" under such a title. And China will not do so simply because this would impose upon the PLA the expectation that it would adopt an American approach to articulating its plans for military modernization, impose U.S. terminology upon the Chinese defense establishment (to include the title "National Military Strategy"), and expound upon a rationale for military modernization that would make sense to foreign readers as opposed to making sense to the PLA itself.

So what has changed over the course of the intervening years to permit a new approach to this topic? The short answer is, quite a bit.

(1) More Data. Throughout the course of the 1990s—especially during the period of the ninth Five Year Plan (1996-2000) and since then—the PLA's military modernization and reform efforts have been accompanied by a veritable explosion of published commentary from credible Chinese military authorities. These publications have been necessary to stimulate discussion among PLA strategists and planners responsible for the reform effort and *equally necessary* to inform the greater PLA, and in some cases other non-PLA PRC government officials, about defense modernization plans, programs, and rationales. Moreover, this data is now increasingly accessible to foreigners.

(2) More Context About the Data. Not only is there more data available to draw upon, but the degree of sophistication that foreign consumers of this literature have developed in carefully vetting and evaluating this data has increased as well. In addition, many in the field of PLA research have been involved in an explication of Chinese military terminology. There is now a much better grasp of the professional lexicon by which the PLA speaks to itself (and others in China) about military issues.

(3) Five Successive PRC Defense White Papers. Since 1998 Beijing has published five official white papers devoted to military and security affairs—*China's National Defense*, in 1998, 2000, 2002, 2004, and 2006. None of these white papers by themselves tells the entire story of the national military strategy under

which the PLA is currently operating. Taken together, however, they provide more official insights into, and explanations of, that military strategy than one might expect or that some give the Chinese credit for.

(4) The "Canonization" of Jiang Zemin's "Military Thought." Finally, circa August 2003, Jiang Zemin's speeches, directives, and guidance to the PLA during his tenure as Chairman of the Central Military Commission (CMC) were elevated to the level of enduring "thought" (*sixiang*) in the hierarchy of CCP political theory. The PLA now adds "Jiang Zemin Thought on National Defense and Army Building" to the canon of "Mao Zedong Military Thought" and "Deng Xiaoping Thought on Army Building in the New Period." The elevation of Jiang's role in military modernization resulted, predictably, in a wave of publishing. And since the national military strategy under which the PLA is still operating was issued on Jiang's watch, these publications, many of which are publicly available, provide new insights. One important example would be the three-volume *Selected Works of Jiang Zemin* (2006).[7]

The approach in this chapter. Given the new sources of data and the new contexts for interpreting this data, it is now possible to offer a preliminary exposition on the essence of China's equivalent of a national military strategy. Drawing upon a body of Chinese literature, both primary source and secondary source, we are now in a position to: (1) identify and explain the most important Chinese terms, constructs, elements, and components that comprise the equivalent of Beijing's national military strategy as a generic framework employed by PLA strategic planners; and (2) discuss the key features and basic content of the current national military strategy, its evolution, and the drivers and assessments associated with it. In short, this chapter discusses China's national military strategy in terms the Chinese use to discuss it.

REVIEWING THE BIDDING: WHAT *ARE* "NATIONAL MILITARY STRATEGIES"?

While there is no need to go into great detail on the question of what, generically, constitutes a national military strategy, a quick review is appropriate in order to set the stage. As a general proposition, national military strategies around the world share certain common features. At the most basic levels, there are six.

- First, of course, is that they represent authoritative guidance. Whether this guidance is publicly released or classified varies from nation to nation.[8] Either way, national military strategies represent a set of policy decisions that set the azimuth for actual planning or action.

- Second, national military strategies usually provide direction on how the military element of national power should support larger national objectives and, in varying degrees of detail, how it will do so.

- Third, the elements that comprise national military strategies are usually articulated in broad directives and large concepts, not minute detail. These broad concepts eventually lead to more detailed planning and programs to implement the larger concepts.

- Fourth, national military strategies usually serve as planning guidelines. These planning guidelines are often associated with specific time frames and provide guidance across a spectrum of near-term, mid-term, or long-term horizons.

- Fifth, while different nations organize the components of their national military strategies in different ways and use different terminology to identify those components, all usually address the issues of "ends, ways, and means." As explained so clearly by Harry R. Yarger, the "ends" represent "what" must be accomplished (objectives). The "ways" (strategic concepts and courses of action) talk to how the objectives will be accomplished. The "means" speak to resources, that is, which resources will be used, or what resources need to be developed. These "means" are not focused just on weapons, technologies, or other "hard" capabilities. They also include "intangible resources" such as "will," "courage," or "intellect" as well as organizational changes, the need for new types of operational units, new doctrines, and other institutional infrastructures or professional paradigms.[9]

- Sixth, national military strategies are usually informed by, and oftentimes articulate, the analytic assessments driving the rationale for the strategy. These analyses can include threat perceptions or likely contingencies, capabilities-based assessments, larger-order strategic assessments, or domestic factors driving or constraining the strategy.

A classic example of the basics of a national military strategy is the 1992 *National Military Strategy of the United States* — *the first time* Washington ever produced a public domain document on this issue. A mere twenty-seven pages long, the 1992 "NMS" addressed four major issues: (1) a review of the security environment, national interests, and national objectives, (2) broad strategic principles and operational guidance for the U.S. armed forces to follow or be able to achieve ("strategic deterrence and defense," "forward presence," "crisis response," etc.), (3) operational planning requirements and deployment postures based on possible contingencies or threats, and (4) the requirements for current and future force structures, capabilities, and supporting infrastructures.

"MILITARY STRATEGIC GUIDELINES" — CHINA'S NATIONAL MILITARY STRATEGY

What Are "Strategic Guidelines"?

As a general practice in China, "strategic guidelines" (*zhanlüe fangzhen*) serve as one type of policy format among many that can be used to articulate national-level directives, policies, or principles that guide action. "Strategic guidelines" delineate the Party-State's fundamental decisions or approaches on major domestic issues, foreign policy issues, or security issues. There can be strategic guidelines for foreign policy, strategic guidelines for economic policy, and strategic guidelines for the reform of this or that sector, etc.[10] While it is unclear as to where, exactly, strategic guidelines fit in the hierarchy of authority as applied to PRC policy pronouncements, it is clear that they are high-level dictates that in terms of supremacy and rank reside generally at or around the strata of the "general line" (*zong luxian*) and "general policies" (*zong zhengce*).[11]

Usually, strategic guidelines are issued by the top

leader of the Party-State on behalf of the collective leadership or on behalf of the Party-State organ responsible for the portfolio under consideration. Strategic guidelines, therefore, provide authoritative direction for action to the members of the particular policy community (*xitong*) to which they are issued, and it is these members who subsequently flesh out the details.

Military Strategic Guidelines.

The highest level of national guidance and direction to the armed forces of China is known as the "Military Strategic Guidelines" (*junshi zhanlüe fangzhen*).

The Military Strategic Guidelines issued to the PLA, and the detailed plans and programs that are subsequently developed to implement them, constitute China's national military strategy. As explained by the Army Building Research Department of the PLA's National Defense University (NDU), "The military strategic guidelines are the fundamental military policies (*junshi zhengce*) of the party and the nation. They are the overall principles (*zong yuanze*) and guiding principles (*zong gangling*) for planning and guiding the development and utilization of the armed forces."[12]

Military Strategic Guidelines are issued to the PLA under the authority of the Military Commission of the Central Committee of the Chinese Communist Party, usually abbreviated CMC.

The issuing of a new set of Military Strategic Guidelines is a significant event. According to a senior PLA strategist, prior to 1993, this has happened only four times since the founding of the PRC in 1949.[13] New guidelines are usually issued under the name of the Chairman of the CMC, who has often also served concurrently as the CCP secretary general and state president. Historically and systemically, significant changes to the Military Strategic Guidelines are initiated by a major speech delivered by the CMC Chairman to the leadership of the PLA at an expanded meeting of the CMC.[14] These expanded meetings (*kuoda huiyi*) include not only the sitting members of the CMC but can include as many as several hundred key PLA leaders from the four general departments, the military regions, the services, and other national-level organizations such as the AMS and the NDU.[15]

These speeches contain a mere handful of core concepts, subsequently considered by the PLA as "strategic guiding thoughts" or "strategic guiding ideology" (*zhanlüe zhidao sixiang*) that serve as the basis for the leadership and planners in the Chinese armed forces to take for implementation: planning, programming, adjustments, acquisitions, resource allocations, priorities, etc. At their heart and at a minimum, the "strategic guiding thoughts" underlying the Military Strategic Guidelines for any particular period provide official judgments that include:

- The ideological and political basis for the Military Strategic Guidelines;
- An assessment of the international environment and its impact on China's security;
- China's overall national security objectives, its domestic objectives, and the relationship of military objectives to other national objectives;
- The most likely type of conflict for which the PLA must prepare (either a capabilities-based assessment, a contingency-based assessment, or both); and

- Broad guidance to the PLA on how it will prepare, reform, or adjust to meet the challenges of the new situation.

All the bullets above except the first are also part of the "strategic assessment" (*zhanlüe panduan*) that informs the Military Strategic Guidelines.

To recapitulate, the "strategic guiding thought" that is the "core" of the Military Strategic Guidelines forms the basis for fleshing out the details of the "ends, ways, and means" of the PRC's national military strategy, and major changes to it can have far-reaching programmatic and operational planning implications for the PLA. For example, and to be quite brief, in the mid-1980s Deng Xiaoping made radical changes to the "strategic guiding thought" underlying the Military Strategic Guidelines inherited from Mao Zedong. Where Mao had characterized the international situation as one of "war and revolution," Deng saw "peace and development" as the "keynote of the times." Where Mao kept the PLA on a wartime footing and directed the PLA to prepare to fight "early wars, major wars, and nuclear wars" that assumed an invasion of the mainland, Deng directed the PLA to enter a period of "peacetime army building" but also to develop the capabilities for dealing with local wars and "incidents" that might erupt on China's periphery.[16]

Clearly, the preceding example does not do justice to the complex analyses of either Mao or Deng, nor to the myriad resulting adjustments the PLA made to China's national military strategy as a result of these differing directives. The point is to provide an example of the type of large-order assessments or directives that are encapsulated in the term "strategic guiding thought" —

which becomes bureaucratic/ideological shorthand for PLA planners to use in developing detailed programs under the Military Strategic Guidelines.

Finally, it is important to point out that major adjustments to the Military Strategic Guidelines must also address (either to revalidate or change) China's domestic objectives and China's larger national security objectives. This is because China does *not* formally generate the equivalent of a "National Security Strategy" (as does the United States) that PLA planners can use as the basis for their subsequent preparations, or which they can use as the supra-institutional basis for coordinating their military plans and programs with larger national objectives and other relevant nondefense organizations.[17]

Key Components of the Military Strategic Guidelines.

When new Military Strategic Guidelines are promulgated, there are several key "strategic issues" (*zhanlüe wenti*), or questions, that must be addressed. Some of these issues are addressed outright in the initial promulgation of new Military Strategic Guidelines; others can only be dealt with or resolved over time and with subsequent military staff work, planning, and resource allocation. Below is a list of the key strategic issues that the Military Strategic Guidelines address as suggested by the body of literature canvassed for this chapter. There may be other factors or strategic issues contained within the Military Strategic Guidelines that are not apparent in the data at hand, and it is not certain at this point that the issues identified below are being presented in the correct sequence. At bottom, however, these issues address two fundamental questions: "What kinds of conflicts must the PLA be prepared to fight?" and "How should PLA modernization and reform programs adjust to comport with the new operational imperatives?" The key concerns or strategic issues are:

- Presenting the Strategic Assessment (*zhanlüe panduan*);
- Adjusting the Content (*nei rong*) of the Active Defense Strategy (*jiji fangyu zhanlüe*);
- Articulating the Strategic Missions (*zhanlüe renwu*) and Strategic Objectives (*zhanlüe mubiao*) of the armed forces;
- Issuing guidance for Military Combat Preparations (*junshi douzheng zhunbei*);
- Identifying the Main Strategic Direction (*zhuyao zhanlüe fangxiang*); and
- Determining the Focus for Army Building (*jundui jianshe*).

The Strategic Assessment.

The "strategic assessment" (*zhanlüe panduan*) is both a political assessment and a military assessment. Politically, it usually provides judgments on the state of international relations and the global order, identifies trends in relations among the major powers or other major groups of nations, and judges China's situation in the global order and that order's impact on larger PRC national objectives—especially domestic objectives. This judgment also assesses the prospects for global security affairs and how they impact China. Is the world at peace or at war? What are the underlying causes of the conflicts that are occurring? How do they affect

China? It was within the context of these larger-order political and strategic assessments that Deng Xiaoping expressed the judgments in the mid-1980s that "peace and development" were the main trend of the times, that a world war between the two superpowers (the United States and Union of Soviet Socialist Republics [USSR]) was not imminent, and that China could look to a period of peace in order to focus on domestic development. From a military standpoint, the strategic assessment issues judgments about the nature of contemporary warfare. It addresses the general forms or types of wars being fought (*zhanzheng xingtai*) and the more specific forms of combat operations (*junshi douzheng xingshi*) by which they are characterized. These judgments are clearly articulated when Military Strategic Guidelines are issued.

Of great import, then, changes to the Military Strategic Guidelines—either adjustments to the guidelines or the issuing of a new set of guidelines—are usually the result of a major change in one or more of the issues addressed in the strategic assessment. To recapitulate, these are (1) changes in the international order; (2) changes in the security environment and China's security situation; (3) changes in China's domestic situation; and (4) changes in the nature of warfare itself.

Adjustments to the Active Defense Strategy.

The relationship between the "Active Defense Strategy" and the "Military Strategic Guidelines" is so intimate—indeed, the two are nearly indistinguishable in the minds of the PLA—that one often encounters the phrase "The Military Strategic Guidelines of the Active Defense." But speaking of the two as being identical is technically incorrect. What, therefore, *is* the "Active Defense" and what is its relationship to the Military Strategic Guidelines?

The "Active Defense" (*jiji fanyu*) or "Active Defense Military Strategy" (*jiji fangyu junshi zhanlüe*) establishes a set of broad strategic concepts and principles, and a set of very general operational concepts, for prosecuting war at the strategic level of conflict. The term itself originates from "Mao Zedong Military Thought" (*Mao Zedong Junshi Sixiang*), specifically Mao's 1936 essay, "The Problems of Strategy in China's Revolutionary War," in which he expounded the concept of fighting a strategically defensive war while at the same time engaging in offensive operations at the campaign and tactical levels of war."[18] It was not until 1956, however, that the "Active Defense Strategy" and the "Military Strategic Guidelines" became intertwined. In that year, Marshall Peng Dehuai linked the two at an expanded meeting of the CMC by adding a political component to the active defense strategy (i.e., strategically, China does not attack until it is attacked), and declaring that China would adopt the "military strategic guideline of the active defense."[19]

Over time, the higher-order strategic-level principles informing the "Active Defense" strategy have remained relatively constant. Briefly, as the PLA would address them, they would include the following tenets:

- Overall, our military strategy is defensive. We attack only after being attacked. *But our operations are offensive.*
- Our counteroffensive will not be limited by space or time.
- We will not put boundaries on the limits of our offensives.

- We will wait for the time and conditions that favor our forces when we do initiate offensive operations.
- We will focus on the opposing force's weaknesses.
- We will use our own forces to eliminate the enemy's forces.
- Offensive operations against the enemy and defensive operations for our own force protection will be conducted simultaneously.
- We will maximize our advantages against the opposing forces.[20]

Obviously, the levels of generality inherent in the tenets bulleted above are not conducive to operational planning; the development of capabilities and doctrines; training; resource allocations; the generation of priorities; or deployment decisions under *specific* scenarios or conditions. Consequently, the basics of the "Active Defense Strategy" as shown are at bottom a framework that must be filled in with details in order for it to become an implementable strategy. True, the major decisions, assessments, judgments, determinations, and policies encompassed in the other components of the Military Strategic Guidelines *inform* the development of the specifics of the "Active Defense Strategy" during any given period of time. Hence, the interconnectedness between the two. In other words, without the major components of the Military Strategic Guidelines to flesh it out, the "Active Defense Strategy" is a near-empty construct. Without the need to flesh out the "Active Defense Strategy," the Military Strategic Guidelines have no higher operational focus. Hence, as a PLA military theorist might say, the relationship is "dialectical."

Strategic Missions and Strategic Objectives.

PLA materials usually employ the terms "strategic missions" (*zhanlüe renwu,* alternately "strategic tasks") and "strategic objectives" (*zhanlüe mubiao*) as part of the same phrase. As best as can be determined, the PLA does not differentiate between the concepts of "missions" and "objectives." The "strategic missions and objectives" of the PLA are usually articulated in the Military Strategic Guidelines, are set down in the broadest of terms, and are derived from the "strategic assessment" as well as the PRC's larger security objectives. Examples of "strategic tasks and missions" would be to "defend sovereignty and maintain internal stability."

Military Combat Preparations.

The term "Military Combat Preparations" (*junshi douzheng zhunbei*) refers to the type of warfare the PLA must be prepared to fight, and therefore also constitutes an official assessment of the next *type of war* that is most likely to be fought by the PLA. This guidance is couched mainly in terms of a capabilities-based assessment, not a contingency-based assessment. It is about the nature of contemporary warfare, not about identifying the next enemy or any specific operational scenario or planning contingency. The guidance contained under "Military Combat Preparations" is closely linked to the issue of "Army Building" because of its obvious programmatic implications for the development of operational capabilities. Examples would be directing the PLA to prepare to fight: (1) total war versus limited or local war; (2) local wars "under normal conditions" versus "local wars under modern high-tech conditions"; or (3) conventional warfare versus nuclear warfare, etc.

The Main Strategic Direction.

Whereas the guidance under "Military Combat Preparations" is mainly geared to a capabilities-based analysis, the concept of the "Main Strategic Direction" (*zhuyao zhanlüe fangxiang*) is a contingency-based assessment. Explaining this concept requires a brief digression.

The term "main strategic direction" (*zhuyao zhanlüe fangxiang*) is a concept from Chinese military science that informs both warfighting (the actual prosecution of a war in progress) and war planning during peacetime.

As a Warfighting Concept. In the case of warfighting, the term is applicable at both the strategic level of war as well as at the "campaign level" (*zhanyi ji*) of warfare in specific theaters of war (war zones, or *zhan qu*).[21] In essence, as a warfighting concept, the "main strategic direction" represents a decision and determination about where (geographically) and against which enemy forces operations must be conducted to achieve the strategic and/or operational results desired. As a warfighting concept, think "theater of operations," "decisive operations," "center of gravity," and "main effort" as described in U.S. doctrinal literature.[22]

Most PLA encyclopedia and military dictionary entries for this term generally focus on its application as a warfighting concept. The example below from the *Chinese Military Encyclopedia* is representative:

> **Strategic Direction (*zhanlüe fangxiang*)** — Refers to the operational direction with an important influence on the overall situation of the war. It directs the strategic objectives and has a defined depth and width including the ground as well as the multidimensional space of air, sea, and outer space. It is often determined on the basis of the military, political, economic, natural geographical,

and demographic factors of the participating sides and other relationships as well as their strategic tasks that must be completed. Whether the strategic direction is selected correctly or not directly impacts the process and the result of the war. No matter if it is dealing with [offensive operations] or [defensive operations], the strategic direction always [distinguishes] between its main direction (*zhuyao fangxiang*) and secondary direction (*ciyao fangxiang*). Within a certain [time frame], there can only be one main strategic direction. The main strategic direction focuses on combat with the enemy, its center of gravity. . . . Because of this, determining the main strategic direction is the most important issue of strategic guidance [for prosecuting a particular war].[23]

As a War Planning Concept. However, in the context of the Military Strategic Guidelines, the main strategic direction speaks to the issue of war planning and preparations in peacetime. It identifies the most likely geographic direction, and usually the most likely potential adversary, that is assessed as posing the highest risk to the PRC as regards the outbreak of a future conflict. Hence, at this level, identifying the "Main Strategic Direction" serves as a "worst case scenario" planning tool for developing forces and capabilities, making force deployment decisions, and making other preparations should conflict erupt.

This larger context for the term comes through very clearly in the excerpt below from the 2003 *Outline for Studying Jiang Zemin Thought on National Defense and Army Building*, produced by the PLA General Political Department:

> Planning for the national defense and modernization of army building, and planning for military combat preparations requires a prominent main strategic direction (*zhuyao zhanlüe fangxiang*). While paying attention to other directions (*qita zhanlüe fangxiang*), the main strategic direction is the impetus for army building (*jundui jianshe*) in other strategic directions.[24]

In other words, if the worst possible case scenario is identified (main strategic direction) and preparations, modernization programs, training, deployments, etc., are focused on being able to counter that threat, then other contingencies considered less pressing will *ipso facto* be taken care of as well.

Since 1949, as China's security situation has changed, the main strategic direction (and other major elements of the Military Strategic Guidelines) has been shifted four times according to one PLA strategist.

- Mid 1950s-Early 1960s: East against the United States and "other invasionary forces";
- Mid 1960s-Early 1970s: To the north and west against the Soviets, *and* east (still) against the United States (causing, obviously, a serious dilemma for PLA planners who, doctrinally speaking, assert that there can be only one main strategic direction);
- Early 1970s-Mid 1980s: North (the "three northern regions") against the Soviets who "became our main target of defensive operations"; and
- Mid 1980s-Early 1990s: The beginnings of a coastal concept with no specific enemy identified. "Under the premise of a stable strategic situation on the northern front, gradually improving the strategic situation of the southern front, strengthening the development of border and maritime defense, attaching importance to managing and maneuvering on the high seas and maintaining our maritime rights and interests."[25]

Giving Focus to Army Building.

The PLA uses the term "Army Building" (*jundui jianshe*) when it speaks of modernization and reform efforts. Under the rubric of "Army Building" within the Military Strategic Guidelines can be found the specific modernization objectives the PLA must pursue, the reforms it must enact, and the capabilities it must develop to enable the armed forces of China to accrue the operational wherewithal it needs and the institutional superstructure it must have to provide for the national defense in any given period of time. Army Building is the programmatic "guts" of China's national military strategy. It covers every aspect of modernization that was briefly enumerated under the Three Pillars posited at the beginning of this chapter— (1) The development, procurement, acquisition, and fielding of new weapons systems, technologies, and combat capabilities; (2) institutional, structural, systemic, and personnel reforms; and (3) doctrinal adjustments, etc.

The Military Strategic Guidelines lend focus to the PLA's modernization efforts (Army Building). They are the basis for the development of more detailed plans, programs, and resource allocations. In some cases, the Military Strategic Guidelines will provide priorities among those programs. Programmatically, the larger-order guidance for Army Building becomes the grist for subsequent documents that are developed by the PLA, such as the very important *Outline of the Plan for Army Building* (*Jundui Jianshe Jihua Gangyao*) that is apparently generated at the beginning of each national Five Year Plan.[26]

Overall, the outputs and deliverables under Army Building answer the question, "*What* is the PLA doing in the realm of modernization?" And to the degree that these outputs are observable or knowable, they constitute the subject of the majority of foreign writings and news media reportage about the PLA. However, it is the Military Strategic Guidelines that provide a larger strategic and programmatic context for answering the question, "*Why* is the PLA doing it?"

These, then, are the key components of the Military Strategic Guidelines as described in a most generic sense. With the preceding as background, the next section moves on to discuss the content of the current guidelines.

THE MILITARY STRATEGIC GUIDELINES FOR THE NEW PERIOD

On January 13, 1993, Jiang Zemin, then CMC Chairman, delivered a speech to an expanded meeting of the CMC in which he promulgated a new set of Military Strategic Guidelines.[27] Known officially as the "Military Strategic Guidelines for the New Period," this document represents the national military strategy under which the PLA has been operating for some 14 years.

Every modernization program, every reform initiative, and every significant change that the PLA has undergone, and which foreign observers have been writing about for over a decade, are the results of some of the fundamental decisions made when the new guidelines were promulgated — especially the ensuing programs the PLA initiated after 1993 to comport with the new guidelines.

Like the Military Strategic Guidelines issued prior to 1993, the "Military Strategic Guidelines for the New Period" is a "rolling national military strategy." This means that while the strategic guiding thought of any iteration of the guidelines continues to serve as the foundation and justification for action over time, it is the concrete programs subsequently developed and implemented that give body to the Military Strategic Guidelines. In the case of the "Military Strategic Guidelines for the New Period," the implementing programs have been "rolled out" over the course of four Five Year Plans: the eighth Five Year Plan (1991-95) when the new guidelines were promulgated, and throughout the ninth Five Year Plan (1996-2000), 10th Five Year Plan (2001-05), and the current 11th Five Year Plan (2006-10). Clearly, this type of "rolling strategy" allows for adjustments along the way. For example, by 1999 PLA professional military literature began to re-characterize the most likely type of future warfare as "Local Wars Under Modern *Informationalized* Conditions" vice "Local Wars Under Modern High-Tech Conditions" (the latter being the initial articulation in 1993). By 2002 the former term, substituting "informationalized" for "high-tech," was officially incorporated into the lexicon of the "Military Strategic Guidelines for the New Period." As such, it represents an *adjustment* to the military strategic guidelines, *not* a new set of military strategic guidelines or a new "national military strategy."[28] It represents, in the words of a September 2006 article in the PLA's official newspaper, *Liberation Army Daily (Jiefangjun Bao)*, an "enrichment and improvement" to the old guidelines, not a new set.[29]

In this section of the chapter, the key elements of the "Military Strategic Guidelines for the New Period" will be presented. Before proceeding, however, some caveats are in order. It is unknown whether there is a sole official document in which the PLA formally commits to paper its Military Strategic Guidelines as does the United States when it publishes its official "National Military Strategy." Therefore, the overview of the main aspects of the "Military Strategic Guidelines for the New Period" that follows is based on data in the public domain: published excerpts of Jiang Zemin's 1993 speech to the CMC, a PLA study guide (*gangyao*) on Jiang Zemin "military thought" that discusses the "Military Strategic Guidelines for the New Period" in one chapter, and various PLA commentaries and articles in professional military literature. Therefore, while the data set used is relatively small, it is considered authoritative as far as it goes. However, the data set could not possibly be considered complete at this point.

The Strategic Assessment.

The perceived need to issue a new set of military strategic guidelines in 1993 was driven by three key assessments. First, of course, was a major change to the international order as a result of the demise of the Soviet Union and the other communist regimes in Eastern Europe. Jiang's original forecast in 1993 that the trend in major power relations would be toward "multipolarity" has since proven overly optimistic, and the PRC now talks about "unipolarity" and a global order dominated by a "sole superpower." Nevertheless, the basic 1993 assessment of China's larger security situation has fundamentally remained in place since that time.[30] At the CMC meeting on January 13, 1993, at which the new guidelines were introduced, Jiang reiterated the Dengist assessment from the late 1980s that the prospects for a world war involving China were slim, and that China was enjoying a window of opportunity for its own economic development and military modernization. Jiang went so far as to opine that China's regional security situation at the time was likely the best it had been since 1949. According to Jiang:

> The contemporary world is in a historic period of momentous change. Overall, the present international situation is beneficial to our country's development. First, for a relatively long time to come, it is probable that the international environment will be peaceful with new world wars being avoided. This is an extremely important strategic assessment [here quoting Deng]: "The increase in the forces for peace in the world is surpassing the increase in the forces for war."
>
> . . . Moreover, compared with other regions of the world, the Asia-Pacific region has maintained a relative degree of stability with economic contacts and cooperation between countries becoming closer by the day with many traditional hot spots either already resolved or in the process of realizing a political resolution. *Our country's peripheral security environment is continuing to improve and friendly relations with neighboring countries have entered their best period since the founding of the nation.* (Emphasis added)
>
> . . . These conditions and factors mentioned above provide a relatively good external environment for us to consolidate our energy on developing the national economy.[31]

At the same time, Jiang pointed out the challenges to Chinese national security, as follows:

- "Ethnic, religious, and territorial disputes that were covered up by the rivalry between the United States and the USSR have become more prominent by the day, with bloody conflicts and local wars continuing to spread."

- "Although negotiations over arms control and force reductions have made some progress, the arms race has transitioned into a new high-tech arena that has produced an impact on the world and in the Asia-Pacific region that cannot be ignored."
- "Although we should not need to fight new world wars and total wars that affect our country for some time, factors giving rise to local wars, armed conflict, and domestic social turmoil (*shehui dongluan*) still exist."
- "Although the competition for economic and technical strength in order to lay a foundation for comprehensive national power has become a leading aspect of international struggle, military measures still play an important role."
- Although Jiang revalidated Deng Xiaoping's assessment that the "keynote of the times" remains "peace and development" (*heping yu fazhan*), he also noted that "hegemony and power politics have already become the main obstacles to world peace and development."
- Moreover, Jiang asserted, "Viewing our country's security environment, we can see that no matter if it is a political or an economic problem, no matter if it is an external military threat or a problem hindering the completion of unification of the motherland and unstable domestic factors, they are all either directly or indirectly related to hegemonism and power politics, and in all cases we can see the shadow of hegemonism and power politics. Regarding this, we must be strategically farsighted. We must resolutely struggle against actions that damage the rights and interests of our people

and national sovereignty. Of course, we must be flexible in taking hold of the methods of this struggle."

- Finally, regarding Taiwan, "Although work towards the great cause of the unity of the motherland continues to make progress, many new complex factors are emerging."[32]

The second assessment revolved around domestic issues: (1) China would continue with the program of "reform and opening up"; (2) economic development was still a paramount objective; (3) China required a stable domestic, international, and peripheral environments to succeed; and (4) PLA modernization would have to be accomplished within the broader context of other national objectives.

Fundamentally, Jiang revalidated the centrality of economic reform in China's search for enhancing its "comprehensive national power" and the critical requirement of maintaining a peaceful and favorable external environment:

> In summary, we must fully evaluate these favorable factors, grasp this rare opportunity, strengthen our foreign affairs work and foreign exchanges, expand our country's latitude in the international situation, and increase our initiative in handling international affairs in order to create even better external conditions for domestic development that are beneficial to our acceleration of the pace of reform, opening up, and the development of modernization, consolidating our energy in handling the national economy, and continuing to enhance our country's comprehensive national power. This is the fundamental essence of guaranteeing the nation's long-term peace and good governance, and the consolidation and development of the cause of constructing socialism with Chinese characteristics.[33]

Jiang's rededication to Deng's domestic line of "reform and opening up" might seem gratuitous from today's vantage point. But it is worth remembering that in the wake of Tiananmen (1989), Deng Xiaoping encountered serious resistance from some CCP elders as to the wisdom of the economic and foreign policies he had put into place. Some elders expressed the view that the crisis of 1989 was the result of the emergence of new socio-economic and political forces attendant to "reform and opening up." It took what is now known as "Deng's Southern Tour" in 1992 to sweep away the last vestiges of post-Tiananmen resistance to the decision to push forward.

Thus, as regards the "Military Strategic Guidelines for the New Period," it has been clear since 1993 that the PLA's modernization and reform programs have been viewed as but one element of China's search for enhanced "comprehensive national power." Although the PLA has unquestionably been the beneficiary of steadily increasing financial resources, especially since 1999, military modernization is not being accomplished in isolation from other PRC national objectives. This fact comes through clearly in a volume authored by the PLA NDU's Army Building Research Department (2004):

> The military strategic guidelines of the new period persists with obeying and serving the development strategy of our nation.... Our national development strategy (*guojia fazhan zhanlüe*) is a strategy of comprehensive national development that employs the strategy of economic development as the core, and is the general strategy (*zong zhanlüe*) for guiding the coordinated development of our nation's economy, politics, military, diplomacy, culture, [etc]. The military strategic guidelines for the new period are a component of the national development strategy so without a doubt they should obey and serve the nation's general strategy.[34]

However, it was the third assessment concerning the changing nature of warfare and the self-recognized inadequacies of the PLA which served as by far the most important impetus for issuing a new set of military strategic guidelines in 1993.

Today, of course, it is almost trite among students of Chinese military affairs to dwell on the impact of the U.S. first Gulf War on the PLA. Even so, it is still useful to remind ourselves occasionally that once the strategic, operational, and tactical implications of Operation DESERT STORM became clear, the leadership of the PLA was forced to confront the disconcerting reality that China's armed forces were woefully inadequate for the demands of modern warfare, and that this inadequacy demanded a major adjustment to China's national military strategy. Here are Jiang's words from his 1993 speech to the CMC:

> Since the beginning of the 1980s, the scope of high-tech competition throughout the globe has intensified by the day. Now, each country is readjusting their own development strategies, making the development of modern and especially high technology a crucial measure of strengthening their comprehensive national power and national defense strength, striving to take hold of the strategic initiative. The facts of the Gulf War have shown that along with the utilization of high technology in the military arena, the enhancement of precision attack weapons and unprecedented operational intensity, the characteristics of sudden, three-dimensional, mobile, rapid, and in-depth attacks, have become more prominent, and the use of high-tech superiority has obviously taken hold of the strategic initiative to an even greater degree. In the present world, if a country does not work hard to strengthen its national defense power alongside its economic and social development, enhancing its military quality and the level of its weapons and equipment, its operational capabilities will not be strong under modern high-tech conditions. As

soon as a war breaks out, it will be in a passive position and suffer attacks, with its national interest, the people's dignity, and its international prestige all suffering greatly. Because of this, many countries in the world are ... readjusting their military strategy in order to adapt to the needs of developments to the international situation and the situation of military combat.

Ten years later, in 2003, the *Outline For Studying Jiang Zemin Thought on National Defense and Army Building* (*gangyao*) still acknowledged the centrality of the Gulf War as a determinant of the need for new guidelines:

At the onset of the Gulf War, [Jiang Zemin] brought up the need to study the special characteristics of modern warfare through this conflict. He personally managed open symposiums and conferences and pointed out that "modern warfare is becoming high-tech warfare." In addition, according to our country's security situation, he promptly brought up [the need] for studying the formation of the Military Strategic Guidelines for the New Period. At the beginning of 1993, the Central Military Commission formulated the Military Strategic Guidelines for the New Period, implementing major adjustments on military strategy.[35]

Consequently, almost everything that the "Military Strategic Guidelines for the New Period" speaks to, and what every subsequently implemented program has been about, is what must be done to develop the requisite capabilities to rectify PLA shortcomings.

Guidance For Military Combat Preparations.

Central to the current Military Strategic Guidelines, therefore, is the question of what *type* of war the PLA must be prepared to fight. As Jiang put it in 1993:

> Since the founding of our country, our military has always implemented the military strategic guidelines of the active defense. *Under the new historical conditions, exactly what kind of military strategic guidelines should we be carrying out?* We believe that we should continue to carry out the military strategic guidelines of the active defense. . . . *At the same time, along with developments and changes to the situation, we should bestow the military strategic guidelines of the active defense with new content at this appropriate moment.* (emphases added)[36]

In this regard, the "Military Strategic Guidelines for the New Period" have been crystal clear from their inception in 1993. The PLA has been told to work towards the ability to fight and win "Local Wars Under Modern High-Tech Conditions."[37] As mentioned earlier, that descriptor was changed in 2002 to "Local Wars Under Modern Informationalized Conditions." That change, however, was basically a variation on the same theme. The important point to make is that the PLA was charged in 1993 to cease focusing its modernization efforts on late industrial age warfare and shift to a long-term program of developing the necessary capabilities for fighting late 20th-century and early 21st-century conventional warfare as exemplified by U.S. forces in 1991. And to the degree that U.S. operations in Kosovo in 1999 and to this day in Afghanistan and Iraq have further defined and refined the nature of 21st-century warfare in the minds of PLA analysts, those operations and capabilities establish the "gold standard" for what the PLA aspires to achieve eventually. As stated by Jiang in his 1993 speech to the CMC:

> In terms of strategic guidance, we have long since transferred the key preparations from being based on fighting early, fighting large, and fighting nuclear wars, to dealing with local wars. *Now, on the basis of developments and changes to the international situation, we must give*

priority to preparations for dealing with local wars under modern high-tech conditions. This is a further development and perfection of our army's strategic guiding thought. (emphasis added)

As explained in 2004 by the PLA NDU's Army Building Research Department:

> ... the CMC with Jiang Zemin as the core clearly pointed out that the focus of military combat preparations in the new period would change from fighting to win local wars under normal conditions to fighting to win local wars under modern, high-tech conditions. . . . Looking at these developing trends, for some time to come in the future, these different characteristics, scopes, and patterns of local wars and armed conflicts will be the main types of warfare. In the local wars our country is likely to face in the future, regardless of whether they are wars to realize the unification of the motherland *or wars to resist and counter a localized invasion by an enemy* (emphasis added), in all cases we are likely to face an enemy that possesses high-tech weapons and equipment. We can see that making the focus of military combat preparations on fighting to win local wars under modern high-tech conditions is a necessary choice based on a scientific analysis of the international strategic framework as well as our national security environment.

It is interesting to note that there is a strain in this assessment geared not just to the necessity for the PLA to gain the capability of conducting offensive operations at the campaign level (operational level) of war, but also to the need for conducting defense at the strategic level of war (see italicized words above).

As early as 1993, the PLA concluded that, while the probability of a full-scale invasion of China was low, the *possibility* of an attack on the PRC mainland could not be discounted, given what it had observed throughout the Gulf War of the new U.S. high-tech precision-guided munitions. In other words, the new face of high-tech warfare left the mainland vulnerable. Once again, Jiang's words in his 1993 speech are relevant:

> At the same time, we must also recognize that local wars under modern conditions are greatly different from past wars. As soon as a conflict or a war breaks out, the likelihood of an enemy first using precision-guided weapons and long-distance operational aircraft to conduct air raids, as well as independent sea and air wars, is high. Under these conditions, we must still persist with implementing people's war. . . .

A 2004 text published by the PLA Academy of Military Science makes the point much more directly:

> Future high-tech local wars are certainly very different from the wars we have fought in the past. Looking at the recent high-tech local wars, the odds of the enemy sending a large number of troops to our national territory to fight at the beginning of the conflict is relatively small. If conflict or a war develops, the enemy will probably use precision attack weapons, and long-distance operational aircraft to launch air raids on strategic targets along our coast or in the interior, carrying out relatively independent maritime and air wars, conducting a so-called "surgical attack operation." Regarding wars conducted to maintain the unification of the motherland in the direction of the Taiwan Strait, these also will be carried out along the coastal areas, and feature both maritime and air operations. Traditional means of mobilization and organizing the masses to participate in the war are already difficult to utilize; however, this does not mean that the ideology of a people's war is already passé.[38]

Strategic Guiding Thought.

Given China's assessment of the larger international situation, its assessment of its own security requirements, its larger national objectives, and the imperatives of modern warfare, what "strategic guiding thought" has been passed on to the PLA for the "new period"? Basically, it consists of four directives derived from Jiang's initial pronouncements at the enlarged CMC meeting:

- Ideologically, continue to adhere to the fundamental military theories first set down in "Mao Zedong Military Thought"; but especially build upon "Deng Xiaoping Theory on Army Building in the New Period" as a means to further research the best ways to construct "a modernized, regularized, revolutionary military that is politically qualified, operationally proficient, possesses a good work style, is strict in discipline, and acts as a powerful safeguard" for China's national security interests.

- Second, the PLA "must obey and serve the nation's development strategy" and in doing so must:

    persist with embarking from the nation's overall situation, carefully guide army building [in conjunction with] preparations for military combat, closely coordinate with the political and the diplomatic . . . in order to safeguard reform, opening up, and the development of the economy, ensuring that army building [does] an even better job of serving the realization of the nation's strategic objectives.[39]

- Third, "we must place the focus of future preparations for military combat on fighting to win possible local wars under modern high-tech conditions," and

- Fourth, when dealing with threats to national security, the military strategic guidelines must be flexibly applied.

This latter point speaks to how China should deal with perceived threats to its national security and to why the military element of national power is only one means, among others, to deal with security issues. In explaining this point, Jiang exhorted the PLA to continue to adhere to the military principle of "striking only after the enemy has struck," and using the diplomatic element of national power to prevent conflict when possible. He rejected wars of aggression as a policy choice and reiterated that conflict can only disrupt other central objectives.

> Militarily, we must strictly guard our stance of self-defense and never invade other countries or take the initiative to cause trouble. Regarding actions that harm our national sovereignty or interests, we must carry out a struggle that is truthful (*youli*), advantageous (*youli*), and controlled (*youjie*). In peacetime, the military must make the containment of wars from erupting as an extremely important duty, actively coordinate with the political, diplomatic, and economic struggle, work hard to improve our country's strategic environment, reduce insecure unstable factors, and work hard to contain local wars and armed conflict from erupting, ensuring that our national economic development is free from the impact of war. Only with a relatively secure and stable environment over a period of decades can our economic power, defense power, and comprehensive national power be able to greatly increase, will our national security be guaranteed, our country's international position even more consolidated and enhanced, and the cause of socialism with Chinese characteristics be enriched with even more vitality and vigor.[40]

Strategic Missions and Objectives for the PLA.

> Maintaining the nation's territorial sovereignty, maritime rights and interests, and social order, as well as a secure and stable internal and external environment for

safeguarding national economic development and reform and opening up are the strategic tasks given to our army by the military strategic guidelines for the new period in order to realize the state's strategic objectives.[41]

The various biennial editions of the PRC defense white papers, entitled *China's National Defense*, have in the past provided lists of missions and objectives for the PLA. Depending upon the larger security context prevailing when each was published (1998, 2000, 2002, 2004 and 2006),the order in which these missions and objectives have been listed, and the length of the list itself, has changed from year to year. The white papers, however, have tended to mix and conflate the PLA's larger strategic missions and objectives with some of its more granular Army building programs and objectives. In this chapter macro goals are treated separately from those associated with Army building, with Army building objectives reserved for their own section.

Reading across various samples of data focused on the "Military Strategic Guidelines for the New Period," we find that the strategic-level missions and objectives assigned to the PLA come down to five major tasks.

- Defending national territory and sovereignty;
- Securing the nation's maritime rights and interests;
- Maintaining the unity of the motherland;
- Ensuring internal stability; and
- Maintaining a secure and stable external environment, especially on China's periphery

These missions and objectives are straightforward. What becomes interesting is how the PLA translates these missions and objectives into larger-order notions of how to enable PLA forces to achieve them.

Basically, PLA commentaries on the "Military Strategic Guidelines for the New Period" suggest that the only way the armed forces of China can play a role in achieving China's larger national objectives, and achieve the missions and objectives given the PLA, is to develop real and credible operational capabilities in order to deter conflict or successfully prosecute conflicts if deterrence fails.

Defending National Territory and Sovereignty. "Mission One" for any military, of course, is the fundamental requirement to defend the nation from attack, to defend its territory, and to preserve the nation's sovereignty. China is no different than any nation in this regard. And, like most nations, the PRC has charged the PLA to develop capabilities that will deter any potential enemy from attacking the nation and to defeat acts of aggression if they cannot be deterred. Jiang makes this quite clear in his 1993 speech: "Properly handling preparations to fight to win local wars under modern high-tech conditions is done in order to possibly prevent or avoid these kinds of wars and is also the fundamental measure to guarantee that as soon as they erupt we are able to achieve victory."[42]

Securing the Nation's Maritime Rights and Interests. Beginning in earnest in the 1980s, China became enmeshed in various disputes in the regional seas off its littoral. By the early 1990s, it became clearer to Beijing that maritime resources would become increasingly important to the PRC's larger modernization objectives, and the Chinese people were told to develop "sea consciousness." It is not surprising, therefore, that the new military guidelines have made securing China's maritime rights a strategic mission for the PLA. And while China ostensibly prefers to handle these disputes by diplomatic means, the PLA has been charged to develop operational capabilities to enhance Beijing's diplomatic leverage.

> ... in the process of building a strong socialist modern nation, there are still many problems concerning securing and safeguarding our country's long-term sustainable development that are becoming more prominent by the day, including how to maintain our maritime rights and interests as well as exploit and utilize maritime resources, how to maintain the security of our strategic and energy shipping lanes.... The essence of our country's guidelines is the principles of common cooperation, equality, and mutual benefit, resolving contradictions and disputes through patient consultations. *However, to do this we also need a certain military force to serve as support. If we have the ability to win high-tech wars, then we can take a position of greater initiative in diplomatic consultations, and more effectively maintain our country's just rights and interests.*[43] (emphasis added)

A strategist from the AMS adds a strategic perspective to this issue that goes beyond merely providing Chinese diplomats the ability to negotiate from a position of strength:

> Our country is a large power close to the sea and it uses vast territorial waters, the continental shelf, and exclusive economic zones. The integrity of our country's sovereign territory includes this maritime area. In order for the Chinese people of the future to exist and develop, we must attach importance to managing and maneuvering in the sea. Securing and maintaining our country's maritime rights and interests is an important aspect of future strategic guidance over a long period of time that must be considered. In terms of developing military power, the standing army must take the ground forces as the main body. However, we must give priority to gradually strengthening the development of the navy and the air force.[44]

Note that as early as March 1993 when this lecture was delivered—within 4 months of the articulation of new guidelines—the need to give more attention to the Navy and Air Force was already beginning to come to the fore in the context of discussions about the newly issued Military Strategic Guidelines.

Maintaining the Unity of the Motherland. The "Military Strategic Guidelines for the New Period" demand that the PLA develop credible capabilities vis-à-vis Taiwan for deterrence and coercion as well as actual military operations, if need be. Without enhanced and credible operational capabilities on the part of the PLA, so the Chinese argument goes, the preferred official strategy for resolving the Taiwan issue (political means) may not be possible, and if force of arms is necessary then actual operational capacities will have to be developed. Either way, the PLA must develop the capabilities to give Beijing serious options, political or otherwise. In discussing the "Military Strategic Guidelines for the New Period," one PLA commentator sums up the logic on the Taiwan issue as follows:

> If our military power cannot defeat Taiwan and cannot deter international interfering forces and Taiwan separatist forces on the island, the Taiwan authorities will not come to the political negotiating table, and international interfering forces will also not abandon their positions. If we have the ability to fight and win high-tech wars, international interfering forces and Taiwan separatist forces on the island will have to think twice, and we can create more opportunities to use peaceful methods to realize unification across the strait. [But as] soon as the Taiwan authorities make a reckless move, we also have the ability to defeat them.[45]

Ensuring Internal Stability. In 1993, a mere 4 years after Tiananmen, concern about internal stability in the strategic guidelines is not surprising. And in the context of the last few years when the Ministry of Public Security continues to report rising numbers of anti-government protests and incidents, the issue of internal stability remains paramount from a CCP and PLA perspective. But as described in the quotation below (from 2004), concerns about internal stability increasingly encompass more than just anti-government incidents. They also include natural disasters as well as transnational security concerns such as pandemics and issues associated with ethnic unrest.

> At this time [2004] where China is a large developing country, particularly at a time of important strategic opportunity for reinvigoration of the Chinese people, stability overrides everything. At a time when China is a large country engaged in transformation, the military plays a crucial role in safeguarding internal stability. Safeguarding internal stability concerns political and economic factors, for example political turmoil or a financial crisis. It also includes . . . factors such as serious natural disasters and epidemics. It also includes taking hold of ethnic and religious factors, for example contradictions between nationalities in the border region or religious problems. Safeguarding internal stability needs the establishment of mechanisms to deal with handling important developing incidents to ensure that important developing incidents can be handled and controlled effectively in a timely fashion in order to prevent them from leading to greater social turmoil.[46]

On the issue of maintaining internal stability, the *Outline For Studying Jiang Zemin Thought on National Defense and Army Building* (2003) makes specific reference to the People's Armed Police (PAP) as a critical component of the armed forces charged with "protecting national security and social stability."[47]

Maintaining a Secure and Stable External Environment, Especially on China's Periphery. This particular objective is commonly viewed by the PLA as a core responsibility for its contribution to the PRC's larger national development goals, and it is often articulated in conjunction with the mission of *"Providing a strong, powerful and secure safeguard for reform, opening up, and modernization."* At bottom, it charges the PLA to develop the requisite capabilities, operational as well as diplomatic-political, to maintain security and stability on China's periphery in order for economic development to proceed at home.

In his 1993 speech to the CMC, Jiang spoke specifically to the issue of improving relations with the nations on China's periphery:

> Only with a relatively secure and stable environment over a period of decades can our economic power, defense power, and comprehensive national power be able to greatly increase, our national security is even more guaranteed, our country's international position is even more consolidated and enhanced, and the cause of socialism with Chinese characteristics is enriched with even more vitality and vigor. . . . Regarding peripheral countries, we must do more work, eradicate anxiety, and promote neighborly, friendly relations in accordance with the guidelines of a stable periphery.[48]

However, from an operational perspective, the Military Strategic Guidelines also direct the PLA to move forward in developing the capabilities and plans necessary to deal with conflicts on the periphery, should they erupt. The PLA, according to one book published by the Academy of Military Science, must

> increase . . . planning in this regard and, on the basis of differing strategic directions and differing possible targets, we must properly handle all kinds of preparations, establishing mechanisms to deal with and handle all kinds of disputes, ensuring that any disputes

that may arise are quickly and effectively resolved, in order to safeguard stability and security.[49]

Identifying the Main Strategic Direction.

In the course of the research for this chapter, no authoritative statement definitively identifying the "main strategic direction" for the "Military Strategic Guidelines for the New Period" was found. There is, however, little question that a "main strategic direction" has been designated. For example, in the *Outline For Studying Jiang Zemin Thought on National Defense and Army Building*, the PLA is exhorted to manage the relationship carefully between the "main strategic direction" and other pressing directions in its "preparations for military combat":

> Another aspect is to take care of the relationship between the main strategic direction (*zhuyao zhanlüe fangxiang*) and other strategic directions (*qita zhanlüe fangxiang*). Planning for the national defense and modernization of army building, and planning for military combat preparations requires a prominent main strategic direction; while paying attention to other directions the main strategic direction is the impetus for military building in other strategic directions.[50]

The *Outline* does not, however, specify what that "main strategic direction" happens to be, and it is likely that it is not identified anywhere in public domain data. However, some of the data reviewed for this chapter would lead one to *speculate* that today, as during the mid-1950s to the mid-1960s, the "main strategic direction" is once again east, off China's coast. One hint comes from Jiang's 1993 speech:

> At present and for a period in the future, our priority in terms of military struggle is preventing Taiwan from

fomenting any great "Taiwan independence" incidents, remaining vigilant against that which harms national sovereignty and territorial integrity. . . .

. . . The military must actively support the Party and government in strengthening its political, economic, cultural, [etc.] attractiveness and influence over Taiwan, giving play to the role of military deterrence, containing "Taiwan independence" separatist forces, working hard to promote peaceful reunification, while at the same time seriously handling emergency military preparations.[51]

The littoral off China's east coast as a potential candidate for the current "main strategic direction," and the western border regions as a potential candidate for today's "secondary direction" (*ciyao fangxiang*), pose a *possibility* rendered even more likely by Major General Peng Guangqian's comments in a lecture delivered in 1993.

> Under the old backdrop of "fighting large," we once were based on large armed groups carrying out decisive strategic battles within our national territory. Because of this, the center of gravity of strategic guidance deviated to the interior. Border and maritime defense only existed as a shield area for the larger war. Relatively speaking, border and maritime defense were relegated to subordinate status. This was determined by the objective environment of the times.
>
> Under the new historical conditions, the strategic position (*zhanlüe diwei*) of maritime and border defense has become more prominent. The borders and coastal areas are not only the frontlines and the window of our country's opening to the outside world in the new period, but also the main battlefield for our country's local wars and military operations. Although the unfavorable situation of border and maritime defense combat cannot immediately constitute a great threat to our country, it directly influences the nation and the

> dignity of the people, influences the nation's territorial sovereignty and rights and interests, and influences popular sentiment and morale, as well as the smooth implementation of national development strategy. In particular, owing to the closely interwoven relationship between problems in border and maritime defense with ethnic problems, religious problems, and territorial problems, they are highly sensitive and complex, and their handling requires true skill. Because of this . . . we need to attach sufficient importance to raising border and maritime defense problems to strategic heights, and must also attach high degrees of historical responsibility to carefully preparing and guiding their command.[52]

Needless to say, the suggestions herein offered about candidates for the main and secondary strategic directions under the current military guidelines must be labeled for what they are at this point: *speculation in the absence of authoritative PRC statements.*

Finally, the sense that one gets from reviewing various PLA materials is that the main strategic direction is an element of the strategic guidelines subject to constant study and review, and likely in need of regular revalidation over time, since the Military Strategic Guidelines for any particular period have a relatively long shelf-life.

Determining the Focus for Army Building.

> Each aspect of army building, together with all jobs — military training, political work, logistics support, national defense research, etc. — will fall under the instruction and management of the Military Strategic Guidelines for the New Period . . . based on the conditions of winning a modern technological, especially high-tech, local war. . . . The Military Strategic Guidelines require each aspect of army building within the entire military to obey and serve the needs of this one strategic guideline. They must all center upon carrying out and the implementing [the guideline] of winning a local

war under modern technological, especially high-tech, conditions.[53]

—Jiang Zemin

The programmatic guts of the "Military Strategic Guidelines for the New Period" center on Army building. The new guidelines announced by Jiang in 1993 revolved around the fundamental decision that the PLA would have to undergo an extended period of significant reform and modernization to cope with the imperatives of the changing nature of warfare. In short, the PLA was told to develop the operational capabilities and the institutional capacity to prosecute "Local Wars Under Modern High-Tech (subsequently "Informationalized") Conditions." And, as mentioned at the beginning of this chapter, the programs and outputs that come under the rubric "Army building" in the guidelines have for many years provided the grist for analyses by scholars and others who follow Chinese military affairs. Consequently, the actual details are already well known. However, it is worth reviewing the key guidance for Army building that was given to the PLA back in 1993 as well as some of the subsequent directives. In retrospect, one cannot but be impressed by how many of the fundamental objectives originally articulated have actually been put in place or have shown significant evidence of progress.

When he issued the new Military Strategic Guidelines to the CMC in January 1993, Jiang Zemin specifically directed the PLA to begin development of modernization programs or institutional reforms in five key areas:[54]

(1) "First we must place the development of science and technology for national defense and

the development of unit equipment in a prominent position." The first order of business for the development of military modernization would have to be the accrual, over time, of modern weapons systems and other supporting technologies, without which the PLA would be unable to prevail in modern warfare. The fighting spirit of the PLA notwithstanding, without modern weapons, said Jiang, "We will have to pay a relatively heavy price to achieve victory."[55] In 1993, Major General Peng Guangqian expressed the same idea without mincing words: "One cannot simply use manpower superiority to compensate for technological inferiority" given the nature of modern warfare.[56] To this end, in order to raise the operational effectiveness of the armed forces, Jiang spoke to the need to "respect science and technology and attach importance to the role of weaponry;" enhance basic research and improve the defense science and technology establishment; and develop new weapons and technologies, while at the same time "improving existing weapons and equipment."

(2) "Second, we must attach high degrees of importance to enhancing the overall quality of officers and enlisted personnel." In this second injunction, Jiang spoke to the need to improve the level of education within the officer corps and enlisted force as well as improving unit training:

> Leading cadre and leading organs at all levels must place their main energies on education and training, act unswervingly, truly grasp, firmly execute, and truly enhance the quality of education and training. [They must] work hard to cultivate and create a batch of talented outstanding personnel with high degrees of political awareness and noble morale who take hold of modern military technology and understand the art of commanding modern warfare. We must recognize that

this is a fundamentally important matter for realizing our army's modernization and winning future wars, as well as an extremely important component of taking the path of crack troops with Chinese characteristics, and strengthening the development of our army's quality.

Continuing, he made the following specific points:

- "We must focus on winning local wars under modern high-tech conditions, strengthening unit training, academy instruction, and military science and research."

- "We must embark from the needs of actual war, further improve unit training, and truly enhance unit tactical and technical levels under modern high-tech conditions" (i.e., train the way we will fight).

- "We must give prominence to properly grasping training for command personnel at all levels, cultivating them into the mature backbone for running the army and taking hold of the art of modern warfare."

- "We must continue to strengthen and perfect the development of academies and schools and improve instructional content and methods in order to develop and cultivate a large batch of qualified, talented personnel to construct our army's modernization and develop science and technology for national defense."

- "We must focus on the characteristics of local wars under modern high-tech conditions, research in-depth and actively explore the rules of the people's war under modern conditions, work hard to inquire about combat methods for using inferior equipment to defeat an enemy with superior equipment, enriching and developing a military theory with Chinese characteristics" (i.e., new doctrine).

(3) "Third, we must further give priority to army building." In this narrower use of the term "army building," Jiang is referring to prioritizing organizational changes. For example,

- "We must have the resolve to properly grasp the development of priority units, ensuring that they possess strong operational capabilities."
- "We must adapt to the requirements of people's war under modern conditions and attach importance to and strengthen the development of the capabilities of the reserve forces."
- "The general departments and the relevant state organs must . . . further improve mechanisms for national defense mobilization and give priority to resolving the problems associated with rapid mobilization for future local wars."

(4) "Fourth, we must truly strengthen and improve the military's ideological and political work."

- "No matter if it is peacetime unit building or fighting to win any local war under modern high-tech conditions that may erupt, none can be divorced from strong and powerful ideological and political work."
- "We must strengthen the development of military Party organizations and leading groups at all levels, guaranteeing the Party's absolute leadership over the military, and guaranteeing the high degrees of stability and collective unity within units."

- "We must educate units in our army [on the PLA's] fundamental duties and honorable traditions, giving play to outstanding traditions and maintaining the essence of the red army. We must vigorously launch education in Chinese contemporary history, modern history, as well as education [about the current] situation and combat readiness. . . ."
- "We must closely center on education and training, properly grasp regular ideological and political work, intensify enthusiasm for training, cultivate a combat work style that does not fear sacrifice and is fierce and tenacious, guaranteeing that units are able to maintain a soaring fighting spirit under any kind of difficult conditions, and satisfactorily complete the tasks accorded to them by the Party and the people."

(5) "Fifth, we must further strengthen the development of military logistics." Specifically,

- "We must fully recognize the role and function of logistics support and continue to strengthen the development of logistics."
- "Upon a foundation of guaranteeing continuing improvements in unit living, [we should] consolidate financial and material resources. . . ."
- "[We should] gravitate towards the development of equipment for priority units and in the important direction" (i.e., main strategic direction).
- "On the basis of operational needs under modern high-tech conditions, we must strengthen the development of logistics and technical support

capabilities and work hard to create multifaceted assistance and support capabilities, particularly enhancing comprehensive emergency support capabilities."
- "We must adapt to the requirements of developing a socialist market economy and improve methods for raising funds for goods and materials and their supply, enhancing the use and benefit of expenditures and materials."
- "We must strengthen logistics mobilization work and gradually form a joint civil-military logistics support system"

Based on these larger-order directives to guide army building, the PLA was told to begin to study, research, and develop the concrete programs that would address the key requirements of "Local Wars Under Modern High Tech Conditions." Hence, the beginning of the 1990s witnessed a deluge of writing, publishing, and "debates" among Chinese military professionals as to how to operationalize that which was handed down to them in the "Military Strategic Guidelines for the New Period."

While work on Army building issues began immediately after Jiang's promulgation of the new guidelines in 1993, it appears that not until 1995 had all the relevant communities within the PLA finalized their initial plans for systemic reforms. In December of that year, the CMC issued the *Outline of the Plan for Army Building During the 9th Five Year Plan* (*"Jiuwu" Qijian Jundui Jianshe Jihua Gangyao*) to give form and structure to the programmatics for retooling Army building to comport with the requirements contained in the "Military Strategic Guidelines for the New Period." This master blueprint—the first issued since the new guidelines were promulgated in 1993 — ostensibly provided a roadmap to key modernization and reform objectives for the period 1996-2000.

Programmatically, the completion of the *Outline for Army Building* at the end of 1995 was timed to coincide with the development of the state's ninth Five Year Plan in order to ensure that the PLA received requisite funding in the defense budget. The new *Outline for Army Building* heralded the inception of the "Two Transformations" program (*liangge zhuanbian*) that directed the PLA to begin the transformation (1) from preparing to fight Local Wars Under Ordinary Conditions to preparing to *fight and win* Local Wars Under Modern High Technology Conditions, and (2) from being a military based on quantity to one based on quality. It called as well for the PLA to transform from being personnel intensive to becoming science and technology intensive.

Not surprisingly, therefore, it was during the years of the ninth Five Year Plan, certainly towards its close in 1999 and 2000, that many of the deliverables under the Three Pillars of PLA reform and modernization broached early in this chapter began to attract notice abroad — reform of the professional military education system; new joint doctrine; new unit field training regimens; countless new regulations on political work, active duty officer management, finance and budgeting procedures; creation of the fourth general department (GAD); establishment of joint logistics bases — the list goes on.

Finally, and worthy of particular note, in addition to these general areas upon which to focus Army building efforts, the "Military Strategic Guidelines for the New Period" also issued new Army building missions to each of the PLA's services (*jun zhong*) and branches (*bing zhong*) based on the forecast that in future high-tech warfare, "control of the air" (*zhi kong quan*), "control of the sea" (*zhi hai* quan), and "strategic strike" (*zhanlüe daji*) capabilities would be operational imperatives. Therefore, according to the *Outline for Studying Jiang Zemin Thought on National Defense and Army Building*, in addition to improving the ground forces, the PLA was directed to "vigorously strengthen the building of the navy, air force, and the Second Artillery."

- The PLA Navy must "truly possesses the comprehensive operational capabilities (*zonghe zuozhan nengli*) to conduct maritime campaigns in the near sea (*jinhai*)."
- The Air Force should "gradually realize the transformation from a national territory air defense model (*guotu fangkong xing*) to an offensive-defensive model (*gongfang jianbei xing*)."
- "The Second Artillery Corp (*erpao*) will have a stronger nuclear deterrent and conventional strike (*changgui daji*) capabilities."

In addition, according to the *Outline* cited above, the new Military Strategic Guidelines called for enhancing the capabilities of "Emergency Mobile Combat Operations Units" (*yingji jidong zuozhan budui*) for dealing with near-term contingencies should they occur.

> Each service and branch will place [army building for] Emergency Mobile Combat Operations Units in a priority position; safeguarding this main objective by providing weapons and equipment, materials and funding, etc; undergoing improved training in order to quickly deal with local warfare and sudden incidents, and complete the military combat tasks for the new period.[57]

Moreover, in addition to extending priority for equipment, funding, and other resources, the PLA was directed to make the "Emergency Mobile Combat Operations Units" the prototypes for developing other future high-tech units throughout the PLA, and to use them as test beds for the conduct of joint operations and "informationalized" warfare.

Finally, each PLA service and branch was given the Army building mission to "establish an intense understanding of joint operations, launch in-depth research into joint operations theory and combat methods [i.e., doctrine], and establish and perfect a joint operations command system that conforms to our country's national situation and military situation."[58]

It is unclear when these latter service and branch missions were actually articulated to the PLA as part of the "Military Strategic Guidelines for the New Period." These very specific mission areas were not included in a recently released version of excerpts from Jiang's original speech in 1993. For all one knows, they may have been there in the manuscript but were edited out as the speech was being cleared for public release in August 2006. What one can state with certainty is that these missions were part of the strategic guidelines as of 2003. They are included in the *Outline for Studying Jiang Zemin Thought on National Defense and Army Building* that was published in 2003. But the *Outline* does not specify whether these mission areas were part of the original promulgation (1993) or were added as part of subsequent adjustments to the strategic guidelines.

Either way, it is highly interesting to note that, with the exception of the mission to enhance "emergency mobile operations units," the Army building objectives set forth above for the services and branches as extracted from the *Outline* (2003) were publicly articulated in the PRC defense paper, *China's National Defense in 2004*. Moreover, that particular version of the white paper provides a wealth of detail on various Army building programs that were launched as a result of the strategic guidelines in 1993. However, for whatever reasons, the PLA does not make the direct linkage between these programs and the "Military Strategic Guidelines for the New Period" for its foreign readership.

Thus, in summarizing the major aspects of China's current national military strategy, let us call upon Jiang Zemin one final time to wrap it all up with the following excerpt from his 1993 speech.

> In summary, for the period in the future the basic content of the military strategic guidelines of the active defense is: employing Mao Zedong Military Thought and Comrade Deng Xiaoping's Thoughts on Army Building in the New Period as the guide, observing and serving the national development strategy, being based on fighting to win local wars under modern high-tech conditions, accelerating the development of our army's quality, working hard to enhance our army's emergency operations capabilities, enhancing strong points and avoiding weaknesses, being flexible in meeting changes, containing the war, winning the war, defending national territorial sovereignty and maritime rights and interests, and maintaining the unity of the motherland and social stability, in order to provide a strong powerful secure safeguard for reform, opening up, and the development of modernization. . . . These guidelines expound on the fundamental guiding ideology that we must persist with in our army building and military combat. They expound on the relationship between our military strategic guidelines and the national development strategy, determine the strategic objectives and strategic tasks of our army, determine the focus of our army building and military combat preparations, and put forth the basic ideology that strategic guidance must take hold of.

CONCLUDING COMMENTS

First, by this time it should be very clear that the PRC does in fact have a "national military strategy" that is guiding the reform and modernization efforts of the Chinese armed forces. Moreover, there should be little question at this point that the Military Strategic Guidelines are the vehicle used to transmit that strategy to the PLA.

- When new guidelines are issued, it is the result of major and significant changes in one or all of the following areas of assessment: (1) changes in the international order; (2) changes in the security environment and China's security situation; (3) changes to China's domestic situation; and (4) changes in the nature of warfare itself.

- The Military Strategic Guidelines tend to have a long shelf-life. The new guidelines issued in 1993 constitute only the fifth instance since 1949 that the PRC has made a major change to its national military strategy.

- The Military Strategic Guidelines are the CMC's authoritative guidance to the PLA to take for further planning and action. As a rolling military strategy, the guidelines are fleshed out in detail over time, and there are various systemic events in the PLA and the PRC to which major programmatic decisions are keyed, such as State Five Year Plans and "Army Building Outlines" (*Gangyao*) keyed to Five Year Plans.

- A major change to the Military Strategic Guidelines, and especially the promulgation of a completely new set of Military Strategic Guidelines, is as much a major political event as it is a strictly military event. Consequently, *it ought to be a knowable event.*

Second, Military Strategic Guidelines must address what appear to be standard strategic issues or questions. At a minimum, these components include:
- The Strategic Assessment (*zhanlüe panduan*);
- The Content of the Active Defense Strategy (*jiji fangyu de zhanlüe nei rong*);
- Strategic Missions and Strategic Objectives for the PLA (*zhanlüe renwu, zhanlüe mubiao*);
- Military Combat Preparations (*junshi douzheng zhunbei*);
- The Main Strategic Direction (*zhuyao zhanlüe fangxiang*); and
- The Focus for Army Building (*jundui jianshe*).

While over time some aspects of the key components of the guidelines may change or be adjusted, such adjustments do not necessarily constitute the promulgation of a new iteration of the Military Strategic Guidelines. It is important to understand the difference between new programs, new "slogans," and new emphases on one hand, and the issuing of a completely new set of guidelines, on the other.

Although some of the terminology applying to the key components of the Military Strategic Guidelines may be new to some, upon reflection, most of the *content* of China's national military strategy as described in this chapter should *not* come across as new. In fact, those who study the PLA on a regular basis have been encountering and writing about many of the concepts, directives, and programs contained herein for many years. Now there is a Chinese framework that can be used to place developments in context, and there is some basis for answering the oft-asked question, "What is driving PLA modernization?"

Third, it is worth reiterating that every modernization program, every reform initiative, and every significant change that the PLA has undergone, and which foreign observers have been writing about for over a decade, are the results of some of the *fundamental decisions* made when the new guidelines were promulgated in 1993—especially the ensuing programs the PLA initiated after 1993 to comply with the new guidelines.

Fourth, the "Military Strategic Guidelines for the New Period" do not fundamentally speak to "numbers"—*it speaks to capabilities.* The number of systems, the number of units, and the number of personnel are all worked out over time. But the "Military Strategic Guidelines for the New Period" are first and foremost about developing strategic and operational capabilities the armed forces of China have hitherto not felt a need to acquire or have not been able for various reasons to develop.

My own reading of the guidelines is that the PLA is being told to develop the capabilities to accomplish the following tasks:

(1) Provide for the defense of strategic assets on the mainland in light of 21st-century precision-guided munitions and other high-tech weapons that could be used to threaten the PRC or actually be employed against the mainland;

(2) Strengthen the deterrent value of its nuclear forces;

(3) Fight and win high-tech joint campaigns in the maritime, aerospace, and electro-magnetic battle spaces off its littoral if need be; and

(4) Field credible operational capabilities to deter potential aggression against the mainland or its interests (political or economic), support the diplomatic element of national power with real "teeth," and provide options to China's leaders across the full spectrum of operations, from "show of force" to the application of force in such a manner that any required escalation control can be managed.

Fifth, slowly but surely—with the very important exception of identifying outright the main strategic direction—the PRC has been releasing information on some of the key components of the "Military Strategic Guidelines for the New Period" into the public domain in China for the past few years. This is certainly the case as regards PLA books, articles, and study materials meant for Chinese military audiences.

As far as materials aimed at *foreign* audiences are concerned, specifically the PRC defense white papers and other materials, the Chinese still do not expound directly the "Military Strategic Guidelines for the New Period." They choose, instead, to repackage aspects of the guidelines, condense them to a stock phrase or two (such as "the military strategic guideline of the active defense"), or talk around the guidelines. Why Beijing chooses not to discuss the guidelines, the components, and the content directly can only be left to conjecture, especially given the release in August 2006 of the three-volume paean to Jiang Zemin that is rich in content on this topic. At the same time, transparency creep does seem to be underway. Especially beginning with the publication of *China's National Defense in 2004*, the PLA is in fact beginning to discuss some, but certainly not all, of the key components of the guidelines. This was evident in the 2006 iteration as well.

Sixth, it is clear that the Chinese approach to crafting a national military strategy is pragmatic, deliberate, and based on the types of calculations that any professional military establishment would undertake. Although the terminology and the organization of the concepts are distinctly a reflection of PLA bureaucratic culture and the intellectual constructs imposed by "scientific" Marxism, there is nothing particularly foreign, strange, exotic, or exceptional about the Military Strategic Guidelines and the national military strategy it transmits. At bottom it is a capabilities-based and contingency-based strategy that sets the azimuth for the development of warfighting capabilities as well as professional and other institutional capacities to provide for the national defense of China that are subject to larger-order national objectives.

The panel reports and papers in the chapters of this anthology which focus on the services—the PLA Navy, PLA Air Force, and Second Artillery—will go into great detail about programs, capabilities, weapons systems, service missions, and other issues that reside at the operational and tactical levels of warfare. But every such chapter has at its root the basic guidance given the PLA as contained in the "Military Strategic Guidelines for the New Period."

ENDNOTES - CHAPTER 3

1. See Wang Wenrong, General Ed., *et al.*, *On The Third Modernization of the PLA*, Beijing: Liberation Army Press, February 2005. The authors argue that the first significant "modernization" of the PLA occurred just prior to the founding of the PRC in 1949, "on the eve of the founding of the New China." The second modernization began in the mid-1980s. The third modernization to which they refer in the title, began, they argue, in the mid-1990s, especially the latter half.

2. For a broad overview of many of the changes evident by the end of the ninth Five Year Plan in 2000, see Kenneth W. Allen, Dean B. Cheng, David M. Finkelstein, and Maryanne Kivlehan-

Wise, *Institutional Reforms of the Chinese People's Liberation Army: Overview and Challenges,* Alexandria, VA: The CNA Corporation, May 2002.

3. For a detailed look at the PLA's new operational level doctrine, see David M. Finkelstein, "Thinking About the PLA's 'Revolution in Doctrinal Affairs'," in David M. Finkelstein and James Mulvenon, eds., *China's Revolution in Doctrinal Affairs: Emerging Trends in the Operational Art of the Chinese People's Liberation Army,* Washington, DC: December 2005, pp. 1-27.

4. *Annual Report to Congress: Military Power of the People's Republic of China* (2006), p. 1.

5. David M. Finkelstein, "China's National Military Strategy," in James Mulvenon and Richard H. Yang, *The People's Liberation Army in the Information Age,* Santa Monica, CA: RAND, 1999, pp. 99-145.

6. The updated version of the original is available on the CNA Corporation website at *www.cna.org*.

7. *Selected Works of Jiang Zemin,* August 2006. (*Jiang Zemin Wen Xuan*) Beijing: Renmin Chuban She, 2006.

8. It is worth pointing out that the very first time the United States published a "National Military Strategy" in the public domain was January 1992.

9. Harry R. Yarger, "Toward a Theory of Strategy: Art Lykke and the Army War College Model," in J. Boone Bartholomees, Jr., ed., *U.S. Army War College Guide to National Security Policy and Strategy,* 2nd Edition, Carlisle, PA: Strategic Studies Institute, U.S. Army War College, June 2006, pp. 79-113.

10. Interviews.

11. Teng Guosui, "National Strategy," in *Chinese Military Encyclopedia,* Beijing: Military Science Publishing House, 1997, Vol. 2., p. 181.

12. PLA National Defense University Army Building Research Department, *Jiang Zemin Guofang he Jundui Jianshe Sixiang Xuexi Duben (Study Guide for Jiang Zemin Thought on National Defense and Army Building)*, Beijing: Chinese Communist Party History Publishing House, April 2004, 2nd Edition, pp. 56-67. Hereafter, *Xuexi Duben.*

13. Peng Guangqian, "The Development and History of Our Country's Strategic Guidelines of an Active Defense Since the

Founding of the Nation," in Peng Guangqian, *Researching Questions of Chinese Military Strategy*, Beijing: Liberation Army Publishing House, January 2006, pp. 86-104, *Zhongguo Junshi Zhanlüe Wenti Yanjiu*. This article, written in 1993, was originally prepared as a lecture delivered at the AMS.

14. Peng, *ibid*.

15. For the different types of CMC meetings, their differing purposes, and types of attendees, see Tai Ming Cheung, "The Influence of the Gun: China's Central Military Commission and Its Relationship with the Military, Party, and State Decision-Making Systems," in David M. Lampton, ed., *The Making of Chinese Foreign and Security Policy in the Era of Reform*, Stanford, CA: Stanford University Press, 2001, pp. 61-90.

16. For a detailed explanation and analysis of these two differing assessments, their implications, and the importance of the concept of "keynote of the times" for China and the PLA, see David M. Finkelstein, *China Reconsiders Its National Security: "The Great Peace and Development Debate of 1999,"* Alexandria, VA: The CNA Corporation, 2000, posted on The CNA Corporation website at *www.cna.org*.

17. That China does not have a formal "National Security Strategy" is explicitly stated in three PLA sources. See (1) Lieutenant General Mi Zhenyu, "The Basic Scope of National Security and the National Security Strategy of the People's Republic of China," in Lieutenant General Mi Zhenyu, *An Exploration of Theories on Warfare and Strategy*, Beijing: Liberation Army Publishing House, January 2004; (2) Chen Zhou, "China's National Defense White Papers and the Development of Defense Policy," speech delivered at the 7th International Symposium on Sun Tzu's *Art of War*, Hangzhou, China, June 2006; (3) Peng Guanquan and Yao Youzhi, eds., *The Science of Military Strategy*, English Ed., Beijing: Military Science Publishing House, 2005, pp. 22-23.

18. See *Selected Military Writings of Mao Tse-Tung*, Beijing: Foreign Language Press, 1967, pp. 77-152.

19. Shan Xiufa, General Editor, *Research on Jiang Zemin Thought on National Defense and Army Building*, Beijing: Military Science Publishing House, March 2004. See Chapter 4: "Formulating and Perfecting the Military Strategic Guidelines for the New Period," Vol. 3, 2004, pp. 72-88.

20. The genesis and explication of these maxims is to be found throughout the body of literature that comprises "Mao Zedong Military Thought."

21. For a full explanation of this term at the campaign (operational) level of warfare, see Wang Houqing and Zhang Xingye, eds., *The Science of Campaigns*, Beijing: National Defense University Press, May 2000.

22. Readers are referred to U.S. Army FM 3-0, *Operations*, June 2001; Joint Publication 1-02, *Department of Defense Dictionary of Military and Associated Terms*, as Amended through April 14, 2006; and U.S. Army FM 100-7, *Decisive Force: The Army in Theater Operations*, May 1995, etc.

23. Lu Mingshan, "Strategic Direction," *Military Science*, Vol. 3, *Chinese Military Encyclopedia*, Beijing: Military Science Publishing House, July 1997, p. 715. Phrases in brackets indicate this author's edits.

24. See Chapter 5: "Using the Military Strategic Guidelines for the New Period to Maintain the Overall Situation of Army Building."

25. These four phases and the quotation come from Peng, *ibid*. Note that Peng's article was written in May 1993. Consequently, subsequent shifts are not accounted for.

26. Yuan Wei and Zhang Zhuo, eds., *History of the Development of Chinese Military Academies and Schools*, Beijing: National Defense University Publishing House, July 2001, pp. 927-934.

27. Excerpts from the actual speech Jiang delivered, entitled "The International Situation and the Military Strategic Guidelines," can be found in the August 2006 3-volume *Selected Works of Jiang Zemin*, Beijing: People's Publishing House, August 2006, Vol. 1, pp. 278-294. Hereafter cited as "JZM Speech 1993."

28. Shan Xiufa, *ibid*. According to this AMS volume, the PLA had been considering recasting this term since 1999. But it was not until Jiang Zemin gave a speech in December 2002, presumably at the annual end of year expanded meeting of the CMC, that this term was officially adopted. According to Shan, Jiang's speech in 2002, *On National Defense and Army Building*, was subsequently published in 2003. Note that this change in terminology is provided in the English edition of *China's National Defense In 2004*.

29. Shou Xiaosong, "Vigorously Implement the Military Strategic Guidelines for the New Period," *Liberation Army Daily*, Internet version, September 21, 2006.

30. Of note, in 1999, the Chinese defense, security, and foreign policy establishment engaged in a heated debate which challenged the Dengist assessment that "peace and development" is the "keynote of the times" and challenged Jiang Zemin's relatively positive assessment of China's regional security situation. The debate was driven by (1) the U.S.-NATO intervention in Kosovo, (2) the errant U.S. bombing of the PRC Embassy in Belgrade, (3) the U.S. BMD program, and (4) regional concerns such as developments in Taiwan and the strengthening of the U.S.-Japanese alliance. For the details, see Finkelstein, *China Reconsiders Its National Security: The "Great Peace and Development Debate of 1999," ibid.*

31. JZM Speech 1993, *ibid.*

32. All quotes in bullets from JZM Speech 1993, *ibid.*

33. *Ibid.*

34. *Xuexi Duben, ibid.*

35. Clearly, the *Gangyao*, published in 2003, was in many ways a paean to Jiang Zemin as he was slowly leaving the stage as CMC Chairman. Therefore, the uninitiated should understand that while Jiang is hailed for his wisdom and foresight, it was actually a very large and capable group of PLA professionals who were behind the assessments and analyses that led to the issuance of the new military strategic guidelines.

36. JZM Speech 1993.

37. This chapter will forgo the discussion of the operational and tactical-level attributes that the PLA ascribes to "Local Wars Under Modern High-Tech Conditions" inasmuch as foreign students of the PLA have written reams of articles over the years on that topic.

38. Shan Xiufa.

39. Regarding the "nation's strategic objectives," Jiang's speech reiterated the decisions made at the 14th Congress of the CCP which stated as goals the advancement of the "socialist market system" throughout the 1990s, thereafter creating a society with an increased standard of living, and by the year 2049 (the centenary of the PRC) having finally realized "socialist modernization,"

thus putting China on a par with middle developed nations. JZM Speech 1993.

40. *Ibid.*

41. *Research on Jiang Zemin Thought on National Defense and Army Building*, Beijing: Military Science Publishing House, March 2004.

42. *Ibid.*

43. Shan Xiufa, *ibid.*

44. Peng, *ibid.*

45. Shan Xiufa, *ibid.*

46. Shan Xiufa, *ibid.*

47. *Gangyao, ibid.*

48. JZM Speech 1993, *ibid.*

49. Shan Xiufa, *ibid.*

50. *Gangyao.*

51. JZM Speech 1993, *ibid.*

52. Peng, *ibid.*

53. Guidance attributed to Jiang Zemin in the *Outline for Studying Jiang Zemin Thought on National Defense and Army Building.*

54. JZM Speech 1993. See Section III of the speech entitled, "On Several Questions Concerning Army Building and Military Combat Preparations."

55. *Ibid.*

56. Peng, *ibid.*

57. *Gangyao, ibid.*

58. *Ibid.*

SELECTED BIBLIOGRAPHY

CHINESE MATERIALS

Reference Materials (Chinese)

Chinese Military Encyclopedia (*Zhongguo Junshi Baike Quanshu*) (Beijing: Military Science Publishing House, 1997, 11 Vols.).

Military Terminology of the People's Liberation Army (*Zhongguo Renmin Jiefangjun Junyu*) (Beijing: Military Science Publishing House, 1982).

PLA General Political Department, *Outline for Studying Jiang Zemin Thought on National Defense and Army Building* (*Jiang Zemin Guofang He Jundui Jianshe Sixiang Xuexi Gangyao*)(July 2003). (This *Gangyao* was produced by the General Political Department of the PLA. While the actual document carries no bibliographical data, the date of publication is fixed at July 2003 based on several meetings the PLA publicized in the PRC media to herald its dissemination throughout the armed forces as a required study guide. See, for example, *Xinhuanet*, 2003-07-31 16:59:02, "Central Military Commission Holds Forum to Mark the Publication of the *'Outline for Studying Jiang Zemin Thought on National Defense and Army Building'*."

Selected Works of Jiang Zemin, Vols. 1-3. (*Jiang Zemin Wen Xuan*)(Beijing: People's Publishing House, August 2006).

Books (Chinese)

Shan Xiufa, General Editor, *Research on Jiang Zemin Thought on National Defense and Army Building* (*Jiang Zemin Guofang He Jundui Jianshe Sixiang Yanjiu*) (Beijing: Military Science Publishing House, March 2004).

Peng Guangqian, *Researching Questions On Chinese Military Strategy* (*Zhongguo Junshi Zhanlüe Wenti Yanjiu*) (Beijing: Liberation Army Publishing House, January 2006). This volume is flagged as a "High Quality Academic Collection for Advanced Scholars of Contemporary Chinese Military Studies" (*Dangdai Zhongguo Junshi Xuezi Shenxuezhe Xueshu Jingpin Congshu*).

PLA National Defense University Army Building Research Department, *Study Guide for Jiang Zemin Thought on National Defense and Army Building (Jiang Zemin Guofang He Jundui Jianshe Sixiang Xuexi Duben)*(Beijing: Chinese Communist Party History Publishing House, April 2004, 2nd Edition). This volume is flagged as a deliverable under the PLA NDU's 10th Five Year Plan for Science and Research.

Wang Wenrong, General Editor, *On the Third Modernization of the PLA (Zhongguo Jundui Disanci Xiandaihua Lun Gang)* (Beijing: Liberation Army Publishing House, February 2005). The English translation of the title of this volume is from the front cover of the book.

Academy of Military Science Strategy Research Department, Peng Guangqian and Yao Youzhi, General Editors, *The Science of Military Strategy (Zhanlüe Xue)*(Beijing: Military Science Publishing House, October 2001). The English language edition of this volume was published by the Military Science Publishing House in 2005. In both the Chinese and English editions the authors chose to translate the title of the book as *The Science of Military Strategy*, consciously adding the English word "military" in English whereas the Chinese character for military (*jun*) is not in the Chinese language title.

Mi Zhenyu, *An Exploration of Theories on Warfare and Strategy (Zhanzheng Yu Zhanlüe Lilun Tanyan)*(Beijing: People's Liberation Army Publishing House, January 2004). This volume is flagged as a "High Quality Academic Collection for Advanced Scholars of Contemporary Chinese Military Studies," *Dangdai zhongguo junshi xuezi shenxuezhe xueshu jingpin congshu*.

Other Secondary Sources (Chinese)

Chen Zhou, "China's National Defense White Papers and the Development of Defense Policy" (*Zhongguo Guofang Baipishu Yu Guofang Zhengce de Fazhan*), Speech delivered at the 7th International Symposium on Sun Zi's *Art of War*, Hangzhou, China, June 2006.

ENGLISH LANGUAGE MATERIALS

Kenneth W. Allen, Dean B. Cheng, David. M. Finkelstein, and Maryanne Kivlehan-Wise, *Institutional Reforms of the Chinese People's Liberation Army: Overview and Challenges* (Alexandria, VA: The CNA Corporation, May 2002).

Tai Ming Cheung, "The Influence of the Gun: China's Central Military Commission and Its Relationship with the Military, Party, and State Decision-Making Systems," in David M. Lampton, Editor, *The Making of Chinese Foreign and Security Policy in the Era of Reform* (Stanford, CA: Stanford University Press, 2001), pp. 61-90.

David M. Finkelstein and James Mulvenon, Eds. *China's Revolution In Doctrinal Affairs: Emerging Trends in the Operational Art of the Chinese People's Liberation Army* (Washington, DC: Beaver Press, December 2005).

Richard Meinhart, *Strategic Planning By The Chairmen, Joint Chiefs of Staff, 1990-2005* (Carlisle, PA: Strategic Studies Institute, U.S. Army War College, April 2006).

Richard Meinhart, "National Military Strategies: 1990-2005," in J. Boone Bartholomees, Jr., ed., *U.S. Army War College Guide to National Security Policy and Strategy*, 2nd Edition (Carlisle, PA: U.S. Army War College, June 2006), pp. 303-312.

Andrew Scobell, *Chinese Army Building in the Era of Jiang Zemin*, (Carlisle, PA: Strategic Studies Institute, U.S. Army War College, August 2000).

Harry R. Yarger, "Toward a Theory of Strategy: Art Lykke and the Army War College Model," in J. Boone Bartholomees, Jr., ed., *U.S. Army War College Guide to National Security Policy and Strategy*, 2nd Edition (Carlisle, PA: Strategic Studies Institute, U.S. Army War College, June 2006), pp. 79-113.

Yao Yunzhu, "The Evolution of Military Doctrine of the Chinese PLA from 1985-1995," *The Korean Journal of Defense Analysis*, Winter 1995, pp. 57-80.

PART II:

PLA STRATEGIC SYSTEMS

CHAPTER 4

"MINDING THE GAP": ASSESSING THE TRAJECTORY OF THE PLA'S SECOND ARTILLERY[1]

Evan S. Medeiros

INTRODUCTION

The Second Artillery is one of the most dynamic branches of an already active and rapidly modernizing People's Liberation Army (PLA).[2] Second Artillery doctrine and force structure have been evolving in the last decade in order to deter new threats and to generate greater coercive military options. There are strong and multiple indications that the Second Artillery will continue to develop in the coming years. The Second Artillery may even assume new missions, such as counterspace operations. These changes are highly consequential for U.S. security interests in Asia and regional stability by dint of the Second Artillery's ability to "reach out and touch" other militaries in East Asia rapidly and accurately, including forward deployed U.S. forces. It is in this sense that the Second Artillery is truly China's *strategic* rocket force and thus worthy of systematic examination and analysis.

This chapter examines the past and current modernization activities of Second Artillery in an effort to generate fresh insights about its future direction. In doing so, it principally examines the relationship between Second Artillery requirements and doctrine on the one hand, and its force structure capabilities on the other. This chapter aims to answer the following questions: What are the roles and missions of the Second Artillery's nuclear and conventional missile forces — and do the two differ; what are the main attributes of its current force structure and related modernization activities; how consistent and congruent are its doctrine and capabilities; how has the Second Artillery adapted to emerging threats, including the possibility of assuming new missions; and, ultimately, what does this doctrine-capabilities comparison indicate about the Second Artillery's future evolution?

These questions provide an analytic construct within which to assess the Second Artillery's trajectory of change. This approach seeks to avoid a classic mistake in strategic analysis: inferring doctrine from capabilities and thereby arriving at worst-case assessments of doctrine because force structure data is often limited.[3] The answers to the questions raised above will provide a baseline useful for assessing the operational military challenges facing the United States as it seeks to deter China from pursuing a military resolution to the Taiwan question and as the United States manages its regional security commitments in the face of rapid PLA modernization.

This chapter is divided into five parts. Following the introduction, the second and third sections analyze the doctrine-capabilities relationship of the Second Artillery's nuclear and conventional missile forces, respectively.[4] The fourth section addresses the possibility of new missions for the Second Artillery. The chapter's concluding section advances several arguments about the degree of coherence within the Second Artillery and discusses the implications of these arguments for the Second Artillery's future force structure; it also enumerates factors which could change current Second Artillery capabilities.

The chapter principally relies on specialized Chinese military writings for information on PLA and Second Artillerey doctrine. These writings are drawn from the wave of military books and journal articles published following the PLA's doctrinal flowering that began during the ninth Five Year Plan (FYP) (1996-2000) and continues today. The chapter treats these professional military writings as *authoritative* but not *definitive*, since they are statements made outside the orbit of official doctrinal documents. Yet, the professional military writings referenced herein were chosen for their consistency of argumentation in order to identify widely-held claims about Second Artillery doctrine and operations. By contrast, the information on Second Artillery missile *capabilities* is drawn from publicly available U.S. data, such as the U.S. Defense Department's reports and those of other U.S. government agencies.

NUCLEAR DOCTRINE AND CAPABILITIES

China's *nuclear* missile forces are the oldest part of the Second Artillery, which was created in 2 years after China's initial nuclear test in 1964. It has been the custodian of China's nuclear-capable missile systems for over 40 years. It is also the youngest branch of the PLA and, as such, for decades received limited attention within a PLA which has long been dominated by Army-centric thinking and leadership. It has been only within the last decade that the Second Artillery has been accorded a more equitable measure of political influence and financial resources, similar to those of the PLA's services.[5] These and other changes in the Second Artillery are reflected in the modernization of its doctrine and capabilities.

The Institutional Development of Nuclear Doctrine.

The institutional context that shaped the development of China's nuclear doctrine is critical to understanding its current content and future direction. For at least 20 years after China's first nuclear test in 1964, Chinese research on nuclear doctrine was highly underinstitutionalized and undertheorized, especially in comparison to U.S. and Soviet doctrine during the Cold War. Few, if any, Chinese research organizations (within or outside the PLA) conducted substantive work on specifying the roles and missions of nuclear weapons. This relative inattention to nuclear issues was particularly acute within the PLA and its Second Artillery. According to PLA sources, military institutions did not begin researching nuclear strategy questions until 1985.[6]

There are at least three reasons for this phenomenon. First, China's nuclear doctrine was largely defined by the general beliefs of Mao Zedong and Deng Xiaoping about the need to possess a basic retaliatory capability to deter an adversary from using or threatening to use nuclear weapons against China. PLA and other research institutions conducted very little work on the requirements of possessing a credible second-strike capability, aside from emphasizing the general notions of survivability and holding at risk some enemy targets. The evolving availability of certain missile and warhead capabilities influenced nuclear doctrine, but this occurred within the above framework.[7]

Second, China's military education and research system was closed for at least 10 years during the Cultural Revolution (1966-76). This resulted in a serious lack of expertise and experience within the PLA and Second Artillery suitable for developing a detailed nuclear doctrine, as well as broader military doctrine. The PLA did not reengage such issues until well into the 1980s.

Third, there was limited political space in China to discuss nuclear doctrine for decades. Such issues were treated with such intense secrecy, internally and externally, that the political environment did not lend itself to broad interagency discussions. A related issue was the political incorrectness of discussing doctrine because it required using highly criticized Western concepts and terminology such as "deterrence."[8]

By the mid-1990s, a critical mass of expertise, experience, and political space began to emerge within China, especially within PLA circles, to discuss China's nuclear strategy and doctrine. PLA strategists and operators began to think more systematically about China's nuclear threat environment, the requirements of credible deterrence, and the capabilities necessary to achieve such deterrence. China's participation in international arms control processes (beginning in the early 1980s), in particular its direct involvement in the Comprehensive Test Ban Treaty (CTBT) negotiations in the early 1990s, prompted internal discussions about nuclear doctrine and force structure issues. More broadly, China had largely rebuilt its military education and research institutions by the mid-1990s. At that time, the Academy of Military Sciences and National Defense University began leading the process of renovating China's entire military doctrine. This culminated in the publication in 1999 of several documents called "campaign outlines" (*zhanyi gangyao*) that collectively detailed a new doctrine for "joint operations," as well as one specific to the Second Artillery.[9]

Although there was some internal discussion and writing about nuclear doctrine from the mid-1980s to mid-1990s, in retrospect this discourse appears to have been largely exploratory and had little impact on doctrine or capabilities. Chinese specialists read Western literature and debated among themselves about how to characterize Chinese nuclear doctrine. This process was manifested in debates about "minimum deterrence" versus "limited deterrence" as China's official doctrine.[10] As one senior AMS strategists told this author, that period in the evolution of nuclear doctrine was one of "let a hundred flowers bloom" (*baihua qifang*).[11]

Chinese writing and deliberations on nuclear doctrine since that time have become far more focused and detailed. New PLA publications offer numerous insights on the content of nuclear doctrine.[12] This development suggests that the locus of expertise on nuclear questions is diversifying from the monopoly of China's nuclear scientific community (i.e., the bomb builders in the "Ninth Academy") to PLA theorists and strategists. This was facilitated by the military's work on renovating its operational doctrine in the 1990s, and its systematic thinking about the requirements of specific military operations. This process included the Second Artillery's elaboration of its "nuclear counterstrike campaign" (*he fanji/baofu zhanyi*), which gave the PLA and the Second Artillery an opportunity to further develop the concepts and actions that collectively constitute nuclear doctrine.

Since the mid-1990s, the Second Artillery has made consistent advances in researching and developing its *guidelines, principles, requirements*, and *tasks* for nuclear missile operations.[13] PLA sources indicate that the Second Artillery has completed a "basic system of military theory" for nuclear operations, which included publication of several internal documents elaborating the elements of such operations. PLA sources also indicate there is still more work to be completed. The Second Artillery, as of 2004, had published the following documents related to a nuclear campaign: *Strategic Studies (Erpao Zhanlue Xue); Campaign Studies (Erpao Zhanyi Xue); Tactics Studies (Erpao Zhanshu Xue); Command Studies (Erpao Zhihui Xue); Logistics Studies (Erpao Houqin Xue); Equipment and Technology Studies (Erpao Zhuangji Xue);* and *Management Studies (Erpao Guanli Xue).* However, despite its intentions, the Second Artillery had by 2004 *not* yet published the following documents: *Nuclear Military Thought (Erpao He Junshi Sixiang); Nuclear Military Academic Research (Erpao He Junshi Xueshu);* and *Intelligence Studies (Erpao Qingbao Xue).*[14]

More broadly than the above work on nuclear doctrine, the Second Artillery has articulated the concepts driving their current and future capabilities — both nuclear and conventional. The Second Artillery has adopted three basic principles *(jiben yuanze)* to guide its future force structure as well as its future operations.[15] They are:

- Maintain conventional and nuclear [missile forces], but put priority on conventional [missile forces] *(hechang jianbei, yi chang wei zhu);*

- Bring out focal points, put priority on quality *(tuchu zhongdian, zhiliang weizhu)* (the latter part of this phrase is about improving survivability of Chinese nuclear forces); and

- Aggressively explore and boldly innovate *(jiji tansuo, dadan chuangxin).*

Analyzing Nuclear Doctrine.

Chinese leaders and PLA strategists continue to view nuclear weapons as primarily political tools to maintain China's freedom of action while minimizing its vulnerability to coercion by other nuclear-armed states. The legacy of Mao Zedong's, Deng Xiaoping's, and Jiang Zemin's views of nuclear weapons as a means to deter both nuclear aggression and blackmail continue to define China's nuclear strategy, doctrine, and policies. These views are reflected in multiple ways in PLA writings *as well as* those of State and Party organs.[16]

Strategic Intent. Chinese publications refer to three missions for its nuclear weapons: deterring nuclear aggression against China, preventing nuclear coercion, and conferring great power status/eliciting deference. Such writings, especially those within the PLA, consistently emphasize possessing nuclear weapons to prevent deterrence failures and to preclude other nuclear powers from issuing nuclear threats to intimidate or coerce China during crises. Chinese military writings have identified only one mission for Second Artillery nuclear forces, i.e., "a nuclear counterstrike campaign," in connection with which they discuss conducting retaliatory nuclear strike operations.[17] This single campaign stands in stark contrast to the numerous conventional missile campaigns and the fact that the latter are far more offensive in orientation, as discussed later.

PLA writings emphasize several themes that further illuminate their conceptualization of nuclear doctrine.[18] First, China's possession of a limited numbers of nuclear weapons is sufficient to deter nuclear aggression against China. China does not seek a capability for *assured destruction*, but rather *assured retaliation*. China seeks to hold at risk enough enemy targets of value with the threat of unacceptable damage such that adversaries are deterred from using or threatening to use nuclear weapons against China. A second but related idea is that China seeks to avoid being drawn into the trap of a nuclear arms race, which most Chinese strategists argue would waste scarce national resources and, ultimately, undermine Chinese security. China will not succumb to the same fate as the Soviet Union in the Cold War, which was bankrupted by trying to keep up with U.S. defense expenditures. How China defines the elements of such a nuclear arms race so as to avoid them remains unclear. Third, the PLA is most focused on improving the survivability, reliability, invulnerability, and penetrability of its nuclear forces so as to bolster the PLA's ability to deter nuclear aggression and coercion. The military's writings are most systematic and technically detailed about these three themes.[19]

These ideas are expressed in both PLA and CCP publications, which have become more explicit in recent years. China's 2006 *National Defense White Paper* publicly outlined for the first time the key principles guiding Chinese nuclear strategy and doctrine. This is China's clearest articulation, to date, in a white paper or other public document of the collective orthodoxy of China's nuclear strategy and doctrine.[20] The 2006 white paper stated that China pursues a "self-defense nuclear strategy" (*ziwei fangyu he zhanlue*). The two principles underlying this nuclear strategy are "self-defensive counterattack" (*ziwei fanji*)) and "limited development of nuclear weapons" (*youxian fazhan*). The report stated that ultimately China seeks to possess a "lean and effective nuclear deterrent capability" (*jinggan youxiao he liliang*). These principles are especially important because they are official ones used within China's intragovernmental deliberations about nuclear strategy and doctrine.[21] To be sure, additional research is needed to understand their precise meaning.

Furthermore, a 2004 report by a Central Committee unit included very similar concepts. It stated that China's nuclear strategy is to possess a "limited, self-defensive counter-strike" capability (*youxian ziwei fanji*). This strategy's primary characteristics are the following: it is defensive (*fangyuxing*), it seeks to counter coercion/intimidation (*kang weishexing*), it is limited (*youxianxing*), and it will be effective (*youxiaoxing*). The report added that China does not seek "to carry out or win a nuclear war," but that rather it "intends to counter and contain nuclear war."[22] The similarity of the ideas in these phrases with the 2006 *National Defense White Paper* suggests relative congruity in beliefs about the role and missions of nuclear weapons across the system in China.

Operational Principles. PLA writings also identify a few operational principles which guide the planning and actual use of nuclear weapons; these concepts are narrower than those mentioned above. Some of these concepts have been widely used in the PLA since its earliest Red Army days, and their use by the Second Artillery should be interpreted as the application of general PLA concepts to Second Artillery missions.[23] The first two principles are persistently and consistently identified by PLA and Second Artillery publications as guiding Second Artillery operations. They are part of the Second Artillery's doctrinal orthodoxy.

- Close defense (*yanmi fanghu*): This concept was one of the first articulated by the Second Artillery in the early 1980s. It is a broad, catch-all concept that embodies the policies and actions used to improve the survivability of China's nuclear forces.

- Key point counterstrikes (*zhongdian fanji*): This concept is about the Second Artillery's conduct of nuclear counterstrikes. In conducting such strikes, the Second Artillery seeks to elicit such a heavy psychological shock that the adversary does not escalate further and ends the nuclear exchange. This concept includes targeting both civilian and military sites in the hopes of causing deep psychological trauma and degrading the adversary's military capabilities. Yet, PLA writings suggest that retaliatory strikes are more about shocking an adversary than inflicting pain.

The following two principles are relatively new and intermittently referenced in PLA publications; it is not clear whether they inform Second Artillery planning, procurement, and operations. In particular, the principle of counternuclear deterrence is the subject of some debate among nuclear strategists as to its meaning and, thus, its applicability to China's nuclear strategy and doctrine.

- Effective nuclear deterrence (*youxiao he weishe*): This concept is a relatively new addition; it has not been part of the historical PLA lexicon. An effective nuclear deterrent is regarded as one that is reliable, survivable, and capable of penetrating an adversary's missile defenses. As long as the PLA is assured that it can retaliate and impose unacceptable damage on an adversary following a nuclear attack, then its deterrent is assessed to be effective. Yet, the level of forces that meets this latter standard is unclear. An idea closely related to effectiveness is sufficiency (*zugou*), which also informs nuclear force structure. A sufficient nuclear force is one sized and scaled to survive an enemy's initial nuclear strike, to execute counterattack and reattack operations, and to penetrate an adversary's defenses. Sufficiency and effectiveness were initially mentioned together in the 2001 version of *The Science of Military Strategy* published by the Academy of Military Sciences, but not in the 1999 version published by the National Defense University.

- Counternuclear deterrence/intimidation (*fan he weishe*): This concept is about military activities that seek to signal China's capability and resolve to resist nuclear coercion or intimidation. It is an expression of China's effort to avoid being blackmailed by an adversary's nuclear threats. It also reflects China's traditional view of "deterrence" by a potential enemy as highly coercive and, thus, as a practice to be opposed. In Western parlance, this concept is an expression of nuclear signaling. There is disagreement among PLA strategists about the accuracy of this term and, thus, its applicability to PLA operations.

External Threats, PLA Responses, and the 'No-First-Use' Bugaboo. Another window into Chinese nuclear doctrine is its responses to changes in PLA threat perceptions. PLA perceptions of its nuclear threat environment have shifted radically since the end of the Cold War. During the 1970s and 1980s, much of China's nuclear forces were oriented to deterring and defeating a large Soviet attack, which included China's possible use of tactical nuclear weapons to defeat a large Soviet land invasion.[24] Following the collapse of the Soviet Union and the rise in U.S.-China tensions over Taiwan, China's nuclear strategists became far more focused on U.S. military capabilities and its forward-deployed forces in Asia. PLA strategists and planners are now heavily preoccupied with threats from U.S. strategic offensive and defensive capabilities.

Currently, PLA strategists view their nuclear security environment as highly complex and deeply uncertain, with U.S. military capabilities as the most dynamic element in their threat assessments. Many Chinese view the U.S. 2001 Nuclear Posture Review (NPR) as lowering the nuclear threshold and validating the development of new "mini-nukes" to use for coercive purposes in regional conflicts. In particular, PLA strategists are very concerned about the threat to China's nuclear forces posed by the combination of U.S. missile defenses, non-nuclear strike options, and related threats from air attacks and special operations forces; it is this combination of capabilities that many in China believe could significantly compromise China's second-strike capability. In addition, Chinese strategists remain concerned about Japan's latent ability to develop nuclear weapons and Indian nuclear modernization. (Interestingly, North Korea's nascent nuclear capability is absent from the PLA analyses surveyed for this chapter.) However, PLA writings strongly suggest that U.S. capabilities are collectively treated as "the inclusive case" against which most PLA planning is conducted.[25]

How has PLA doctrine evolved in the context of these changing threat perceptions? At a minimum, China's nuclear doctrine has received more attention among strategists, which has resulted in a fuller theoretical development of the beliefs and concepts comprising nuclear doctrine. As discussed above, this development was facilitated by many factors such as the PLA's improving institutional capabilities to address nuclear questions. Perhaps most significant, China's doctrinal discussions and debates have remained focused on using nuclear weapons for deterrence, countering coercion, and signaling great power status, despite China's increasingly dire perceptions of its nuclear security environment.

Another place to look is China's long-standing policy on the no-first-use (NFU) of nuclear weapons. China has engaged in a debate in recent years about eliminating or qualifying its NFU commitment as a means to bolster China's deterrent. This debate was resolved in favor of *not* altering it, due in part to the expected damage to China's international image and reputation. Some discussion and writing about NFU and broader nuclear doctrine continued after that decision, but as of at least June 2006 the government banned further internal debate or public commentary about NFU.[26]

PLA sources suggest the continued application of NFU to its planning and operations. PLA internal writings consistently treat China's NFU pledge as a structural feature of and constraint on Second Artillery nuclear operations. A senior Second Artillery officer, for example, described nuclear strike planning as guided by the principle of "first resist, then penetrate" (*xiankang, houtu*), in which the nuclear forces survive a first strike and then retaliate by puncturing the adversary's missile defenses.[27] This idea is repeated in numerous other PLA and Second Artillery writings about nuclear counterstrike operations. To be sure, there is some questioning of NFU and related beliefs within the PLA, especially within the Second Artillery; yet, such reservations are expressed by indirection and "straw-man arguments," indicating the continued existence of political constraints on criticizing long-standing aspects of China's official nuclear policy, such as NFU.[28]

The occurrence of this internal debate about NFU does not appear to be indicative of a qualitative shift in China's nuclear doctrine. Rather, PLA strategists are exploring means to improve the credibility of China's nuclear deterrent in the face of the dual threats of America's missile defense systems and its non-nuclear strike weapons (e.g., conventional *Tomahawks*). Many Chinese fear the United States will use this combination of strategic defense and offense to neuter China's nuclear capabilities, making China vulnerable to coercion in a crisis. In other words, the viability of China's deterrent faces threats that some Chinese argue could be countered by injecting a degree of ambiguity into official doctrine, such as by conditionalizing NFU.

In assessing the implications of China's NFU debate, a far more pressing issue is the precise conditions under which China would adhere to its NFU pledge. It remains unclear what foreign military actions constitute "first use" for Chinese leaders, and thus what would trigger nuclear retaliation. Some PLA writings state that an enemy intention to carry out a nuclear strike on China is grounds for retaliation.[29] Other Chinese sources suggest that conventional strikes on Chinese "nuclear assets" *or* conventional strikes that produce weapons of mass destruction (WMD)-like effects could justify nuclear "retaliation" by China. Chinese officials and analysts are intentionally ambiguous on this point, probably to bolster the PLA's ability to deter such actions.[30] These conclusions about Chinese views on NFU are reinforced by the author's discussions with Chinese strategists during a U.S.-China conference on nuclear affairs held in June 2006.[31]

The apparent absence of other debates within the PLA also sheds light on the evolution of China's nuclear doctrine. In other words, there are numerous nuclear dogs that are not barking. There is no evidence of a Chinese discussion of using nuclear weapons as a cost saver or substitute for conventional weapons; indeed, China's intense conventional buildup in recent years belies that very notion. There is very little, if any, substantive discussion about using tactical nuclear weapons to deter major conventional aggression against China.

Moreover, there is little indication in PLA and Second Artillery writings that, for the purposes of escalation control or damage limitation, they seek primarily to use nuclear weapons to deny an adversary specific military gains. In other words, there appears to be little interest in Cold War-style "nuclear warfighting" strategies in which nuclear exchanges can be calibrated and finely managed. While some Second Artillery *operators* have *hinted* at such notions in internal military publications, their ultimate conclusions about adhering to existing policy and strategy suggest a political environment that is unwilling to engage concepts of nuclear warfighting.[32] PLA writings about nuclear counter-strike operations do not extensively and systematically discuss the conduct of nuclear warfare and the modalities of nuclear escalation. These writings do not indicate that PLA strategists are "thinking the unthinkable" nor do such publications possess the game-theoretic character of U.S. and Soviet writings during the Cold War. The PLA's most detailed, systematic, and technologically sophisticated writings focus on improving the survivability, invulnerability, and penetrability of Second Artillery nuclear missile forces.

Nuclear Missile Capabilities.[33]

The main attributes and overall direction of China's nuclear force structure modernization have been known to PLA observers for decades. Since the mid-1980s, China has been expanding the size and improving the quality of the Chinese missile forces in an effort to enhance their reliability, survivability, response time, and — most recently and urgently — their penetrability. This has been a gradual effort, not a crash program; many of the missile systems coming online have their roots in decisions that date back to the mid-1980s and, in some cases, even earlier. To be sure, once these modernization programs are finished and operationally implemented, perhaps by the end of this decade, China will have significantly upgraded the quality of its nuclear forces. When complete, China will have revolutionized its nuclear capability, providing it with a highly credible deterrent against major nuclear powers for the first time since 1964.

In terms of quality, China's nuclear modernization efforts are focused on diversification. China's nuclear forces are moving beyond their decades-long reliance on land-based, liquid-fuel, silo-based missile systems — the DF-3A (2,790+ kilometers [km]), DF-4 (5,470 + km), and DF-5A (12,900 + km) systems — to far more complex mobile missile systems such as the DF-31 (7,250 + km) and its extended range-version, the DF-31A (11,270 + km). For over a decade, China has fielded one nuclear-capable road-mobile, solid-fuel, medium-range missile known as the DF-21 (1,770 + km) and a sea-based variant known as the JL-1 (1,770 + km). The JL-1 is expected to be replaced by a longer range JL-2 (8,000 + km) by the end of the decade, which will be deployed in a new *Jin*-class ballistic missile nuclear submarine (SSBN) under development.[34]

In addition, the PLA has been engaged in a military-wide effort to modernize its command, control, communications, computers, intelligence, surveillance, and reconnaissance (C^4ISR) capabilities, which may include improvements to its nuclear command and control and missile-related early warning capabilities. Public data on the latter two types of capabilities are limited, but such modernization is critical to possessing a credible second strike capability. Understanding China's progress in improving such capabilities is essential in assessing accurately the strength of China's overall nuclear force structure.

In terms of quantity, China currently deploys some 20 DF-5A intercontinental ballistic missiles (ICBMs), 16-24 DF-4 intermediate-range ballistic missiles (IRBMs), 14-18 DF-3A and 40-50 DF-21 missile systems. These numbers are expected to grow by 10s in the next decade as China adds several new systems to its arsenal. Many of the older, land-based missile systems (such as DF-3A and DF-4) will likely be phased out as their solid-fuel successors are deployed. China will likely retain its DF-5As in service to add to its ICBM capabilities. The new DF-31s are just now being deployed but in unknown numbers.[35]

The U.S. Defense Department's *China Military Power Report* (2006) noted a highly dynamic and uncertain dimension of Second Artillery nuclear modernization, stating that China "will deploy several new conventional and nuclear variants of medium range ballistic missiles (MRBMs) and IRBMs for regional contingencies and augment its long-range missile forces. China is also developing air- and ground-launched cruise missiles that could have a nuclear capability."[36] Another area of significant Second Artillery procurement is that of capabilities to counter missile defenses such as decoys, penetration aids, and possibly multiple warheads. China is likely pursuing several options at once but an emphasis on multiple warheads could have destabilizing consequences depending on their number and operational capabilities. All such activities require watching to see which option receives the most serious investment.

Assessing the Doctrine-Capabilities Nexus.

In assessing the relationship between Second Artillery doctrine and capabilities, three main trends are discernible. First, China's nuclear doctrine (regarding both strategic intent and operational principles) have become more developed and explicit while remaining focused on the long-standing missions of deterrence, counter-coercion, and great power status. This has occurred despite the perceived emergence of numerous new threats to China's nuclear security, mainly those related to U.S. capabilities. The PLA is now playing a greater role in the process of doctrinal development as well.

Second, there is a broad degree of consistency between China's stated objectives for its nuclear forces and its modernization activities. Neither the size, scope, nor pace of China's nuclear force structure modernization is indicative of new or hidden missions. The Second Artillery's procurement is focused on deploying systems that improve the reliability, survivability, and penetrability of Chinese nuclear forces. This is consistent with a doctrine that seeks to use nuclear weapons to deter nuclear aggression and prevent coercion.

Absent evidence that China is significantly expanding the size of its missile forces *and* developing highly accurate and lethal nuclear-capable missile systems (i.e., ones that can destroy hard targets), there is little basis to conclude that China's nuclear strategy and doctrine are fundamentally changing. China would have to make substantial and sustained improvements to all aspects of its nuclear force structure—missiles, warheads, command and control and early-warning—to indicate a doctrinal shift that supported Cold War-like nuclear warfighting strategies. China would have to develop and deploy a significantly different force structure if it sought to adopt a more aggressive nuclear doctrine. If anything, China's nuclear forces have more work to do to ensure their survivability in perpetuity; in this sense, the PLA's longstanding nuclear doctrine-capabilities gap is closing but is not yet there.

This is not to say that China's nuclear modernization does not raise numerous implications for U.S. military planners. China may very soon possess a sufficiently invulnerable second strike capability, such as when it deploys a real sea-based nuclear capability, that it will affect U.S. calculations and limit U.S. options during a crisis. To draw a rough analogy from history, the Soviet Union's deployment of its first *Yankee*-class SSBN in the late 1960s convinced President Nixon that the United States no longer possessed a viable damage limitation option against Soviet nuclear forces. A disarming first strike was no longer conceivable. This accelerated a shift in U.S. thinking towards escalation control options in the U.S.-Soviet nuclear competition. As China's deploys its new SSBNs equipped with several 8,000 km range JL-2 missiles, the United States will confront similar challenges.

Third, even China's reported development of new variants of nuclear-capable MRBM and IRBM systems is arguably consistent with the strategic intent of China's current nuclear doctrine. Such systems improve China's ability to deter *a broader range of threats*, which is a weakness of the *composition* of its current arsenal. Such MRBMs and IRBMs, if eventually deployed as nuclear systems, would allow China to hold at risk a greater variety of regional targets and do so in a more credible manner than its currently deployed systems. This is especially important for China as it confronts the possibility of either conventional strikes on its nuclear assets or limited nuclear threats by potential enemies against China in a regional crisis. China's ability to credibly deter these two eventualities is currently constrained by the large and blunt nature of its present arsenal, which thereby reduces the credibility of China's threats to retaliate. Thus, the deployment of new and more accurate theater nuclear strike capabilities provides China with the ability to credibly threaten retaliation without raising the immediate specter of an all-out nuclear "wargasm," as Herman Kahn so oddly characterized it 4 decades ago.

CONVENTIONAL MISSILE DOCTRINE AND CAPABILITIES

The Second Artillery's conventional missile forces differ in numerous ways from their nuclear brethren. They are far younger, having been formed in the mid-1990s as a relatively cheap and effective means to threaten Taiwan. The Second Artillery has accorded them a higher priority than that of the nuclear forces.[37] Their numbers are far greater than China's nuclear missiles, and the numbers are growing at a more rapid pace. Their doctrine is far more offensively oriented, as addressed below. Overall, China's conventional missile force is by far the most dynamic leg of the Second Artillery. The PLA's emerging conventional missile strike capabilities have several implications for regional security and stability, irrespective of Beijing's initial intentions in acquiring them.

Conventional missile doctrine and capabilities are also converging, albeit in a different manner than that of the nuclear forces. The doctrine-capabilities relationship with regard to conventional missile forces is highly dynamic because both aspects are simultaneously evolving. Beginning in the mid-1990s, the Second Artillery's acquisition of conventional missiles outpaced the development of a corresponding doctrine for close to a decade; the completion of a comprehensive doctrine for conventional missile strikes *may* have been finalized in recent years—or it may not be complete yet. Doctrine lagged behind force structure capabilities because the latter were developed on the basis of technological availability and the leadership's search beginning in the mid-1990s for a quick and relatively inexpensive way to threaten Taiwan and thus deter actions to a formalize independence.

However, it now appears that capabilities may be lagging doctrine. As the Second Artillery completes development of its conventional missile doctrine, the missiles themselves remain limited in their ability to conduct the types of precision strikes called for by the new doctrine. Therefore, rather than talking in terms of a classic "doctrine-capabilities gap" in assessing the Second Artillery's conventional missile forces, it may be more appropriate to talk in terms of a "doctrine-capabilities dialectic." This latter characterization may help readers better appreciate the dynamism and evolving interrelationship between these two aspects of the Second Artillery's conventional leg.

Conventional Missile Doctrine.

As with nuclear doctrine, there are two central concerns in assessing PLA doctrine for conventional missile operations: the doctrine's institutional development and its content as expressed in various principles. One of the most striking results of my research for this chapter is the finding that the development of conventional missile doctrine is in a nascent stage and thus potentially incomplete. In fact, according to PLA sources, internal "theory development work" on conventional missile operations is far from finished. A PLA study in 2004 stated that "work has just begun" and that "little to no work" has been completed on a "basic system of military theory" for conventional missile operations. This source notes that a Second Artillery goal for the 10th FYP (2001-05) was to *establish* the discipline of conventional missile theory research.[38] The 2004 study stated,

> The development of conventional missile theory has just recently begun in earnest; little to no work has been done in this theoretical research area. It is urgent that we quickly fill in these research gaps. . . . In addition, the units that are responsible for researching conventional missile theory are not yet on a sound foundation; they have insufficiently strong research abilities. This is also a major factor inhibiting the development of Second Artillery military theory.[39]

The expected completion of such work is 2010 (at the end of the 11th FYP). At that point, the task will enter the stages of "refinement and advancement," which are similar to the Second Artillery's characterization of the status of nuclear doctrine during the 10th FYP.

As a further indication of the nascent level of doctrinal development for conventional missile strike operations, the following precursor studies had not yet been completed as of 2004: Conventional Missile Strategy; Conventional Missile Campaign; Conventional Missile Tactics; Conventional Missile Command; Conventional Missile Equipment and Technology; Conventional Missile Logistics; Conventional Missile Systems and Organizations; Conventional Missile Management; Conventional Missile Intelligence; and Conventional Missile Targeting. The first six topics (strategy, campaigns, tactics, command, military hardware, and logistics) were noted as particularly important and would receive top priority in doctrinal development.[40] Completion of these documents would be highly indicative of a fully developed doctrine for conventional missile operations.

Doctrinal Principles. Institutional weaknesses aside, China's professional military writings detail the evolving roles and missions of the Second Artillery's conventional missile forces. The main operational activity of these forces as discussed in PLA literature is conducting "joint firepower attacks" (*lianhe huoli daji*), which appear to be the basic unit of analysis in conventional missile doctrine.[41] Such joint firepower attack" can support at least three types of joint PLA campaigns: (1) firepower campaign (or other independent campaigns), (2) island landing campaign, and (3) blockade campaign.[42] Thus, the Second Artillery conventional missile forces, in contrast to its nuclear ones, contribute to a joint campaign involving coordination with other PLA services.

PLA publications identify six "main combat operations" or applications for the Second Artillery's conventional missile forces: (1) deterrence combat; (2) countermissile attacks; (3) blockade attacks; (4) disturbance attacks; (5) mobile force combat; and (6) firepower combat.[43] These six types of operations can be employed, individually or collectively, in a variety of PLA campaigns. The *Science of Military Campaigns* details several of the "main tasks" of conventional missile strikes, which further underscore the requisite coordination with other PLA services:

- Conducting a combined ground campaign together with the Army and the Air Force to attack key enemy targets in deep areas.
- Supporting the Army, Navy, and Air Force in conducting a sea blockade, an island blockade, or a landing campaign — by attacking key enemy targets such as naval bases, air force bases, and C^{4}I systems, and seizing local campaign control over the sea and sky.
- Conducting a combined air attack campaign together with the Air Force against enemy airports, air defense systems, C^{4}I systems, and other key targets to seize local control over the sky.
- Carrying out other special combat tasks when needed.[44]

The dominant theme in these writings is the *offensive* nature of conventional missile operations, that is, conventional missiles are *not* just for deterrence and retaliation. The PLA emphasizes using conventional missiles to strike first, strike hard, strike precisely, and strike rapidly. The aim of this approach is to "seize the initiative" and quickly gain "campaign control" in order to speed up the process of warfare leading to the adversary's quick capitulation. PLA writings state that the goals of such attacks are to "smash or weaken the enemy's military strength, to politically shock the enemy, to shake the [enemy's] willpower [to wage] war, to check the escalation of war, and to speed up the progress of war."[45] The conceptual importance of preemption and striking critical targets to joint firepower attacks is reflected in the PLA's "guiding ideology" for conventional missile operations—"forestalling the enemy and striking with focus" (*xianji zhidi zhongdian tuji*)—which is repeatedly stressed in PLA publications.[46]

As an indication of the continued evolution of conventional missile doctrine, Li Tilin, then-head of the Second Artillery's Command Academy, relied on the phrase "three strikes and one resistance" (*san da yi kang*) in outlining in 2004 the goals for the development of China's conventional missile force. Li's phrase means to "strike accurately, strike quickly, strike ferociously, and mount a successful resistance."[47] Li Tilin explains these elements as follows:

> Strike accurately means carrying out a precise attack against the enemy's strategic points and vital links, and quickly paralyzing the enemy's operational system. The significance of this approach lies in its ability to gain the most operational interest at minimum cost, reduce collateral damage, avoid falling into a state of diplomatic passivity, and seize the initiative in the battle.
>
> Strike quickly means that in the midst of the constantly changing conditions of the battlefield—and the fleeting moments when an opportunity for victory appears—one must gather intelligence on the enemy's position as soon as possible, manage information in real time, organize command and control with the highest efficiency, quickly launch the attack, and go all out to apply sufficient pressure to enemy forces within the shortest possible time frame.
>
> Strike ferociously means that one must meticulously select key strategic targets; adopt a method such as convergent

strikes, sustained attacks, or multi-wave attacks; fully bring into play the superiority of 'information plus firepower'; and carry out sabotage operations, paralytic strikes, and devastating attacks against the enemy.

Mounting a successful resistance means that in the informationized warfare of the future, when countering enemy attacks against key sectors of our command and control system, we must utilize the survivability and reconstitution capabilities of an informationized command and control system, effectively counter soft and hard casualties inflicted by the enemy, and maintain the continuity and stability of command and control without interruption.[48]

These terms can be expected to evolve toward greater specificity as the PLA and Second Artillery complete the process of formulating a complete doctrine for conventional missile strike operations.

One large area that remains unclear is the range of contingencies to which Second Artillery joint firepower attacks could be utilized. Current PLA writings specify the aims of joint firepower attacks and the three types of joint campaigns to which they could contribute (i.e., firepower, island landing, and blockade). Yet, it is possible that "joint firepower attacks" could be applied to new types of campaigns as PLA needs grow, as the PLA improves its ability for joint multiservice operations, and as the Second Artillery's capabilities for long-range precision strikes improve. This is an area worth watching as conventional missile doctrine is completed and then used to organize, equip, and employ Second Artillery forces.

Conventional Missile Capabilities.

The Second Artillery's conventional missile capabilities appear to be receiving the lion's share of the political and financial resources within the Second Artillery. This force is large (compared to those of regional militaries), rapidly growing, and increasingly accurate and lethal with its armaments. The Second Artillery is also diversifying its conventional capabilities with the development and deployment of new SRBMs, MRBMs, IRBMs, and land attack cruise missiles (LACMs). In terms of its collective capabilities, this missile force is clearly focused on acquiring the capability for precision strike and preemption, as called for in its doctrine. Thus, there is an impressive consistency between existing doctrinal concepts and ongoing force structure modernization.

Most of these capabilities have been developed for employment during a Taiwan conflict, specifically to deter and counter both Taiwan independence and third-party involvement in such a conflict. Yet, as the Second Artillery's conventional leg grows, modernizes, and diversifies, it will acquire capabilities that have broader implications for U.S. security interests in Asia and regional military balances. In particular, the Second Artillery's evolving conventional missile capabilities raise the possibility of America's eventual loss of its operational sanctuary in the Western Pacific.[49]

China's current conventional missile force structure is dominated by two families of SRBMs and one type of MRBM. China currently deploys the CSS-6/DF-15 SRBM (600 km/500 kg) and the CSS-7/DF-11 (300 km/500kg), most of which are garrisoned opposite Taiwan. Both classes of SRBMs use solid fuel and are road-mobile; both also have variants possessing improved accuracy and greater ranges. In terms of MRBMs, China has developed a conventional version of the CSS-5/DF-21 missile, which previously was deployed only with a nuclear warhead. This conventional missile, reportedly known as the DF-21C, has a range of 1770-2500 km and is gradually replacing the liquid-fuel DF-3A as it is phased out of service.[50]

The most pronounced feature of China's conventional missile forces is the sheer rate of growth of its SRBM force. Beginning in 1995, the total CSS-6 and CSS-7 force grew from 30-50 missiles to well over 900 by 2007. The average annual growth rate increased from 50/year in the late-1990s to roughly 100/year by 2004. In 2006, the rate of expansion grew to over 100/year. These trends are detailed in Figure 1.

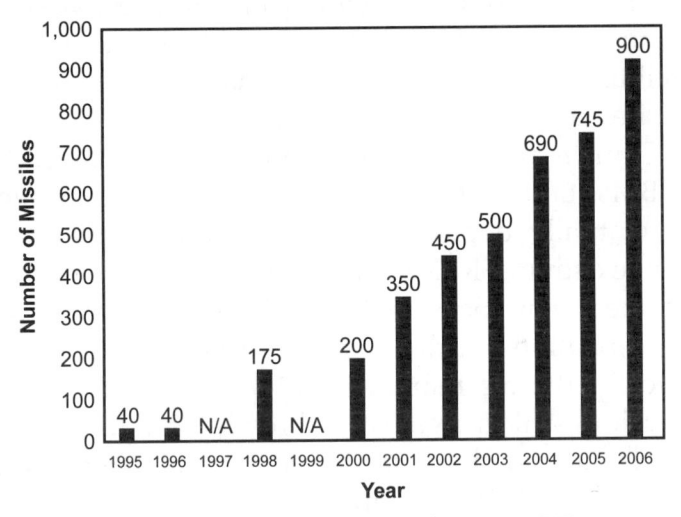

Figure 1. Growth in China's SRBM Force (Average Numbers).

These quantitative trends are matched by qualitative ones. China has developed new variants of both the CSS-6 and CSS-7 with improved range, accuracy, and lethality. China can use these newer variants to strike a broader range of regional targets. For example, a CSS-6 variant can now reach U.S. airbases on Okinawa when forward-deployed near China's eastern coastline. These variants importantly provide the Second Artillery with a true precision-strike capability, which the first-generation CSS-6s and CSS-7s lacked. China has been actively using both terminal homing technologies and satellite-assisted navigation to improve the accuracy of these classes of missiles. China supplements the CSS-6 and CSS-7 onboard guidance systems with global positioning systems (GPS) and indigenous satellite navigation technologies to archive circular error probable (CEP) accuracies reportedly below 50 meters. China is also being assisted by Russia with the application of its GLONASS GPS system to missile guidance and control systems. China's collective access to the trinity of U.S. GPS, GLONASS, and its own satellite systems (like BeiDou) may further improve missile accuracies over time.[51]

China is also developing new conventional IRBMs, MRBMs, and LACMs.[52] These new missile systems, if eventually deployed, offer numerous advantages over existing SRBMs, suggesting that they could become a new focus of Second Artillery procurement and operations in the coming years. First, and most basically, these capabilities will help the PLA conduct precision strikes against a broader range of land-based and, possibly, sea-based targets. China could use IRBMs, MRBMs, and LACMs to target—with increased stand-off distances—foreign military assets located far from the mainland, such as U.S. bases on Guam. These new missile systems—depending on their ultimate range, accuracy, and numbers—could provide China with the capability to threaten all of the U.S. operational sanctuaries in the Western Pacific, further complicating U.S. power projection.

Second, IRBMs, MRBMs, and LACMs provide China with multiple deployment options for attacking targets it can already reach, such as those on Taiwan,

Okinawa, and the Philippines. This provides the PLA with greater operational flexibility. To strike such targets, the Second Artillery missile launch sites would no longer have to hug China's coastline (as required by existing SRBMs) and could operate with relative impunity from inland provinces—potentially creating an operational sanctuary for the PLA.

Third, such missile systems present far more difficult interception targets for U.S. short-range missile defense capabilities such as those deployed in Taiwan and at U.S. airbases in Japan. IRBM and MRBM reentry speeds are often too great for interception by such point-defense systems. Therefore, China could use an initial wave of MRBM and LACM strikes to heavily degrade or even eliminate such missile defense systems; this approach would help to ensure that follow-on waves of SRBM strikes hit their targets with maximum destructive effect.

Moreover, China is reportedly developing an MRBM with a maritime strike capability to target a U.S. aircraft carrier operating in the vicinity of Taiwan. According to a 2006 U.S. Department of Defense report, "One area of apparent investment involves the pursuit of MRBMs, an extensive C^{4}ISR system for geo-location of targets, and onboard guidance systems for terminal homing to strike surface ships on the high seas or their onshore support infrastructure."[53]

China would have to overcome significant challenges related to locating, tracking, and targeting U.S. naval vessels in order to acquire an effective maritime strike capability. However, *if developed*, such a capability would severely complicate the U.S. ability to establish and maintain a 24-hour combat air patrol over Taiwan and the Taiwan Strait during a conflict. Research by retired U.S. Navy Admiral Eric McVaden provides details in China's progress on developing such a missile and its efforts to address the related targeting challenges:

> China is trying to move rapidly in developing ballistic missiles that could hit ships at sea at MRBM ranges—in other words, to threaten carriers beyond the range at which they could engage Chinese forces or strike China. Among its other advantages for China, this method of attack avoids altogether the daunting prospect of having to cope with the U.S. Navy submarine force—as anti-submarine warfare is a big Chinese weakness. . . .[The Chinese are] working diligently to perfect the means to locate and target our carrier strike groups (CSGs). In that regard, an imperfect or rudimentary (fishing boats with satellite phones) means of location and targeting might be employed even earlier than the delay of several more years likely needed to perfect more reliable and consistent targeting of ships. Chinese missile specialists are writing openly and convincingly of MIRV'd ballistic missiles (missiles with maneuverable reentry vehicles) that maneuver both to defeat defenses and to follow the commands of seekers that spot the target ships. . . .[54]

An area of PLA modernization highly relevant to the Second Artillery's conventional capabilities is its ability to locate, identify, track, and target an adversary; such a capability directly contributes to greater missile accuracy. This is a high priority for the PLA, which is making substantial progress. According to the DOD's 2006 report, "China has accorded building a modern ISR [intelligence, surveillance, and reconnaissance] architecture a high priority in its comprehensive military modernization, in particular the development of advanced space-based C4ISR and targeting capabilities." China is taking a number of steps to improve its [ISR], including using constellations of satellites, unmanned aeronautical vehicles (UAVs), and special operations forces to gather targeting data for long-range precision strikes. China's development of a robust and reliable C⁴ISR system will be a critical link in acquiring the capacity for precision missile strikes.⁵⁵

Limitations of Conventional Missile Capabilities. Although the advances in Second Artillery conventional missile capabilities have been steady and substantial, the Second Artillery still faces operational constraints on its ability to effectively carry out precision strikes against a range of regional targets. Some PLA writings have highlighted such limitations; many of these stem from a lack of adequate C⁴ISR systems. It is not clear how widely held these beliefs are within PLA or Second Artillery circles, however. A 2004 article in *Junshi Xueshu* argued that conventional missiles "have their own flaws" and are not weapons with "do-it-all" capabilities (*baodatianxia*).⁵⁶ The article noted several limitations of conventional missiles, as follows:

- "The preparation time for launching conventional [missiles] is lengthy and not suited for instant surprise attacks *after the campaign starts* . . . . They are only appropriate for surprise attacks at the initial stage of a campaign." [Emphasis added.]

- "The launching of missiles is heavily affected by natural conditions and thus [our] lack of [an] all-weather launch capability. Conventional [missile launches] must be conducted under certain weather conditions or their precision is compromised and may even lead to failure. Furthermore, at this stage, our reconnaissance and communication systems are underdeveloped and heavily affected by climate factors. Terrain and climate factors also need to be taken into consideration for troop mobility and camouflage. Objectively speaking, therefore, our conventional missile troops are not yet equipped with all-weather operational capabilities."

- "The launch of conventional missiles is limited by various logistics elements, and thus produces less than ideal surprise attack results on certain targets. Currently, we are unable to precisely position the fixed spot targets or provide high resolution target imaging. Consequently, it is not appropriate to use conventional missiles to attack those spot targets near sensitive areas, such as residential areas, schools, and churches. Concerning small moving targets, effectively capturing, tracking, and providing instant information for moving targets is a capability that has not yet been established. Hence, without guided warheads (*mozhi daodan tou*), conventional missile attacks are unfit."

- "Given their high costs, conventional missiles are not suited for large-scale 'blanket' attacks. Conventional missiles are manufactured with the combination of highly advanced electronics and mechanical technologies. Compared with other weapons, they are rather costly. Therefore, it is only sensible to deploy them based on scientific selection rather than using them as artillery or air bombings as one would in 'blanket' attacks."

Assessing the Doctrine-Capabilities Nexus.

The above claims suggest several arguments about the relationship between conventional missile doctrine and capabilities. First, there is a broad compatibility between doctrine and capabilities in which both are focused on giving the Second Artillery the capability to rapidly and accurately strike a growing variety of targets for the purposes of deterrence (of Taiwan and the United States), escalation control, and escalation dominance. These goals likely include new applications of conventional missiles to missions such as counteraircraft carrier operations. Yet, based on the analysis above, there is little indication to date that the Second Artillery's conventional missile force possesses broader regional missions. This force is strongly preoccupied with Taiwan scenarios. For the time being, the Second Artillery has its hands full completing its doctrine and developing the requisite panoply of conventional strike capabilities.

Second, the doctrine-capabilities relationship is highly dynamic, with both elements steadily evolving. It is not the case of capabilities catching up with doctrine — as with China's nuclear forces. Conventional missile doctrine is just now becoming fully developed as China continues to acquire a range of increasingly precise conventional missiles and related ISR assets. China's conventional missile capabilities still face range limitations. In this sense, force structure is growing into congruence with the evolving doctrinal requirements for conventional missile operations. Given their uncertain future direction, the simultaneous evolution of both doctrine and force structure is worthy of continued attention.

Third, such trends have serious implications for American military planners and U.S. regional security interests. As China improves the range and accuracy of its conventional missiles (especially MRBMs and LACMs), the PLA will increasingly be able to hold at risk in various ways a greater number of U.S. military assets in the Western Pacific, including possibly large naval combatants on the high seas. This will further complicate the U.S. ability to deploy and operate from multiple locations in Asia because such forces could, to a degree, become chronically vulnerable to Chinese missile strikes. Thus the U.S. military might eventually lose its operational sanctuaries in this part of the world. In addition, China's growing conventional missile capabilities could threaten regional military balances. The PLA may eventually possess the capability for conventional counterforce strikes on numerous regional states flowing from to the preemptive and coercive options inherent in China's possession of precision SRBMs, MRBMs, and LACMs. Such capabilities could be used to facilitate coercive diplomacy or contribute to adoption of a broader sea control strategy beyond the waters surrounding Taiwan and China's immediate periphery.

NEW MISSIONS FOR THE SECOND ARTILLERY?

Beyond missiles, the Second Artillery could diversify into new areas of military activity related to other strategic capabilities. The Second Artillery could assume responsibility for China's antisatellite (ASAT) weapons, its computer network attack capabilities, or radio frequency (RF) and laser weapons. China is actively acquiring the former two types of weapons. Such capabilities, if inherited by the Second Artillery, would create a PLA branch having a mandate similar to that of the U.S. Strategic Command. To be sure, such an expansion of the Second Artillery's missions remains decidedly uncertain. Such capabilities remain under development and no part of the PLA has yet been given responsibility for them.

Regardless of whether the Second Artillery eventually gains control of such assets, there are multiple indications that China is working on acquiring a range of such strategic capabilities. In January 2007, China conducted its first successful test of a direct ascent ASAT weapon; it used a ballistic missile with a kinetic kill vehicle to destroy an old Chinese weather satellite stationed in low-earth orbit. News media reports indicated that this test was one of many in the PLA's ASAT development program.[57] According to the Pentagon's 2006 report on Chinese military power, China has initiated a "major effort" to develop RF weapons including high-power radio frequency sources, prime power generators, and antennas to radiate RF pulses. These weapons could be used to incapacitate guided missiles, C^{4}ISR assets, computer networks, and even carrier battle groups. Also, the Pentagon report noted that China will eventually possess the technological capability to produce low- and high-energy lasers, given its commercial work in these areas; these technologies could be weaponized in the future if the PLA so decided. Finally, the PLA clearly sees information warfare and offensive computer network operations as critical to "seizing the initiative" in a crisis. It has been devoting significant resources to developing such capabilities, including development of specialized units and their integration into military exercises.[58]

Such new missions, however, would also create an organizational burden for a Second Artillery that is already redesigning its command structures to better manage its larger and more sophisticated nuclear and conventional missile forces. The command and control demands of mobile nuclear forces, in particular SSBNs, are both new and onerous. Any future control of China's

ASAT, RF, laser, and information warfare capabilities by the Second Artillery would serve as an important indication of its evolving role within the PLA as well as the broader ambitions of the PLA itself.

CONCLUSION: FUTURE SECOND ARTILLERY ROLES AND MISSIONS

In returning to the original mandate of this volume, the preceding analysis suggests several conclusions about the future direction of the Second Artillery. At present, there is a broad degree of congruency between doctrine and capabilities within the Second Artillery's nuclear and conventional forces. While the relationship between doctrine and capabilities in these two legs of the Second Artillery is decidedly different, the degree of overall compatibility suggests that the Second Artillery remains focused on existing missions as expressed in its current doctrine. At a minimum, the Second Artillery is still in the process of acquiring nuclear and conventional missile capabilities to meet existing doctrinal requirements. This process may include new applications of emerging capabilities, but within the context of existing missions. To be sure, there appears to be greater room for doctrinal evolution or "mission creep" within the Second Artillery's conventional forces than within its nuclear forces.

The Second Artillery's nuclear missile forces and related capabilities are trying to catch up rapidly with an increasingly explicit strategy and doctrine premised on using nuclear weapons to deter nuclear aggression and to preclude nuclear coercion. With the imminent deployment of the DF-31 and the looming deployment of a real sea-based nuclear retaliatory capability, China will have substantially reduced its vulnerability and upgraded the quality of its nuclear deterrent. Substantial improvements in ballistic missile early warning and nuclear command and control would contribute materially to accomplishing these goals. It is highly significant that even as China confronts an increasingly complex nuclear threat environment, neither its nuclear doctrine nor its nuclear force structure has radically changed. Rather, the PLA has responded by developing new missile systems (with better accuracy, survivability, and penetrability) in order to hold at risk a greater variety of regional targets. These actions will inject a degree of flexibility into China's retaliatory options so as to bolster the overall credibility of its deterrent. In other words, capabilities are being acquired within existing doctrinal requirements. There are few indications to date that the PLA plans to move beyond the acquired wisdom and principles comprising China's nuclear strategy and doctrine. This is not to say that the future will resist such changes, but rather that several ideological, institutional, and technological constraints persist.

The Second Artillery's conventional missile forces are its most dynamic leg. Both doctrine and force structure are readily evolving. The PLA may just now be completing a comprehensive doctrine for conventional missile strike operations, though for at least the last 5 years the missions for such weapons have been fairly evident. The evolution of conventional missile doctrine bears watching. Joint firepower attacks could become relevant to a range of regional campaigns lying beyond the existing doctrinal focus on a Taiwan contingency. In terms of capabilities, the Second Artillery is deploying increasingly sophisticated SRBMs for launching precision strikes around the mainland's immediate periphery and is procuring LACMs and MRBMs to provide greater operational flexibility, potentially including maritime strikes.

China's acquisition of these latter capabilities bears the closest watching. Given that both conventional missile doctrine and capabilities are evolving at the same time, the potential exists for the PLA to contemplate using increasingly precise, accurate, and lethal LACMs and MRBMs for regional missions lying outside the boundaries of a Taiwan contingency. A mutually reinforcing dynamic could emerge in which new capabilities enable more coercive missions or a broader geographic application of existing missions. This could include using such strike capabilities for anti-access missions in areas other than the Taiwan Strait or missions to facilitate coercive diplomacy for resolving off-shore territorial disputes. Such developments would have serious implications for U.S. force projection in the Western Pacific, the security of U.S. allies and security partners, and regional stability in the Asia-Pacific.

ENDNOTES – CHAPTER 4

1. This chapter benefited significantly from the comments of Roger Cliff, Ron Christman, Taylor Fravel, Bonnie Glaser, Catherine Johnston, Alastair Iain Johnston, Franklin C. Miller, Forrest Morgan, Brad Roberts, and Larry Wortzel. Research assistance was provided by Luke Armerding and Timothy Cook. Of course, any mistakes are my own.

2. The Second Artillery is technically a PLA *branch* (or *bingzhong*), whereas the Army, Navy and Air Force are all higher-level *services* (or *junzhong*). Even though the Second Artillery commander is a member of the Central Military Commission, it is not clear that this distinction is relevant to assessing intra-PLA bargaining for influence and resources. For an excellent overview of the Second Artillery, see Kenneth Allen and Maryanne Kivelhan-Wise, "Implementing PLA Second Artillery Doctrinal Reforms: Emerging Trends in the Operational Art of the Chinese People's

Liberation Army," in James Mulvenon and David Finkelstein, eds., *China's Revolution in Doctrinal Affairs*, Alexandria, VA: The CNA Corporation, 2005, pp. 159-200.

3. To be sure, the "doctrine-capabilities" analysis used in this chapter possesses its own limitations. First, there is a "chicken and egg" type of problem: Is doctrine driving force structure or is doctrine being fitted to existing capabilities? Second, the analytical approach in this chapter begs the question of what Second Artillery doctrine is and how it can be conclusively identified. There are often numerous differences between public doctrine, internal doctrine, and doctrine by dint of operational military planning. This is not an issue unique to China; these various layers existed within U.S. nuclear policy during the Cold War.

4. The distinction between the nuclear and conventional missile forces is treated as important to the analysis in this chapter because of the multiple historical, doctrinal, and operational differences that distinguish these two legs of the Second Artillery.

5. This is likely due to the Second Artillery's deployment of conventional missile forces and their centrality to the PLA's Taiwan strategy.

6. "Dier Paobing Junshi Xueshu" ["Second Artillery Military Studies"], *Junshixue Yanjiu Huigu yu Zhanwang* [*Military Academic Research Review and Prospects*], Beijing, China: Junshi Kexue Chubanshe, 1995, pp. 358-371.

7. John Wilson Lewis and Xue Litai, *China Builds the Bomb*, Stanford, CA: Stanford University Press, 1988; and John Wilson Lewis and Xue Litai, *China's Strategic Seapower*, Stanford, CA: Stanford University Press, 1994, pp. 209-244.

8. The striking disarray and politicization of the Second Artillery in the 1960s and 1970s are chronicled in John Wilson Lewis and Xue Litai, *Imagined Enemies: China Prepares for Uncertain War*, Stanford, CA: Stanford University Press, 2006, pp. 173-214. The author is grateful to Iain Johnston for highlighting the point about the political incorrectness of discussing Western concepts related to nuclear strategy and doctrine.

9. The full name of this document is *Zhongguo Renmin Jiefangjun Dierpaobing Zhanyi Gangyao*. In 1999, *gangyao* were also issued for the Army, Navy, Air Force, and joint operations. They are listed in Xue Xinglin, ed., *Zhanyi Lilun Xuexi Zhinan* [*Campaign Theory Study Guide*], Beijing, China: Guofang Daxue Chubanshe, 2001, p. 636.

10. These debates are detailed in Alastair Iain Johnston, "China's New 'Old Thinking': The Concept of Limited Deterrence," *International Security*, Vol. 20, No. 3, Winter 1995/96, pp. 5-42.

11. Conversation with senior PLA strategist, Beijing, June 2006.

12. Important sources include Wang Hongqing and Zhang Xingye, eds., *Zhanyixue* [*Science of Military Campaigns*], Beijing, China: Guofang Daxue, May 2000; Xue Xinglin, *Zhanyi Lilun Xuexi Zhinan*; "Dier Paobing Junshi Xueshu," *Junshixue Yanjiu Huigu yu Zhanwang*; Zhao Xijun, ed., *Shezhan: Daodan Weishe Zonghentan* [*Deterrence Warfare: A Comprehensive Discussion of Missile Deterrence*], Beijing, China: Guofang Daxue Chubanshe, 2005.

13. These terms accord with a classic conceptual hierarchy in PLA thinking; they are *zhidao sixiang, jiben yuanze, zongti yaoqiu*, and *zhuyao renwu*.

14. All of these Second Artillery documents are listed as a set in this source, with indications of which had been published and which ones were not. This is the basis for my claim that the non-published sources are forthcoming. See "Ershiyi Shiji Chu Erpao Junshi Lilun Fazhan yu Chuangxin" ["The Development and Innovation of 2nd Artillery Military Theory in the 21st Century"], as published in the National Military Philosophy and Social Science Planning Office, eds., *Ershiyi Shiji Chu Junshi Xue Xueke Jianshe yu Chuangxin* [*Development and Innovation of Military Science in the 21st Century*], Beijing, China: Junshi Kexue Chubanshe, 2004, pp. 342-348. This source is hereafter referred to as *EPJLFC*.

15. *EPJLFC*, p. 345.

16. This claim is based on the military sources referred to in this chapter as well as a Central Committee publication. See Pan Xiangchen, ed., *Zuigao Juece: 1989 Zhihou Gongheguo Zhongda Fanglue* [*Highest Decisions: Key State Strategies after 1989*], Beijing, China: Zhonggongdangli Chubanshe, 2004, pp. 740-743.

17. This campaign is detailed in Xue Xinglin, *Zhanyi Lilun Xuexi Zhinan*, pp. 384-393. The issue of what constitutes retaliation for China is addressed in subsequent sections of the chapter.

18. These themes are drawn from several sources including Xue Xinglin, *Zhanyi Lilun Xuexi Zhinan*; Wang Hongqing and Zhang Xingye, eds., *Zhanyixue*; Wang Wenrong, ed., *Zhanluexue* [*Science of Military Strategy*], Beijing, China: Guofang Daxue Chubanshe,

1999; Yao Youzhi and Peng Guangqian, eds., *Zhanluexue* [*Science of Military Strategy*], Beijing, China: Junshi Kexue Chubanshe, 2001; *Zhanyixue Yanjiu* [*Campaign Studies Research*], Beijing, China: Guofang Daxue Chubanshe, 1997, pp. 278-86; Gu Dexin and Niu Yongjun, *Heyouling De Zhengdong—Ershishiji Hewenti Huihu Yu Sikao* [*Rumblings of the Nuclear Specter: Looking Back at and Considering the Nuclear Problem in the 20th Century*], Beijing, China: Guofang Daxue Chubanshe, 1999, pp. 274-88; Hu Guangzheng, *Zhongwai Junshi Zuzhi Tizhi Bijiao Jiaocheng* [*Teaching Materials on a Comparison of Chinese and Foreign Military Organizational Systems*], Beijing, China: Junshi Kexue Chubanshe, 1999, p. 223. For excellent Western analyses of Second Artillery nuclear doctrine see Larry M. Wortzel, *China's Nuclear Forces: Operations, Training, Doctrine, Command, Control, and Campaign Planning*, Strategic Studies Institute, Carlisle, PA: U.S. Army War College, May 2007; and Evan S. Medeiros, "China's Evolving Nuclear Doctrine," in Albert Wilner and Paul Bolt, eds., *China's Nuclear Future*, Boulder, CO: Lynne Rienner Publishers, Winter 2005, pp. 39-78.

19. Dai Siping, "Daodan Budui de Shengzai yu Fangyu" ["The Survivability and Protection of Missile Forces"], *Junshi Xueshu*, February 2003, pp. 67-68.

20. The last discussion of nuclear questions was in the 2000 *Defense White Paper*. This document did not use the term nuclear strategy (*he zhanlue*); it simply stated that "China possesses a small number of nuclear weapons entirely for self-defense" (*chuyi ziwei de xuyao*). The discussion of nuclear issues did not even warrant its own subsection; it was part of a section on "safeguarding world peace, and opposing aggression and expansion."

21. See *www.csis.org/media/csis/events/060620_china_nuclear_report.pdf*; interviews with senior PLA strategists, Beijing, China, June 2006.

22. See Pan Xiangchen, *Zuigao Juece*, p. 742.

23. These concepts are addressed in two PLA sources. See Xue Xinglin, ed., *Zhanyi Lilun Xuexi Zhinan*; and Wang Hongqing and Zhang Xingye, eds., *Zhanyixue*.

24. Charles Ferguson, Evan S. Medeiros, and Phillip C. Saunders, "Chinese Tactical Nuclear Weapons," in Alistair Miller and Brian Miller, eds., *Tactical Nuclear Weapons: Emergent Threats in an Evolving Security Environment*, London: Brassey's, 2003, pp. 110-128.

25. These arguments are part of most of the PLA analyses of nuclear questions cited in this chapter.

26. Interviews with PLA strategists, Beijing, June 2006.

27. Li Tilin, "Dui Dier Paobing Xinxihua Jianshe de Sikao" [*Thoughts on the Development of 2nd Artillery Informationization*], *Junshi Xueshu*, December 2004, p. 55; Zhang Peimin, "Ruhe Fazhan Zhanlue Weishe Shouduan" [*How to Develop Strategic Deterrence Methods*], *Junshi Xueshu*, February 2004, pp. 33-34; Liu Zhenwu, "Lun Guojia Anquan yu Jiji Fangyu Zhanlue" [*A Discussion of the National Security Strategy and Active Defense*], *Junshi Xueshu*, April 2004, p. 8; Dai Siping and Gong Yunchang, "Guanyu He Zhanzheng Wenti de Jidian Sikao" [*A Few Thoughts on the Issue of Nuclear Warfare*], *Junshi Xueshu*, January 2003, p. 15-19; Dai Siping, "Daodan Budui de Shengzai yu Fangyu" [*The Survivability and Protection of Missile Forces*], *Junshi Xueshu*, February 2003, pp. 67-68.

28. A good example of this is Dai Siping and Gong Yunchang, "Guanyu He Zhanzheng Wenti de Jidian Sikao."

29. Xue Xinglin, *Zhanyi Lilun Xuexi Zhinan*, p. 398; Guo Shunyuan and Wang Heng, "Zhanyi Zhong de Dier Pao Zuozhan Zhihui Juece" ["2nd Artillery Command & Control Decision-Making During a Joint Campaign"], *Junshi Xueshu*, July 2004, p. 65.

30. Shang Yi, "Zhu Chenghu: Foreign News Agency 'Cites Out of Context'," *Ta Kung Pao*, July 17, 2005 as translated by *Foreign Broadcast Information Service* (FBIS) in *CPP20050718000113*. In this media report, "military commentator" Li Yunsheng stated, "The Chinese military and society are considering how China should react at a critical juncture if national security comes under substantial threat." He added that "some countries' precision guided weapons have similar power to that of a small-scale nuclear weapon, thus if China's nuclear facilities are attacked by such weapons, it actually implies the outbreak of a nuclear war, hence the country will consider the use of nuclear weapons under such circumstances."

31. See *www.csis.org/media/csis/events/060620_china_nuclear_report.pdf*.

32. On this latter point, see Dai Siping and Gong Yunchang, "Guanyu He Zhanzheng Wenti de Jidian Sikao"; these Second Artillery operators used Gorbechev's "nuclear thinking" as

a straw-man argument to criticize the general propositions that nuclear war is unwinnable, and that nuclear escalation is uncontrollable. Yet, following such analysis, they did not call for changes to China's nuclear doctrine, which is a natural implication of their arguments. Rather, they merely confirmed current CCP nuclear policies related to NFU and using nuclear weapons for retaliatory purposes. This approach was strongly suggestive of writings that seek to "push the envelope" about current beliefs about nuclear warfare, while also not questioning prevailing doctrinal orthodoxy.

33. This section draws from the data included in the following reports: U.S. Department of Defense, *Annual Report to Congress on the Military Power of the People's Republic of China 2006,* Washington, DC: Office of the Secretary of Defense, 2006; U.S. Department of Defense, *Annual Report to Congress on the Military Power of the People's Republic of China 2007,* Washington, DC: Office of the Secretary of Defense, 2007; U.S. Department of Defense, *Annual Report to Congress on the Military Power of the People's Republic of China 2005,* Washington, DC: Office of the Secretary of Defense, 2005; Shirley Kan, *China: Ballistic and Cruise Missiles,* Washington, DC: Congressional Research Service, August 2000.

34. These missile ranges are taken from *Annual Report to Congress on the Military Power of the People's Republic of China 2006,* pp. 27, 50.

35. U.S. Department of Defense, *Annual Report to Congress on the Military Power of the People's Republic of China 2007,* Washington, DC: Office of the Secretary of Defense, 2007, p. 42.

36. *Annual Report to Congress on the Military Power of the People's Republic of China 2006,* pp. 27, 50. It is likely that if China deploys nuclear-capable ALCMs, these would be controlled by the PLA Air Force, and the nuclear-capable LACMs would be controlled by the Second Artillery.

37. On this point, see *EPJLFC,* p. 345, as noted earlier in the chapter.

38. The corresponding task for the Second Artillery nuclear forces was to *refine* nuclear theory during the 10th FYP.

39. *EPJLFC,* p. 344.

40. This list of documents is taken directly from *EPJLFC,* pp. 344-345.

41. Earlier PLA writings talked about the "conventional missile strike campaign" as the unit of analysis. More recent writings frequently mention "joint firepower attacks" as a core activity of conventional missile forces and as contributing to a variety of PLA joint campaigns.

42. Zhang Qihua and Lui Xiangfei, "Lienhe Huoli Daji Bu Tong Jieduan de Changgui Daodan Huoli Yunyong" ["Applications of Conventional Missile Firepower on Different States of Joint Firepower Attacks"], *Junshi Xueshu*, February 2003, pp. 29-30.

43. Wang Hongqing and Zhang Xingye, *Zhanyixue*.

44. *Ibid.*

45. Xue Xinglin, *Zhanyi Lilun Xuexi Zhinan*, p. 394.

46. *Ibid.*, p. 395.

47. Li Tilin, "Dui Dier Paobing Xinxihua Jianshe de Sikao," p. 55. As noted in the previous section of this chapter, Li Tilin's concept for nuclear force development is "first resist, later attack."

48. *Ibid.*

49. Following the demise of the Soviet air and naval threat at the end of Cold War, the United States has since possessed the unchallenged ability in the event of a crisis to impose air and sea dominance in the regions around its military bases in the Western Pacific (e.g., Japan, Okinawa, and Guam). And as China and other potential adversaries have not had ballistic missiles in sufficient numbers with the ranges and accuracies needed to seriously impede operations at those bases, U.S. commanders in this theater have enjoyed a condition of "operational sanctuary." The latter term describes a situation in which U.S. commanders could act with confidence that, in the event of war, they could conduct high-tempo military operations with minimal threat of interruption by enemy attack. China's possible future deployment of accurate SRBMs, IRBMs, and MRBMs would significantly undercut such a condition.

50. See Eric A. McVadon, "Recent Trends in China's Military Modernization," testimony before U.S.-China Economic and Security Review Commission, September 15, 2005, at *www. uscc.gov/hearings/2005hearings/written_testimonies/05_09_15wrts/ mcvadon_eric_wrts.php.*

51. On this point, see *Annual Report to Congress on the Military Power of the People's Republic of China 2006*, p. 33. This report suggests

that China's BeiDou satellite system alone is not sufficient for the tracking and targeting required for higher missile accuracies.

52. *Ibid.*, pp. 11, 25, 26, 29.

53. *Ibid.*, pp. 25-26.

54. On the threats to U.S. maritime forces posed by MRBMs, see McVadon, "Recent Trends in China's Military Modernization."

55. *Annual Report to Congress on the Military Power of the People's Republic of China 2006*, pp. 31, 27-29.

56. Ren Mingyong, "Bianzhengdi Renshi Changgui Daodan Budui Zai Lianhe Huoli Daji Zhong de Zuoyong" ["Dialectically Recognize the Roles of Conventional Missile Troops in Joint Firepower Attacks"], *Junshi Xueshu*, February 2003, pp. 33-34. Ren Mingyong is identified as a research student in the Second Artillery's Command Academy located in Wuhan, Hubei province.

57. Edward Cody, "China Confirms Firing Missile to Destroy Satellite," *Washington Post*, January 27, 2007, p. A8. For two intriguing analyses of the test, see Kevin Pollpeter, "Motives and Implications Behind China's ASAT Test," *The China Brief*, Jamestown Foundation, Vol. 7, No. 2, January 24, 2007; and James C. Mulvenon, "Rogue Warriors?—A Puzzled Look at the Chinese ASAT Test," *China Leadership Monitor*, No. 20, Winter 2007. For an analysis of Chinese writings about ASATs, see Michael P. Pillsbury, *An Assessment of China's Anti-Satellite and Space Warfare Programs: Policies and Doctrines*, report prepared for the U.S.-China Economic and Security Review Commission, January 2007, at *www.uscc.gov/researchpapers/2007/FINAL_REPORT_1-19-2007_REVISED_BY_MPP.pdf.*

58. See *Annual Report to Congress on the Military Power of the People's Republic of China 2006*, pp. 31-36.

CHAPTER 5

PLA COMMAND, CONTROL, AND TARGETING ARCHITECTURES: THEORY, DOCTRINE, AND WARFIGHTING APPLICATIONS

Larry M. Wortzel

This chapter examines contemporary Chinese People's Liberation Army (PLA) military publications, military literature, reports on exercises, and equipment to determine how the PLA is incorporating new information technology in its force and how the technology will be integrated into China's warfighting architecture. I find that at the intellectual level, the PLA understands the way technology has driven a "revolution in military affairs" (RMA) affecting how commanders organize forces and how those forces coordinate on the battlefield. I argue that, for the most part, PLA military theorists are learning to apply technology to war by watching how the U.S. armed forces have experimented with technology and performed in combat.

Senior PLA leaders and military strategists consider the United States to be the most advanced military force on which to model their own military development. They also see the United States as the most advanced and likely potential enemy; to counter this enemy, they may need to employ the latest means of command, control, communications, computers, intelligence, surveillance, and reconnaissance (C4ISR). In the view of many in the PLA, it is the demonstrable power of the United States, and their concern that the United States has the potential to use that power to coerce or dominate China and its interests, that requires the PLA to follow U.S. military developments more carefully than those of other nations. The *White Paper on National Defense* released by China's State Council in December 2006 acknowledges that "a revolution in military affairs is developing in depth worldwide," noting also that "military competition based on 'informationalization' is intensifying."[1] "Hegemonism and power politics" are seen as intensifying, a code phrase often used as an indirect way to characterize the United States. In response to these concerns, the authors express the belief that the PLA must be prepared to fight or counter American forces. Senior leaders and military strategists have developed both theory and doctrine for the employment of information warfare concepts by the PLA. More importantly, China's military forces have developed a networked warfare architecture that is effective on a limited scale.

The long-term goal of the PLA is to create a more modern force that can challenge (or deter) the best military forces in the world.[2] Therefore, PLA military thinkers use the United States as the model for the force they must train to counter. However, China's military today is still not a uniformly high-technology force. A number of systems are able to work at sophisticated levels, but across the spectrum of its military systems, the PLA cannot field or operate a fully digitized force. The PLA understands and is working to apply "network-centric warfare" concepts, but lacks a comprehensive set of data transfer systems necessary to field and maintain a modern force that employs these concepts in warfare in a uniform way. It may be 2 to 5 years until, in the Asia-Pacific region, the PLA achieves anything close to the level of networking that U.S. forces can apply globally today.[3]

PLA military theorists are convinced that to be successful in battle in the information age, any commander must be able to use integrated C4ISR systems.[4] In the theoretical realm and in doctrine development, the PLA has been aggressive and quite successful in grasping the RMA. Senior PLA leaders do more than merely discuss information operations, they incorporate them in force-on-force field exercises.[5] New purchases of equipment from Russia and technology from Europe are part of a limited warfighting architecture that depends on C4ISR technology. Moreover, they are applying the technologies and weapon systems to platforms that may be decades old.

As the PLA studies aspects of network-centric warfare and the C4ISR systems that such warfare requires, its theorists see the U.S. armed forces as "the gold standard" on how to apply information technologies and automated electronic data exchange to war.[6] The major works on the subject in PLA military literature are drawn from American military manuals or scholarship on modern war. There are no explicit calls from senior Chinese leaders to prepare for war against the United States, but it is clear that the PLA sees American forces as presenting the greatest challenge China's military could face.

Perhaps the most authoritative long-term guidance to the PLA on the subject of C4ISR and networked warfighting architectures is from General Zhang Wannian. He tells the PLA that "command and control systems must be 'networked' to increase the effectiveness of combat units . . . which will naturally be accompanied by a reduction in the number of layers of command and control."[7] Zhang was chief of the General Staff Department of the PLA from 1992 to 1995 and vice chairman of the Central Military Commission (CMC) after that. While he was CMC vice chairman, he edited the treatise *China's National Defense and Contemporary World Military Affairs*, published by the PLA Academy of Military Science. General Zhang, citing the experience of the U.S. armed forces, says that the process of digitization and networking reduced the number of layers of higher command from five to three in American command and control practice. He predicts that the PLA can expect similar results to produce a "comprehensive system of networked forces and command and control."[8]

Some of the PLA literature is not altogether realistic about what changes digitization and information technology will produce in the conduct of war. One author, a veteran of the fighting on the Sino-Vietnam border, believes that creating a high-technology force able to engage in sophisticated information operations has the potential to make warfare "more limited, less bloody, and less destructive."[9] The examples he uses are from the war in Kosovo and the former Yugoslavia. Writing in 1997, however, the author clearly did not foresee the way that urban guerrilla operations, improvised explosive devices, and suicide attacks tend to negate the blood-sparing potential of high-tech systems through the killing and maiming of so many troops and civilians. In a similar idealistic assessment, a PLA armor officer opined that "electronic warfare operations can be conducted without violating another country's sovereignty"; therefore, he believes that any enemy response is likely to be in kind.[10] This armor officer clearly has not spent much time thinking about kinetic responses to information or electronic attacks.

THE U.S. MILITARY AND ITS INFLUENCE ON CHINA'S C4ISR PROGRAMS

The PLA follows foreign military developments closely, paying special attention to what is going on in the United States. The lessons of the Falklands War, the first Gulf War, and Kosovo awakened China's military theorists to what technology does in the realm of war, as did American debates about the RMA and information warfare.[11] The performance of American forces also convinced the PLA that if it had to confront the United States, it faces a formidable enemy. General Zhang Wannian, then chief of the General Staff Department of the PLA, argued that "modern limited warfare under high-technology conditions is conducted under a cloud of a threat of becoming a nuclear war" and that China must therefore pay special attention to the great nuclear powers.[12] In a discussion of the first Gulf War, he suggests that the "forces of hegemony in the world will use nuclear weapons to dominate other nations," a clear reference to the United States as a potential enemy. Moreover, Zhang suggests that the United States is of special interest to the PLA because China's nuclear weapons can be used to "deter moves to split the sovereign state," a reference to Taiwan.[13] Therefore, it is the power of the United States, and the potential to use that power to coerce or dominate China and its interests, that requires the PLA to follow U.S. military developments more carefully than those of other nations.

One of the most respected PLA strategists and leaders, Lieutenant General Li Jijun, makes it clear why the PLA spends so much of its efforts preparing its forces to confront the United States. Li commanded a Group Army in Manchuria and was responsible for the ground warfare experiment that validated combined arms group armies in the PLA. Later he was the director of Deng Xiaoping's military office. In his evaluation of contemporary world security threats, General Li Jijun sees the major problem facing China as being "large countries" that create "threat theories, including the countries that espouse the 'China threat theory'."[14] This is a clear, albeit indirect, reference to the United States as the nation with the most capability to threaten China because of its policies, its military power, and its alliances.

Li draws historical parallels between England in the Napoleonic age and the United States today. He says,

> like England (in the Napoleonic age), the U.S. is the world's strongest power; the United States has the greatest number of international interests and "colonial" [-like] relationships; U.S. military power is dispersed widely throughout the world; the wide range of interests and military deployments mean that U.S. forces are over-committed and stretched thin; and there is a great need to work with allies and coalition partners to achieve security goals.[15]

Major General Wang Baocun of the Academy of Military Science summarized his view of the United States this way:

> The new military transformation has led to the rise of a United States possessed of overwhelmingly dominant military might. The United States is also an arrogant country with strong ambitions for hegemonism. The United States will take advantage of its absolute superiority in supreme military might in order to pursue power politics and hegemonism, seek to maintain its position as the world's only superpower, and slow down the process of multipolarization for the world's strategic structure.[16]

Views of this type seem to represent official policy at the highest levels in China. The 2006 *White Paper on National Defense* complains that "a small number of countries . . . have intensified their military alliances and resorted to force or threats." The same paper expresses concern that Japan and the United States "are strengthening their military alliance in pursuit of operational integration," and that "hegemonism and power politics remain key factors undermining international security." These views, which put the United States and its alliances at the center of China's threat perceptions, fuel the PLA's efforts to build a modern, information-based, digitized military force. Indeed, even if the PLA did not envision seeking a direct confrontation with the United States, an awareness that the two countries could clash in the event of a Chinese attack on Taiwan is enough to drive PLA modernization. The general tendency in Chinese security thinking to be prepared in the event that a larger power seeks to coerce or dominate China also flows from this sort of analysis. Such concerns drive the PLA to modernize itself.

PLA researchers are quite aware of the data links that support combat systems for the U.S. military, and they have created a catalogue of the knowledge necessary to replicate, counter, or attack them. Two PLA Air Force authors, Sun Yiming and Yang Liping, have built a virtual roadmap for attacking joint U.S. data control systems and communications. They have carefully consulted dozens of corporate websites and tactical data link operator guides, as well as North Atlantic Treaty Organization (NATO) and U.S. military tactical and technical manuals, to produce a guidebook for electronic warfare and jamming to disrupt critical U.S. cooperative target engagement and C4ISR data links. Moreover, the two authors have produced other books and manuals on how to disrupt tactical- and campaign-level missile operations and U.S. electronic systems.[17]

In a PLA National Defense University text on nuclear warfare and nuclear strategy, researcher Wang Zhongquan notes that strategic command and control networks "have multiple uses and systemic effects."[18] Such networks, Wang concludes, "can contribute to command and control systems, strategic warning systems, and intelligence organizations when linked together in a network. The parts of a network of this type include defense communications networks, satellite communications systems, national military command and control networks, and networks of strategic or regional command and control centers."[19] Wang goes on in the book to provide a sophisticated analysis of the U.S. strategic warning system and nuclear command and control network based on a review of published literature in the United States.[20]

This subject as seen through Western eyes is relevant here. The U.S. effort to "harness the revolution in military affairs" was a way to take advantage of "technological leaps in surveillance, command and control, and longer range precision guided munitions" in order to make joint military forces more effective in war.[21] The RMA required the United States to explore a range of force structure issues and changes that revolve around advances in technology and weapons requiring "information-empowered, dominantly knowledgeable forces" that fight in "flattened, less hierarchical organizations."[22] The U.S. Navy may well have led the way in linking C4ISR with the concept of "timely, sensor-to-shooter information direct to the warfighter."[23] All the services caught on, however, in an effort to link command and control systems, information technologies, dissemination systems, and space assets to "strike targets with an accuracy of feet from standoff distances."[24]

These efforts were not lost on the PLA. One PLA Academy of Military Science researcher expressed the view that to engage in modern war the PLA must be able to "attack the enemy's knowledge systems and such high value targets as communications, carrier battle groups, and aviation warfare units."[25] The goal set for the PLA by this researcher was to "destroy the enemy's ability to fight and control war."[26] Moreover, the PLA's information warfare battle doctrine was largely drawn from U.S. manuals, such as U.S. Army Field Manual 100-6, *Information Warfare Doctrine*.[27]

PLA generals working on military transformation have mined the literature and experience of Western military forces for ideas on incorporating information technology into military doctrine and how to build forces that can function in the information age.[28] In fact, in an interview with a *Liaowang* reporter, one military analyst, Major General Zhang Ling, expressed the view that "informationized war of the future will be second only to nuclear war in terms of firepower" when modern weapons are linked to technology.[29]

Addressing how the RMA has affected warfare, Li Bingyan, a major general on the editorial staff of the PLA's newpaper, *Jiefangjun Bao*, pointed out in a recent book that new forms of warfare involve more than massing troops or massing fires against an enemy. Instead, the introduction of high-technology warfare means that to wage modern war, the PLA must be able to "use precision guided missiles" instead of massing traditional fires, and also be able "to use viruses to attack enemy computer systems, and to carry out electronic warfare to attack enemy command and control systems."[30] A significant focus of Li's book is to encourage PLA officers to think in terms of traditional Chinese strategies and classics of military theory, such as *The Romance of the Three Kingdoms* and *The Thirty-Six Stratagems,* but Li encourages them to apply the lessons from the Chinese classics to the modern battlefield.[31] Thus, any Western military force facing the PLA must be prepared for *adapted* applications of technology, somewhat different from those they might expect from a contemporary Western armed force.[32]

THE PLA AND C4ISR IN MODERN WAR

Military theory in China focuses on warfare across a battlefield of five dimensions, or "domains" (or "realms") of war, as they are called in PLA military literature. These five dimensions are land, sea (including undersea), air, space, and the electromagnetic spectrum (some authors refer to the "information realm or domain" instead of the electromagnetic spectrum).[33] PLA military science experts believe that new technology and the development of automated systems have made strategic cues and warning, intelligence, communications, and command and control more critical in all of these dimensions of warfare.[34] Moreover, PLA authors express the view that "information age warfare has broken down the traditional levels and structure of command."[35] Some believe that "military forces must structure themselves around the latent capacities of information."[36] Senior American officers, like Admirals William Owens and Jeremy Boorda, also concluded a few years earlier that the RMA and information systems would generate a restructuring of forces.

Among the "domains" of war, the PLA particularly emphasizes space, with some strategists arguing that the PLA must be ready to conduct warfare in that domain. One of the PLA officers interviewed in the *Liaowang* article cited earlier in this chapter, Major General Zhang Ling, expressed the belief that "control of space will be of tremendous significance in future information warfare [with] the primary combat operation in future war [being] the struggle for space control."[37] Zhang opined that militaries will engage in "soft strikes" against space-based information systems to neutralize enemy satellites and "hard" strikes to destroy enemy space systems with anti-satellite weapons.[38] Addressing rules of engagement in space, he was clear that in space warfare over 120 kilometers above the earth's surface, there are no restrictions related to national sovereignty on military combat operations. Two researchers, Song Yongxin and Guo Yizhing, make similar points in an aeronautics electronics countermeasures journal published in Nanjing. They argue that warfare in space will be part of the information warfare battlefield and that "whoever controls space will have the initiative in war."[39]

Senior PLA officers take a view toward the effects of the RMA on a military's force structure similar to that of American military thinkers such as Owens and Thomas Mahnken. Owens and Mahnken believed that the RMA and the advances in C4ISR would have a radical effect on force structure and warfare, changing organizational structures and even modes of war. Despite the emphasis on automation and electronic systems, however, PLA writers still believe there must be a "man-in-the-loop" in information age warfare with a "strong will and a clear mind" because even "advanced computer systems are no substitute for the strength of the human brain."[40] However, the discussion of the importance of human control and decisions does not explain exactly what that means and how military planning will integrate the "man in the loop" in modern war. There is no explicit recognition that automation will speed engagement decisions. In any event, it is unlikely that the PLA will ever opt for permissive automated action links in place of redundant human systems in making firing decisions.

In an interview with a correspondent from *Qingnian Cankao*, Major General Li Deyi of the Academy of Military Science stated firmly," "It would be inconceivable [today] if a commander in the PLA did not know how to operate a command automation system."[41] But Li opined that the PLA had fallen behind both Russia and the United States in developing an automated command and control system, with the current system being "plagued by inadequate integration and coordination, as well as incompatible [foreign] imports."[42] The PLA therefore understands its problems and envisions eventually correcting them with indigenous systems.

Xin Qin, in his book, *Warfare in the Information Age*, argues that the side with the most comprehensive command and control system in a modern war will also have the strongest maneuver capability and be able to concentrate the greatest combat strength against the enemy.[43] According to Xin, good command and control systems, including sound communications, facilitate maneuver and thus the capability of a nation's military forces to exploit the strengths of mobility and weapon systems in war. This is not a new concept for the PLA. In a 1994 book on information warfare, PLA authors argued that "information technology is the core connecting link for high-technology command and control."[44]

In exercises, PLA commanders challenge their staffs through simulations of extended periods of combat. In one exercise scenario, they intentionally created a "highly informationized" Blue Force that overwhelmed a PLA Red Force operating at a C4ISR disadvantage.[45] The exercise was designed to demonstrate to a PLA division (the Red Force) the advantageous effect of sophisticated reconnaissance and networked command and control systems. These networked systems supported a Blue Force long-range precision strike on the Red Force. The exercise scenario timed the effects of the strike to disrupt the Red Force in its assembly areas as they were forming for maneuver operations.[46] The exercise planners included scenarios of imitative communications deception (e.g., fraudulently joining the enemy's net) and jamming as part of electronic warfare play to confuse Red Forces. Senior PLA leaders were able to demonstrate to subordinate leaders and troops the disadvantages under which the PLA operates in facing a sophisticated enemy with advanced C4ISR systems. The effort reportedly convinced junior PLA leaders and staff officers of the need to field and master such systems for use at the divisional level of combat.[47]

Moreover, at the highest levels of the PLA, senior officers understand that to increase the effectiveness of combat units, the Chinese military must digitize and network its command and control systems.[48] A decade ago, Zhang Wannian emphasized the importance of decisive action in warfare, aided by C4ISR systems that could locate the enemy, control attacks on that enemy, and ascertain the effectiveness of those attacks.[49] The speeches of various PLA leaders at the All-PLA Military Training Conference in June 2006 reflect this broad understanding of the way that C4ISR and information systems affect the battlefield. Jinan Military Region (MR) commander Lieutenant General Fan Changlun made the point that an integrated combat capability requires scientific and technical training, the aim of which should be winning a war. He stressed "informationization, real war simulation, and field training" as the focus of the MR's training efforts.[50] Zhu Wenquan, commander of the Nanjing MR, also discussed the importance of networked training systems, information systems, and electronic databases in creating a modern military force.[51]

Layers of Command and Control.

The PLA as an institution is relatively flexible in layering its command and control structure. Many of its elements still reflect back on the doctrine of "people's war." For example, contemporary military command and control systems routinely involve political, government, and Communist Party organizations inside the fronts or military regions in the command group organization.[52] The structure of a "command and control joint campaign warfighting coordination organization," however, varies according to the "objectives of the campaign, the scale of the campaign, and the actual conditions on the battlefield."[53] Command and control structures, therefore, are both pre-planned and task-organized when needed.

The "supreme command headquarters" (*tongshuaibu*) is the joint command and control organization for a campaign.[54] This level of headquarters may be at the MR or war front level in a single-front or MR war. However, a higher headquarters may be established on the decision of the General Staff Department and CMC for a large-scale war of two or more fronts.[55] The command and control structure and task organization are laid out reasonably well in Xue Xinglin's *A Guide to the Study of Campaign Theory*. The "supreme command headquarters" includes command group representatives from the CMC, the General Staff Department and other General Departments, the PLA Navy, the Air Force, and the Second Artillery. It is a "higher command headquarters with great power and responsibilities."[56]

The next echelon of command and control down from the *Tongshuaibu* is the "War Zone" or "frontal" joint command and control organization. In cases where a campaign is limited to a single war zone or front and the forces assigned to the front are sufficient for the campaign, then the military and political leadership in the war zone will form the War Zone Joint Warfighting Command and Control Organization Headquarters. The commander of the war front can draw from local political, military, and Communist Party organizations. This headquarters "executes orders from the higher supreme command headquarters, the Central Military Commission, and the General Staff Department."[57]

As a third echelon of command and control, in large-scale operations, the PLA may form Army Groups that include more than one Group Army and command groups from the PLA Air Force, Navy, and Second Artillery. In a major front on a large scale, there may be two or more Army Groups subordinate to a war zone headquarters. Representatives from the local political, military, and Communist Party organizations needed to support the Army Group would be assigned to this level of headquarters as well.

Headquarters at all levels may include representatives from other control centers, and, as needed, the PLA may task organize the main command and control center, an alternate command and control center, a forward command center, and rear area command centers for logistics purposes.[58] All of these command centers could include local political, military, and civil defense representatives, Communist Party representatives, and representatives from other PLA arms and services. The propensity to draw on the local populace and use personnel from local universities demonstrates the continued tradition of employing certain vestiges of people's war on the informationized battlefield.

This structure has been implemented in the past at the levels described. For the 1979 attack on Vietnam, the PLA established a major supreme command headquarters at Duyun that included a forward command element from the General Staff Department and the CMC. It controlled two war zones, one centered in the east on Guangzhou MR and one in the west that included forces from the Kunming and Chengdu MRs.

For the purposes of this chapter, several points bear emphasis. First, the inclusion of local forces and local political and Chinese Communist Party (CCP) organizations means that concepts of "people's war" still have a place in PLA doctrine. Second, the PLA is very flexible in task organizing. A frontal headquarters commander in a war zone can draw from educational institutions, reserve units, towns, or industries in the zone as required for the support of his forces.[59] In addition, at least the conventional and short-range missile forces of the Second Artillery are included in the structure. Whether they have any nuclear weapons with them is not clear, and how this command and control structure relates to Second Artillery firing orders needs more research. In any case, it is not known whether the Second Artillery or CMC cell or representative in a military region or frontal headquarters has the authority to approve or countermand firing orders. Nor is it clear how free a frontal commander may be to initiate a firing order for Second Artillery units in the war zone. In *Jiefangjun Bao,* articles have referred to the PLA Navy headquarters as the Navy's *tongshuaibu,* thus reinforcing the possibility that operational firing orders at frontal or military region level could come from the local commander.[60]

Nuclear Command and Control.

Despite the lack of clear definition on the degree of control exercised by frontal commanders over assigned Second Artillery firing orders, a number of PLA sources make it clear that command and control for missile forces is highly centralized. Two PLA officers addressing strategic systems in the book, *Missile Combat in High-technology Warfare,* describe Second Artillery command and control this way: "The nodes in a ballistic missile command and control network are (1) the commander in chief [or supreme command authority] (*tongshuaibu*),[61] (2) the command organizations of the military departments, (3) the missile bases, and (4) the firing units."[62] Furthermore, they emphasize that "where it concerns strategic missiles, the ability of the supreme command authority to control firing orders must be executed quickly, and firing orders must be encrypted (encoded)."[63] Finally, PLA manuals specify that "the war positions of the Second Artillery are established by the supreme command authority (*tongshuaibu*) in peacetime and are dispersed over a wide area for strategic reasons."[64]

On the 40th anniversary of the founding of the Second Artillery, Hu Jintao spoke to an assemblage of people that included Xiang Shouzhi, first commander of the organization, and a number of previous leaders. Hu was present in the combined capacity of President of China, Chairman of the CCP, and Chairman of the Communist Party CMC.[65] He wore a PLA uniform without insignia or rank. In the account of Hu Jintao's speech published by *Xinhua News Service*, Hu is quoted as saying that "the Second Artillery Corps is a strategic force directly commanded and used by the Party Central Committee and the Central Military Commission and is our core force for strategic deterrence."[66] In the case of strategic systems, it is clear that the supreme command authority for the PLA is the CMC.[67]

Second Artillery command orders are centralized, encoded, and protected, and require human authentication. As we noted earlier, PLA military writers do not endorse completely automated command and control systems. The PLA's preference for human control of decisions and a "man in the loop," even in modern, information age warfare, comes out clearly in the literature on the subject. The guiding mantra for the Second Artillery is to "strictly protect counterattack capability and concentrate [nuclear] fires to inflict the most damage in the counterattack."[68] Authorities emphasize that the Second Artillery's strategic warning system is closely tied to the General Staff Department, and that the Second Artillery must continually keep current an estimate of whether the enemy will use other forms of weapons of mass destruction.[69]

NETWORK-CENTRIC WARFARE AND OFFENSIVE ACTION

At the theoretical level, at least, the PLA seems to have grasped the implications of "a knowledge infrastructure for Network-Centric Warfare."[70] In other words, China's military leaders believe that communications and electronic data exchange are the core of an integrated warfighting capability. Researchers at the PLA's National Defense University of Science and Technology in Changsha are clear that the evolution of information technology and its incorporation into weapons and strategies will make a networked military force more effective. In arriving at their conclusions, these researchers have drawn on writings by U.S. and European scholars on web technology and computer languages as well as U.S. Department of Defense publications on network-centric warfare. They have followed all of the published literature in the United States on advanced warfighting experiments and battle laboratories. Important questions remain, of course, as to how deeply this theoretical knowledge has penetrated into the PLA and how widely it is applied across the PLA.

At other PLA academic institutions, sophisticated efforts have been under way for some time to improve joint operations and increase the effectiveness of attacks on ground targets by air and naval forces.[71] Two graduate students at the PLA Naval Engineering Institute published a paper analyzing ways to apply C4ISR systems in network-centric warfare more effectively.[72]

Younger officers can be quite aggressive about the potential for using C4ISR systems to improve the PLA's ability to wage offensive operations. One officer from the Navy Command Academy is clear that "the Second Artillery is the major factor in successfully attacking an enemy naval battle group."[73] To accomplish such an attack, the PLA must use all of its electronic warfare and reconnaissance assets properly, must neutralize enemy anti-missile systems and missile sensor systems, and should use electronic jamming on the enemy fleet. Such combined kinetic and electronic attacks help the PLA attack an enemy fleet or naval base with a combination of explosive, anti-radiation, and fake warheads to deceive enemy radar and sensor systems and defeat a deployed battle group or one in port.[74]

For some time, American naval officers have dismissed the idea that China could conduct an attack on a deployed naval battle group as being beyond the grasp of the PLA. They reasoned that since China does not have the space sensor systems to detect warships at sea or the maneuvering warheads required to execute such an attack, there was no credible threat from China in this area. However, PLA officers seem convinced that using ballistic missiles to attack naval battle groups is a viable concept, and they obviously are actively pursuing the capability.

Two officers from the Second Artillery Engineering College have studied how to modify the trajectory of a maneuverable warhead on its reentry into the atmosphere to determine the effective range for attacking an enemy aircraft carrier with ballistic missiles.[75] They conclude that providing terminal guidance will allow up to 100 kilometers of maneuverability on reentry during the terminal phase of a missile attack. They believe that a carrier "cannot effectively escape an attack within a short period of time."[76] Simulations to predict how the final attack ranges against moving targets at sea will affect maneuvering reentry vehicles are also part of the research agenda for Second Artillery engineering officers.[77] They have concluded that since a carrier battle group can project force out to about 2,500 kilometers, the PLA must reduce its missile warhead circular error probable (CEP) to attack maneuvering targets at sea from outside the carrier's strike range.

For a military force like the PLA, lacking a well-developed, long-reach naval air arm and newer air platforms, this approach makes sense. Three PLA officers from the Second Artillery Command Academy advance the idea that "guided missile forces are the trump card *(sa shou jian)* in achieving victory in limited high-technology war."[78] The keys to achieving such capabilities, in the argument of other PLA officers, lie in three areas: the use of countermeasures, the ability to achieve precision targeting, and the use of space platforms to support the effort.[79]

Analogous concepts are getting serious consideration in the United States today. Senior officers of the U.S. Strategic Command argue that the United States needs a conventional intercontinental or intermediate-range, submarine-launched ballistic missile capable of attacking terrorist or special weapons targets accurately in response times as short as 60 minutes.[80] This concept, called "precision global strike," is treated in the Bush administration's nuclear posture review. Proponents of the capability believe that such missiles would be "uniquely capable" if the United States had to attack promptly, i.e., within hours, of the start of an approaching conflict. Moreover, they could launch such speedy attacks anywhere while to accomplish similar attacks with bombers or cruise missiles might take hours or days.[81] Therefore, for a nation like China, possessing limited force-projection capabilities, no aircraft carriers, limited air-to-air refueling, and a Navy that is not yet fully capable of large-scale blue water operations, the ballistic missile concept must truly look like a "trump card."

Building Knowledge-Based Warfighting Architectures.[82]

Military theory is a grand thing if it is captured in doctrine that is assimilated by military forces and can be effectively employed in battle. However, the mere intellectual exploration of these capabilities is nothing but smoke and mirrors if a military does not have the forces, equipment, and systems to use the theory and doctrine in battle. The PLA has those requisites, albeit on a limited scale.

In general, the PLA is transforming itself into a modern force able to take full advantage of C4ISR technologies and the network-centric warfighting concept. Given the state of affairs in 1996 when the sudden appearance of two U.S. aircraft carrier battle groups in the Western Pacific during the Taiwan missile crisis embarrassed China's senior political and military leaders, the PLA has done remarkably well in its modernization effort.[83] There is a basic data-exchange and target-acquisition locating architecture to support the PLA Navy and Air Force, even if the platforms have limited range. There are national-level and regional C4ISR networks, and the PLA will have a near real-time regional intelligence collection capability from space in a few short years, if it does not already have it.

The PLA theater-level automated command and control capability is embodied in the *Qu Dian* system. It is a redundant, military region or frontal (war front) system linking the General Staff Department headquarters and the PLA's arms and services with regional combat headquarters and their subordinate major organizations. However, the system requires satellite data-exchange support and airborne radio and communications relay.

China's first defense communications satellite, the *Fenghuo-1*, was launched in January 2000. Originally designated *Zhongxing-22 (Chinasat-22)*, it provided C-band and ultra high frequency (UHF) communications for the integrated military command, control, communications, computer, and intelligence system known as *Qu Dian*.[84] China launched a second such satellite in 2003.[85] The *Qu Dian* system uses fiber-optic cable, high frequency and very high frequency (VHF) communications, microwave systems, and multiple satellites to enable the CMC, the General Staff Department, and commanders to communicate with forces in their theater of war (*Zhan qu*) on a real-time or near real-time basis.[86] The system also permits data transfer among the headquarters and all the units under its joint command.[87] The system has been compared to the U.S. Joint Tactical Information Distribution System (JTIDS), a secure network used by the United States and some allies.[88]

Discussing the potential threat posed to United States forces by a functional tactical data, communications, and intelligence distribution system like *Qu Dian*, Congressman Bob Schaffer of Colorado told the House of Representatives:

> Accurate ballistic missiles and the ability to observe U.S. forces from space will give China the potential to attack U.S. ships at sea and in port. Thus, capability is being enhanced by China's development of an integrated command and control system called *Qu Dian*, which relies on its *Feng Huo*-1 military communications satellite launched on January 26, 2000. *Qu Dian*, considered a major force multiplier, is similar to the U.S. Joint Tactical Information Distribution System, or JTIDS, and boasts a secure, jam-resistant, high capacity data-link communications system for use in tactical combat.[89]

Other PLA combat systems have a more limited capability to act as an airborne command post and assist with combat data exchange. The enhanced *Sukhoi* Su-30MKK2 fighter under development for China will improve long-range power-projection for the PLA. According to *Janes's Defense Weekly*, when equipped with a sensor system including side-looking airborne radar, the Su-30MKK2 will be capable of "tasking and controlling up to 10 other aircraft on a common [communications] net."[90] The model already delivered to the PLA, the Su-30MKK, controls up to four Su-27s and, like the more advanced model under development, functions as an airborne command and control system with data exchange to facilitate cooperative targeting.[91]

Of course, the PLA already has an airborne warning and control system (AWACS) built around the Russian *Beriev* A-50.[92] The Russian aircraft (with a NATO reporting name "Mainstay") is designated the *Kong Jing-2000* (KJ-2000) by China. It is equipped with Chinese-made phased-array radar and has a data link capability; a data processing system; friendly, hostile, and unidentified Identification Friend-or-Foe system; and a C3I capability. The KJ-2000 can exchange data with other aircraft and naval ships equipped with compatible data links. The aircraft loiter time on station, however, is only about 90 minutes.

China's own Y-8, a four-engine turboprop, will be equipped with an Ericsson ERIEYE AWACS system, increasing China's airborne early warning and command and control capabilities.[93] The original Y-8 based AWACS system apparently relied on the French firm, Thales, for its airborne early warning radars, and incorporated British Racal technology.[94] China has several of these in its inventory, although one was apparently lost in a training accident earlier this year. The PLA Air Force configured other special versions of the Y-8 (along with the Tu-154) for signals intelligence collection.[95]

The AWACS systems have been data-linked to the F-8 *Finback* fighter, produced by the Shenyang aircraft factory, and to the *Zhi-9* helicopter. The *Zhi-9* is a Chinese version of the French *Dauphin 2* Eurocopter, the AS 365N, produced under license.[96] In the case of the *Zhi-9*, a data-link passes targeting information to ship-based helicopters, thus some of China's indigenously produced destroyers presumably also have a data-link capability.[97] These helicopters are standard equipment on the *Sovremenny* destroyers and elsewhere.

The system also permits data and communications transfer to at least some PLA Navy surface ships. In fact, according to *Jane's Fighting Ships*, the *Sovremennys* are "the first Chinese warships to have a data systems link," which *Jane's* analysts believe is a PRC version of the NATO-designated *Squeeze Box*.[98] They also have the *Band Stand* data link for the C-802 antiship missile[99] as well as a data link for the SS-N-22 *Moskit* supersonic antiship missile.[100] Certain other destroyers can take advantage of these data links. For example, the *Luda* Type-51 destroyers have been fitted with Thompson-CSF data link systems as well as Chinese developed systems, as have the *Luhai* destroyers. These systems will link with the *Zhi-9* helicopters and the surface-to-surface missiles on the destroyers.

According to the Armed Forces Communications Electronics Association (AFCEA) journal *Signal*, China's destroyers are all now capable of data linking with AWACS systems, each other, their on-board helicopters, and their antiship missiles.[101] The *Sovermenny* Ka-25 helicopters are equipped with the

A-346Z secure data link, and other Chinese ships have the HN-900 data link, which incorporates other foreign technologies.

The bad news for the United States and other navies in Asia is that today, the PLA Navy's *Luhu, Luhai, Luda,* and *Sovremenny* destroyers are equipped with systems that function like the U.S. data link combat information transfer systems to support battle management and coordinated strikes on time-sensitive targets.[102] Chinese destroyers and most Chinese frigates have a system that works like the JTIDS, and they can pass data for targeting to the Su-30MKK for over-the-horizon targeting and attack vectors.[103] According to an AFCEA analyst, in some areas the Chinese ships are limited to "1940s era radar tasks of detecting and tracking air and surface targets for their own ship weapons." However, the Chinese have managed to get foreign technology, primarily from France and Russia, that will allow integrated battle management and the integration of sensors, ship guns, and missiles, as well as data management of information from other ships and aircraft.[104]

Space Support for C4ISR.

To reach and support deployed naval forces or air forces at a distance from the coast, the *Qu Dian* system needs a constellation of satellites, including tracking and data-relay satellites, as do other intelligence collection systems and sensors in the PLA.[105] Space, therefore, is increasingly critical to the PLA for the conduct of war. PLA headquarters can support deployed forces with remote sensing from space and airborne platforms and process "remotely captured images of the battlefield" in real time.[106]

Digitized military mapping is part of the space architecture needed for these capabilities. Digitized mapping supports all types of analysis, information networks, and targeting. At present, China's capability in this area is nearly real-time, according to PLA Major General Wang Xiaotong, writing in *Guangming Ribao*.[107] Digital mapping also supports sophisticated combat simulations. Although the PLA is apparently not yet in a position to provide real-time battlefield mapping and information, it is close to that point. PLA experts expect that as new integrated "space-ground military remote sensing survey and mapping technology" comes on line, the military's processing, handling, and distribution "will be more automatic, more intelligent, and more real-time." Such improvements increase the size of the battle area in which the PLA can operate, and the PLA is indeed working to manage forces and information in this new expanded battle space.[108]

Over the mainland and in close proximity to China's borders, the PLA already is able to provide real-time support for joint military operations with communications and data relay satellites. Indeed, China's military forces and command organs exercise this capability. An article in *Jiefangjun Bao* details exercises in the Guangzhou, Chengdu, Shenyang, and Beijing "war theaters" (*zhanqu*) using networked forces supported by satellite communications.[109] In Guangzhou, an exercise reportedly relied on a satellite-supported C4ISR network and fiber-optic systems to "integrate deployed military units in field locations and fixed locations." The Shenyang MR exercise described in the *Jiefangjun Bao* article integrated reserve units and regular PLA forces. To accomplish this, the Shenyang MR Commander established communications networks with local military departments, transportation bureaus, and meteorological bureaus. During an exercise in Chengdu MR, the "war zone" incorporated the General Staff Department's Communications Academy in Chongqing to support satellite communications requirements.

As noted earlier, to make its C4ISR network operational on a real-time basis, China needs tracking and data relay satellites. Space forces cannot function in today's combat environment without such an architecture. The PLA can support manned space activities, reconnaissance, and other military missions with a common platform placed in geosynchronous earth orbits. It is also possible that the PLA could rely on mini or micro satellites and constellations of relay satellites in low earth orbit for the same purpose. However, the PRC is working on a satellite system, the DJS-2, that will function like U.S. tracking and data relay satellite systems. This satellite will have a lifetime of about 15 years, likely operating in the Ku and C bands, making it capable of relaying communications and imagery data.[110]

The *Dongfanghong-IV* satellite, the product of a project announced in 2001, will meet these requirements. With a 15-year life span, it has 50 communications transmitters and is capable of multiple loads of large-capacity communications, data, and broadcast relay.[111] The *Dongfanghong-IV* was developed for military and civilian use in a program directed by the Commission of Science, Technology, and Industry for National Defense (COSTIND).[112] According to *China Defence Today*, this satellite can "distribute information to the lowest echelon in a battlefield, potentially transmitting data (maps, pictures, and enemy deployments) on demand to small units, each using a . . . device to receive orders and situational information."[113]

The first satellite for launch in this *Dongfanghong-IV* series carries the commercial name, *XinNuo-2*. It was set for launch in 2005 as part of a constellation of satellites, but its launch was delayed.[114] The capacity to launch a constellation of small (mini or micro) satellites is also within the capabilities of the PLA. China will launch a constellation of earth environmental monitoring satellites—the *HJ-1, HJ-1A, HJ-1B and HJ-1C*—in the second half of 2007.[115] The *1A* and *1B* are small optical satellites, while the *HJ-1C* is a radar satellite. In the area of military imaging and reconnaissance, China has launched a series of *Jianbing* satellites with recoverable photo packages. It has other packages that provide near real-time electro-optical images. Finally, there is now in space China's first military remote imaging satellite using synthetic aperture radar, the *Jianbing-5*.[116]

China can use the signals from the U.S. Global Positioning System (GPS), the European Galileo, and the Russian Glonass satellites for precision navigation. These signals support military requirements, including directing precision weapons and warheads. However, the CMC is concerned that the United States might interrupt China's ability to use the GPS system if hostilities looked imminent. Therefore, China has developed and launched its own *Beidou* navigation satellites.[117] Clearly, in the near term the PLA and China's defense infrastructure are willing to rely on foreign partners or technology, but as in most other areas, they seek to develop indigenous capabilities for the long term. China can also relay electro-optical imagery back to earth from its remote sensing satellites, which support a military reconnaissance capability similar to that of Western commercial sensor systems in the 1990s.[118]

Without these space systems, China will not achieve a networked, integrated C4ISR architecture to support the military operations it conceives or plans. Moreover, its space reconnaissance architecture must have the necessary tracking and data relay satellites to be able to function on a real-time basis. If the PLA is going to achieve its goals of tracking deployed naval task forces and hitting them with ballistic missile warheads, let alone with air and ship-launched cruise missiles, it will need to collect and transmit radar returns, images, and electronic intelligence reliably over extended distances beyond the mainland. Also, satellite relay systems support logistics communications necessary to ensure that deployed military forces get supplies on a timely basis.[119]

CONCLUSIONS

The PLA of today is not the force that U.S. and United Nations forces fought in Korea in 1950. In some cases, it may be armed with some of the same weapons, but it has modernized significantly. At the theoretical level, PLA academicians, strategists, and senior military leaders have grasped the lessons of the RMA. In the operational arena, PLA officers and leaders at all levels are being educated in these lessons in units and at command academies. In military doctrinal affairs, PLA units can now turn to manuals and a range of publications that outline how to use C4ISR systems in war. In training exercises, the PLA practices using these systems. In the area of offensive and defensive information operations, the PLA is heavily involved. In addition, the PLA is building a space architecture to support real-time information operations.

The thrust of the conference for which this chapter was written was to gauge the "right sizing" of the PLA. The question to be addressed was: "Does the PLA need the capabilities it is developing?"

All of the command, control, and targeting architectures already fielded or under development by the PLA are necessary and appropriate responses for a major military power in the information age if that nation desires to keep pace with improvements in armaments and technology. Thus, the dilemma confronting American military planners is not whether China's military needs these capabilities, it is rather to anticipate the uses to which the capabilities will be put. The problem for the United States (and its allies) is that there is no clear roadmap or outline of the intentions of the CCP or how its Politburo Standing Committee will use such military power and technology. The major straw in the wind regarding China's intent is that many of China's military strategists and senior leaders seem to conceptualize the United States as the target of this new military force. Moreover, when Chinese strategists talk about "comprehensive national power," they want the combination of economic, political, diplomatic, military, and cultural strength to equal "the power to compel" other nations to do China's bidding.

The PLA has solved the over-the-horizon targeting problem conceptually. It has solved it mathematically and in simulation. It has built much of the hardware necessary to underpin a modern military force. It is also very close to fielding the full C4ISR architecture to fight a campaign out to about 2,000 kilometers from China's coast. However, it is not clear how the PLA will put such a system together, engineer it, or use it. For the United States, this means that we must continue to develop and stay ahead in the areas of kinetic and electronic energy weapons, electronic warfare and countermeasures, and information warfare.

China's military forces are developing some potentially dangerous capabilities, certainly more dangerous than they were a decade ago; but they are still not "peer competitors" to the forces of the United States. The PLA's battlefield applications of net-centric warfare concepts still heavily depend on foreign technology. Without the AWACS and data link systems supported by Russian and French technology, the PLA Navy and Air Force would be relegated to the levels of sophistication prevailing in the American and NATO militaries of the 1960s. Even with the architecture the PLA has built, its ability to apply the systems with deployed forces at long distances from its borders is limited. The duration on station of its AWACS aircraft is short (90 minutes), their range limited, and not all of them are capable of inflight refueling. Most of the PLA's combat ships and aircraft can engage in networked operations, but can handle only a limited number of targets. In addition, not all of the weapons they carry can receive the networked combat data.

All this said, the PLA has made significant strides in less than 2 decades in transforming itself into a force that can engage in a modern war along its periphery out to a range of about 1,500 miles. When it achieves its goals of deploying satellite tracking and data relay systems and fielding new long-range missiles with multiple (maneuverable) warheads, it may well achieve its goal of targeting an enemy's deployed naval battle groups. This equates roughly to the capability to defend against and deny access to enemy forces inside the "second island chain" that Liu Huaqing in 1984 conceived that China must dominate. Thus, China is close to achieving a viable anti-access strategy that, at a minimum, would impede U.S. and Japanese military operations. This capability may be only 2 to 5 years away. If China is not a peer competitor to the United States today, it is certainly turning itself into a dominant regional power.

Moreover, with the exceptions of Japan and Australia, it is perhaps the only power in the region able to fight a "knowledge-based" war.

Much of what the PLA has achieved relies on the technical assistance of foreign defense companies, primarily Russian, French, and British. Because China's long-term intent is not clear, because it continues to threaten Taiwan, and because it has violated the sovereignty of Japan, a U.S. ally, some policy responses are required. The "hedging" in the last U.S. *Quadrennial Defense Review* with a shift of forces and priorities to Asia is a military-diplomatic response.

Other responses are necessary. Former Deputy Secretary of State Robert Zoellick's "responsible stakeholder" formulation is one type of diplomatic policy response, as are the renewed military contacts between the PLA and the U.S. armed forces. Effective policy responses to Russian assistance to China are limited, but diplomatic and economic pressure should aim to discourage this military cooperation on the ground that it is not in Russia's interest to see the military balance in Asia changed through weapons or technology transfers. Only recently have European Union (EU) states accepted that the United States has security interests in the Western Pacific, and that their technology sales to China can threaten American forces. Legislation by Henry Hyde and Duncan Hunter got the EU's attention when EU nations were considering lifting the Tiananmen-based arms sanctions on the PLA. This legislation would have excluded European firms from participation in U.S. defense cooperation programs if they sold certain technologies to China. This type of legislative response is useful against long as PLA intentions are unclear, and China's military actions or declarations work against U.S. security interests.

It is also important to remember that as the PLA becomes more dependent on the electromagnetic spectrum for military operations, it is more susceptible to interference in that spectrum. Over the last decade or so, PLA warfare experts have concentrated on exploiting the weaknesses inherent in the American dependence on space and information. That dependence is becoming a two-way street. As the PLA modernizes, it also cannot function without access to space and the electromagnetic spectrum. Strong competition in space control and information warfare will characterize the future military development of China and the United States for some time to come.

ENDNOTES - CHAPTER 5

1. "China Publishes White Paper on National Defense," CPP20061229701005 *www.opensource.gov*.

2. Zhang Wannian, ed., *Dangdai Shijie Junshi yu Zhongguo Guofang (China's National Defense and Contemporary World Military Affairs)*, Beijing: Military Science Press, 1999, p. 25. Zhang points out that even though the Soviet Union has broken up, "hegemonism still looms on the international stage" and that the United States is "primary among Western hegemonist nations."

3. In an article on the theory of applying and managing sensor systems in battle, two researchers from the Aeronautical Radio Electronic Institute in Shanghai see the fusion of radar, infrared, laser, video, and electro-optical sensor systems as future goals to be used by the PLA with technologies like pilotless aircraft. See Mou Zhiying and Wu Jianmin, "Duo pingtai xinxi fuhe yu zhinenghua chuanganqi guanli jishu zongshu" ("Survey of Multi-Platform Information Fusion and Intelligence Sensor Management Technologies"), in *Hangkong Dianzi Jishu (Avionics Technology)*, Vol. 36, No. 4, December 2005, pp. 20-21. See also Peng Guangqian and Yao Youzhi, eds., *The Science of Military Strategy*, Beijing: Military Science Press, 2005, pp. 340-341. China's goals and needs in this area are also discussed in Pan Youmu, "Zhuoyan Kongtian: Yitihua Tansuo Kongtian Anquan Zhanlue" ("Focus on Air-Space Integration and Study National Air-Space Security Strategy"),

Zhongguo Junshi Kexue (China Military Science), Vol. 19, February 2006, pp. 60-66.

4. Min Zengfu, *Kongjun Junshi Sixiang Gailun (An Introduction to Air Force Military Thinking)*, Beijing: PLA Press, 2006, pp. 376-377.

5. "Zhongguo lujun yuancheng jidong yanlian liu kandian" ("Six Points about Chinese Army's Long-Distance Mobility Exercise"), *Ta Kong Pao*, Hong Kong, September 6, 2006, www.takongpao.com.

6. Min Zengfu, *Kongjun Junshi Sixiang Gailun*, pp. 379-380.

7. Zhang Wannian, ed., *Dangdai Shijie Junshi yu Zhongguo Guofang (China's National Defense and Contemporary World Military Affairs)*, Beijing: Military Science Press, 1999, p. 80.

8. *Ibid.*, pp. 80-81.

9. Shen Weiguang, *Xin Zhanzhenglun (The New "On War")*, Beijing: People's Publishing House, 1997, pp. 5-6.

10. Shang Jie, *Xiandai Dimian ZuoZhan Lliang de Yunyong he Fazhan (The Use and Development of Modern Ground Warfighting Power)*, Beijing: People's Liberation Army Press, 1994, pp. 58-59.

11. Fu Quanyou, "Make Active Explorations, Deepen Reform, Advance Military Work in an All-Around Way," *Qiushi*, No. 6, March 6, 1998, in Foreign Broadcast information Service (FBIS)-CHI-98-093, April 3, 1998; Shen Weiguang, "Xinxizhan: Mengxiang yu Xianshi" ("Information Warfare: Dreams and Reality"), *Zhongguo Guofang Bao (China Defense News)*, February 14, 1997, p. 3; Wang Huyang, "Five Methods in Information Warfare," *Jiefang Junbao*, November 7, 1995, p. 6, in FBIS-Ch1-96-194; Deng Raolin, Qian Shi Xu, and Zhao Jincun, eds., *Gao Jishu Jubu Zhanzheng Lilun Yanjiu (Theory and Research in High Technology Warfare Under Limited Conditions)*, Beijing: Military Friendship and Literature Publishers, 1998; Li Jijun, "Xin Shiqi Jundai Jianshe de Zhidao Gangling" ("Guiding Doctrine on Military Building in the New Age," *Junshi Lilun yu Zhanzheng Shijian (Military Theory and Practice in War)*, Beijing: Military Science Publishers, 1994.

12. Zhang Wannian, ed., *Dangdai Shijie Junshi yu Zhongguo Guofang*, p. 107-108.

13. *Ibid.*, p. 202.

14. Li Jijun, *Junshi Zhanlue Siwei (Thinking about Military Strategy)*, Beijing: Military Science Press, 1996, 2000, 2002, pp. 188-189.

15. *Ibid.*, pp. 210-212.

16. Yun Shan, "Zhimian Xin Junshi" ("Facing New Military Transformation Squarely"), *Liaowang Xinwen Zhoukan (Clear Lookout News Weekly)*, No. 28, July 14, 2003, p. 20.

17. Sun Yiming and Yang Liping, *Xinxihua Zhanzheng Zhong de Zhanshhu Shuju Lian (Tactical Data Links in Information Warfare)*, Beijing: Beijing Post and Telecommunications College Press, 2005, pp. 5, 276-314.

18. Wang Zhongquan, *Meiguo He Liliang yu He Zhanlue (American Nuclear [Weapons] Strength and Nuclear Strategy)*, Beijing: National Defense University Press, 1995, p. 73.

19. *Ibid.*

20. *Ibid.*, pp. 69-74.

21. William A. Owens, "JROC: Harnessing the Revolution in Military Affairs," *Joint Forces Quarterly*, Summer 1994, pp. 55-58. On the RMA, see William A. Owens, "The Emerging System of Systems," *U.S. Naval Institute Proceedings*, Vol. 121, No. 5, May 1995, pp. 35-39; Paul Bracken, "The Military After Next," *The Washington Quarterly*, Vol. 16, No. 4, Autumn 1993; Antulio J. Echevarria and John Shaw, "The New Military Revolution: Post-Industrial Change," *Parameters*, Vol. 22, No. 4, Winter 1992-93; and James R. FitzSimmonds and Jan M. van Tol, "Revolutions in Military Affairs," *Joint Force Quarterly*, No. 4, Spring 1994.

22. William A. Owens, "The American Revolution in Military Affairs," *Joint Forces Quarterly*, Winter 1995-96, p. 38.

23. Jeremy M. Boorda, "Leading the Revolution in C4I," *Joint Force Quarterly*, Autumn 1995, p. 16.

24. Thomas G. Mahnken, "War in the Information Age," *Joint Forces Quarterly*, Winter 1995-96, p. 40.

25. Guo Wujun, *Lun Zhanlue Zhihui (On Strategic Command and Control)*, Beijing: Military Science Press, 2001, p. 226.

26. *Ibid.*

27. *Ibid.*, p. 248.

28. Yun Shan, "Zhimian Xin Junshi," *Liaowang Xinwen Zhoukan*, pp. 10-20.

29. *Ibid.*, pp. 15, 17.

30. Li Bingyan, *Da Moulue yu Xin Junshi Biange* (*Grand Strategy and the New Revolution in Military Affairs*), Beijing: Military Science Press, 2004, p. 52.

31. *Ibid.*, pp. 155-156, 180-183.

32. Liang Biqin, *Junshi Zhexue* (*Military Philosophy*), Military Science Press, 2004, p. 526-537.

33. Han Xiaolin, ed., *Gao Jishu Zhubu Zhanzheng Lilun Yanjiu* (*Research and Theory on Limited War Under High Technology Conditions*), Beijing: Military Friendship and Literature Press, 1998, p. 183. See also Chen Yong, Xu Guocheng, and Geng Weidong, eds., *Gao Jishu Tiaojian xia Lujun Zhanyi Xue* (*The Study of Ground Forces Campaign Theory under High Technology Conditions*), Beijing: Military Science Press, 2003, p. 71. Chen and Xu refer to the fifth domain of warfare as information, whereas Han sees it as in the electromagnetic spectrum. Han's description is more inclusive, since it includes electronic warfare and information warfare.

34. Han, *Gao Jishu Zhubu Zhanzhang Lilun Yanjiu*, p. 184.

35. Xin Qin, *Xinxihua Shidai de Zhanzheng* (*Warfare in the Information Age*), Beijing: National Defense University Press, 2000, pp. 1, 10.

36. Shen Weiguang, *Xinxi Zhangzhen* (*The New "On War"*), Beijing: People's Publishing House, 1997, p. 178.

37. Yun Shan, "Zhimian Xin Junshi," *Liaowang Xinwen Zhoukan,* pp. 15-16.

38. *Ibid.*, p. 16.

39. Song Yongxin and Guo Yizhong, "Research into Ground Radar Station Countermeasures for Satellite Reconnaissance," *Nanjing Hangtian Dinazi Duikang,* July 19, 2006, pp. 37-39, 64, in Open Source Center CPP20060720276001, *www.opensource.gov*.

40. Shen Weiguang, *Xinxi Zhanzheng Lun,* pp. 16, 20.

41. Jian Yun, "Academician Views National Defense: Future War is Focused More on 'Command' than Forces," *Beijing Qingnian Cankao,* February 6, 2002, in Open Source Center CPP 20020221000271, *www.Opensource.gov*.

42. *Ibid.*

43. *Ibid.*, p. 41.

44. Zhu Youwen *et al.*, eds., *Gai Jishu Tiaojian Xia de Xinxi Zhan (Information Warfare Under High Technology Conditions)*, Beijing: Military Science Press, 1994, p. 3.

45. Open Source Center, Wu Chao, Jia Zhaoping, and Chu Zhenjiang, "Getting close to the mysterious 'informationized' Blue Force," *Beijing Zhanyou Bao,* February 23, 2006, p. 3.

46. This is a lesson designed to reinforce the critique of Iraqi military decisions in the first Gulf War.

47. *Ibid.*

48. Zhang, *Dangdai Shijie Junshi yu Zhongguo Guofang,* p. 80.

49. *Ibid.*, pp. 111-112.

50. Open Source Center, "Jiefangjun Bao: Excerpts of Speeches by Leading PLA Officers on Military Training," CPP 20060627710007, *Beijing Jiefangjun Bao,* June 27, 2006, p. 2.

51. *Ibid.*

52. Xue Xinglin, *Zhanyi Lilun Xuexi Zhinan,* pp. 161-162.

53. *Ibid.*, p. 161.

54. Section III, "China's Leadership and Administrative System for National Defense," in *White Paper on National Defense,* December 2006, contains an equally clear explanation of the levels of command and control and their relationships.

55. Xue Xinglin, *Zhanyi Lilun Xuexi Zhinan,* pp. 161-162.

56. *Ibid.*

57. *Ibid.*

58. *Ibid.*, p. 162.

59. This process is described in "Second Artillery Stages Simulated 'Missile War' on Screen," *Beijing Zhongguo Xinwen She,* Open Source Center CPP20060524004001, 0753 GMT, May 24, 2004.

60. See *www.chinamil.com.cn/jbzlk/outline/, Jiefangjun Bao,* April 21, 2006, and May 25, 2006.

61. The use of the term *tongshuaibu* is uncommon but not unheard of in explanations of Chinese command and control systems. *Tongshuai* can mean commander or commander in chief. The Nationalist, or Kuomintang (KMT), forces used the term to refer to a couple of major frontal headquarters during the civil

war. In the *Huaihai* Campaign in 1949, for instance, the KMT combat headquarters for the campaign was called the *Tongshuaibu*. PLA military histories also refer to Eisenhower's headquarters for Overlord and the Supreme Headquarters, Allied Powers Europe as the *tongshuaibu*. Clearly, this use is meant to designate more than the command authority of the General Staff Department Operations Department.

62. Liu Mingtao, Yang Chengjun *et al.*, eds., *Gao Jishu ZhanzhengZhong de Daodan Zhan (Ballistic Missile Battles in High Technology Warfare)*, Beijing: National Defense University Press, 1993, p. 107.

63. *Ibid.*, p. 108.

64. Xue, *Zhanyi Lilun Xuexi, Zhinan*, p. 386.

65. *Jiefangjun Bao*, June 30, 2006, p. 1, FBIS Open Source Center, CPP20060630718002.

66. *Xinhua News Service*, July 2, 2005, FBIS Open Source Center, "Account of Party Central Committee's Care and Concern for Strategic Missile Units," rccb.osis.gov, July 5, 2006, 12:30 AM.

67. See Michael S. Chase and Evan Medieros, "China's Evolving Calculus: Modernization and Doctrinal Debate," in James Mulvenon and David Finklestein, eds., *China's Revolution in Doctrinal Affairs: Emerging Trends in the Operational Art of the Chinese People's Liberation Army*, Arlington, VA: Rand Corporation and the Center for Naval Analysis, 2006, p. 147; Ken Allen and Maryanne Kivlehan-Wise, "Implementing PLA Second Artillery Doctrinal Reforms," in Mulvenon and Finklestein, eds., *China's Revolution in Doctrinal Affairs*, pp. 159-200; and Bates Gill, James Mulvenon, and Mark Stokes, "The Chinese Second Artillery Corps: Transition to Credible Deterrence," in James Mulvenon and Andrew N. D. Yang, eds., *The People's Liberation Army as an Organization*, Santa Monica, CA: Rand Corporation, 2002, pp. 510-586. The author is also grateful to M. Scot Tanner of the Rand Corporation and David Cowhig of the the U.S. Department of State for their insights on the term as it is used by the Second Artillery.

68. *Ibid.*, p. 387.

69. *Ibid.*, p. 388.

70. Song Junfeng, Zhang Weiming, Xiao Weidong, and Tang Jiuyang, "Study of Knowledge Infrastructure Construction for

Network Centric Warfare," in *Beijing Xitong Gongcheng Lilun yu Shijian (Systems Engineering Theory and Practice)*, August 1, 2005, in Open Source Center CPP20060822424002.

71. Si Laiyi, *"Lun xinxi zuozhan zhihui kongzhi jiben yuanze"* ("On Basic Principles for Command and Control Information Warfare"), in Military Science Editorial Group, *Wo Jun Xixi Zhan Wenti Yanjiu (Research on Questions about Information Warfare in the PLA)*, Beijing: National Defense University Press, 1999, pp. 245-251.

72. Wang Lu and Zhang Xiaokang, "Analysis of C4ISR System in NCW and Analysis of Its Effectiveness," *Lianyungang Zhihui Kongshi yu Fangzhen*, April 1, 2006, Open Source Center CPP20060814424008.

73. Nie Yubao, "Daji haishang di da jian jianting biandui de dianzi zhan zhanfa" ("Combat Methods for Electronic Warfare Attacks on Heavily Fortified Enemy Naval Formations"), in Military Science Editorial Group, *Wo Jun Xixi Zhan Wenti Yanjiu (Research on Questions about Information Warfare in the PLA)*, Beijing: National Defense University Press, 1999, pp. 183-187.

74. Li Xinqin, Tan Shoulin, and Li Hongxia, "Model, Simulation Actualization on Threat Maneuver Target Group on Sea," in *Qingbao Zhihui Kongzhi Xitong yu Fangzhen Jishu (Information Command and Control Systems and Simulation Technology)*, August 1, 2005, in Open Source Center, www.opensource.gov.

75. Tan Shoulin and Zhang Daqiao, "Effective Range for Terminal Guidance Ballistic Missile Attacking Aircraft Carrier," in *Qingbao Zhihui Kongzhi yu Fangzhen Jishu (Information Command and Control Systems and Simulation Technology)*, Vol. 28, No. 4, August 2006, pp. 6-9.

76. *Ibid.* p. 9.

77. Li Xinqi, Tan Shoulin, and Li Hongxia, "Precaution Model Simulation Actualization on Threat of Maneuver Target Group at Sea," in *Qingbao Zhihui Kongzhi yu Fangzhen Jishu, Information Command and Control Systems and Simulation Technology*, August 2005, in Open Source Center, November 25, 2005.

78. Ge Xinliu, Mao Guanghong, and Yu Bo, *"Xinxi zhan zhong daodan budui mianlin de wenti yu duici"* ("Problems Faced by Guided Missile Forces in Information Warfare Conditions and Their Countermeasures"), in Military Science Editorial Group, *Wo Jun Xixi Zhan Wenti Yanjiu*, pp. 188-189.

79. Min Zengfu, *Kongjun Junshi Sixiang Gailun* (*An Outline of Air Force Military Thought*), Beijing: PLA Press, 2006, pp. 377-378.

80. Amy F. Woolf, *Conventional Warheads for Long-Range Ballistic Missiles: Background and Issues for Congress*, Congressional Research Service Report for Congress RL 33067, Washington, DC: Library of Congress, March 13, 2006.

81. *Ibid.*, Summary and pp. 1-5.

82. Luke Amerding of the National Bureau of Asian Research provided the research support for this section. Melanie Mickelson Graham of Johns Hopkins University, SAIS, also assisted with information on PLA order of battle and weapon systems.

83. Regarding how the U.S. deployment surprised China, see Arthur S. Ding, "The Lessons of the 1995-1996 Military Taiwan Strait Crisis: Developing a New Strategy Toward the United States and Taiwan," in Laurie Burkitt, Andrew Scobell, and Larry M. Wortzel, eds., *The Lessons of History: The Chinese People's Liberation Army at 75*, Carlisle, PA: Strategic Studies Institute, U.S. Army War College, 2003, pp. 379-402. Mark Stokes discusses the effort to use ballistic missiles and maneuvering warheads to attack U.S. naval formations in Mark A. Stokes, "Chinese Ballistic Missile Forces in the Age of Global Missile Defense: Challenges and Responses," in Andrew Scobell and Larry M. Wortzel, eds., *China's Growing Military Power*, Carlisle, PA: Strategic Studies Institute, U.S. Army War College, 2002, pp. 107-168.

84. See *cns.miis.edu/research/space/china/mil.htm*. See also Bill Gertz, "China's Military Links Forces to Boost Power," *Washington Times*, March 16, 2000, in *taiwansecurity.org/News/WT-031600.htm*.

85. See *Jane's Space Directory 2006-2007* and Union of Concerned Scientists' "UCS Satellite Database" *www.ucsusa.org/global_security/space_weapons/satellite_database.html*.

86. *Chinese Defense Today, www.sinodefence.com/strategic/spacecraft/fenghuo1.asp*.

87. K. K. Nair, "China's Military Space Program," in *Promoting Strategic and Missile Stability in Southern Asia*, Report of the MIT/IPCS/CAPS Conference, New Delhi, March 28-29, 2006. New Delhi: Institute of Peace and Conflict Studies Special Report 17, April 2006, pp. 9-10, at *www.ipcs.org*.

88. On the description of the JTIDS system, see Sun Yiming and Yang Liping, *Xinxihua Zhanzheng Zhong de Zhanshu Shuju*

Lian (*Tactical data links in Information Warfare*), pp. 48-50. The comparison between JTIDS and *Qu Dian* is in the Honorable Bob Schaffer, "Remarks on China" in the House of Representatives, March 14, 2002, *Congressional Record*: March 14, 2002, Extensions, pp. E-360-E361. See *www.fas.org/irp/congress/2002_cr/h031402.html*.

89. Schaffer, "Remarks on China," March 14, 2002.

90. Robert Hewson, "More Details on China's Su-30MKK2," *Jane's Defense Weekly*, September 10, 2003, in *search.janes.com*.

91. *Jane's All the World's Aircraft, 2005-2006*, p. 443, at *jawa.janes.com*.

92. See *www.sinodefence.com/airforce/specialaircraft/kj2000.asp*. See also Jane's Electronic Mission Aircraft, *0-www8.janes.com.usmalibrary.edu/search/documentview.do?docId+/content*.

93. *Jane's C4I Systems: 2005-2006*. 17th Ed., Surrey, UK: Jane's Information Group, 2005, p. 274.

94. *0-www8.janes.com.usmalibrary.edu/search/documentview.do?docId+/content*.

95. See *0-www4.janes.com.usmalibrary.usma.edu/K2*.

96. *Jane's All the World's Aircraft: 2005-2006*, 96th Ed., Surrey, UK: Jane's Information Group, 2005, pp. 86-87, 93, 95-96.

97. *Ibid.*, p. 93.

98. Stephen Saunders, ed., *Jane's Fighting Ships: 2005-2006*, 108th Ed., Surrey, UK: Jane's Information Group, 2005. p. 123.

99. *Ibid.*, p. 124.

100. *Ibid.*, p. 123.

101. James C. Bussert, "China Debuts Aegis Destroyers," *Signal*, July 2005, at *www.afcea/signal/articles/anmviewer.asp?a=992*.

102. Zhang Kaide and Zhao Shubin, "Shimin Daji Zhihui Kongzhi Jishu Chutan" ("The Command and Control Technology of Time Critical Strikes"), in *Zhihui Kongzhi yu Fangzhen* (*Command, Control and Simulation*), Vol. 28, No. 2, April 2006, pp. 1-5.

103. Bussert, "China Debuts Aegis Destroyers," *Signal*, July 2005.

104. James C. Bussert, "China Builds Destroyers Around Imported Technology," *Signal*, August 2004, at *www.afcea.org/*

signal/articles/anmviewer.asp?a=252. See also "New-Generation Warships for the PLA Navy," *Military Technology*, February 2004, Vol. 28, Issue 2, pp. 90-91.

105. See *www.txsj.com/www/asp/news/list.asp?id=1242*.

106. Major General Wang Xiaotong, "Military Survey and Mapping: The Three-Dimensional Expansion of Combat Operations Support," *Beijing Guangming Ribao*, January 4, 2006, PRC S&T, Open Source Center, at *www.opensource.gov*.

107. *Ibid*.

108. Tan Yanqi, Yan Jianbo, and Ding Jianmin, "Operating on Network that Measures the Sky and Maps the Ground—On Site Report About a Certain Jinan Military Region Mapping Information Center Enhancing Informationization Building," *Jiefangjun Bao*, October 31, 2005, at *www.Chinamil.com.cn*.

109. "China: Four Major Commands Launch Joint Air, Land, and Sea Military Exercises," *Jiefangjun Bao*, March 2, 2006, p. 2, in Open Source Center CPP20060303502001.

110. The author depended heavily on James Howe, of Vision Centric, Inc., for help with this section of the chapter. Dean Cheng of the Center for Naval Analysis also provided assistance in developing these assessments.

111. Du Minghua, "China Launches Dongfanghong-IV Satellite Project," *People's Daily Online*, December 28, 2001, *english. peopledaily.com.cn/200112/27/eng20011227_87589.shtml*.

112. See *www.globalsecurity.org/space/world/china/dfh-4.htm*.

113. See *www.sinodefence.com/strategic/spacecraft/dongfanghong4. asp*.

114. "China to launch new communications satellite," at *www.chinaview.cn*, 2006-09-05 09:20:12, at *news.xinhuanet.com/english/2006-09/05/content 5049526.htm*.

115. "China to launch 1st environment monitoring satellite," *www.chinaview.cn*, 2006-08-11 21:51:38, *news.xinhuanet.com/english/2006-08/11/content 4951600.htm*.

116. See *www.ucsusa.org/global_security/space_weapons/satellite_database.html*.

117. James A. Lewis, "China as a Military Space Competitor," Center for Strategic and International Studies," August 2004. Three or four satellites are now part of the *Bei Dou* system.

118. *Ibid.*, p. 4.

119. Huang Yanru, "Aiming at Future Warfare, Constructing Informationalized Platforms for Military Transportation in the Harbor and Shipping Industry," Guangzhou *Zhanshi Bao,* January 16, 2006, p. 3, Open Source Center at *www.opensource.gov.*

PART III:

THE PLA GROUND FORCE

CHAPTER 6

"PRESERVING THE STATE": MODERNIZING AND TASK-ORGANIZING A "HYBRID" PLA GROUND FORCE[1]

Cortez A. Cooper III

Rapid defense modernization is a logical priority for a technologically challenged, combat-inexperienced military focused on a mission that might bring it into conflict with the world's most powerful armed force. Faced with the potential for such a conflict in the Taiwan Strait and backed by the strong conviction that use of force in certain cross-Strait circumstances would not only be justified but legally required, China's central leadership has set in motion over the past decade an Army building program of impressive scope and scale. The progress of this peacetime modernization effort, particularly given the low baseline from which it was launched, is perhaps exceeded in the past century only by the rise of the *Wehrmacht* in the 1930s and the transformation of the U.S. military between 1980 and the first Gulf War.

Despite a modernization effort covering to some extent every aspect of military force structure and posture, the Chinese have clearly prioritized development of capabilities to severely damage Taiwan in the event of a conflict over the island's stance on perpetual separation from the mainland—and to deter or slow U.S. responses to such a conflict. As such, the People's Liberation Army (PLA) ground force has played fourth fiddle to missile, air, and naval forces in terms of modernization priority. It has not, however, been forgotten.[2] The PLA remains a Party Army in an era when the Chinese Communist Party's (CCP) grip is less than firm—PLA ground forces remain the primary arbiter of Party control throughout the country, key contributors to Beijing's foreign policy initiatives, and protectors of a 22,000-kilometer land boundary adjacent to a number of current and potential flash points.

The ground force is also preparing for a Taiwan contingency, albeit with a lower profile than its sister services.[3] While China appears to be avoiding for now telltale programs to greatly increase amphibious and airlift capacity to project ground forces onto Taiwan, the PLA continues to train and equip task-organized brigades and divisions to fight an island landing campaign. Beijing wants to avoid alarming U.S. and regional neighbors with an overt preparation for force projection operations but has positioned a defense industrial base to provide, when needed, the projection platforms for a force trained and organized to attack Taiwan and occupy, at least temporarily, key terrain.

The PLA ground force is tasked to support domestic stability operations; defend borders across mountain, jungle, and desert terrain; conduct military diplomacy abroad; and prepare for a local war with significant power projection requirements. The approach of China's Central Military Commission (CMC) appears to be to task-organize specific units for specific campaigns or local missions, rather than to modernize the force across the board and expect each unit to conduct myriad missions. For this reason, the PLA ground force likely will retain well over a million soldiers for at least the next decade having collectively a wide range of modernity and warfighting expertise. Outside analysts who seek to assess PLA ground force mission capabilities based on a homogeneous force will thus probably miss the mark. Addressing the central question of whether or not the PLA ground force is "right-sized," China's 2006 *National Defense White Paper* expresses the belief that the force is very close to being "proper in size" and "optimal in structure" for the missions at hand. The paper indicates that ground force reductions begun in 1985, 1997, and 2003 were instrumental in achieving this goal—but does not mention any plans for further reductions.[4]

Analysts debate the real level of Chinese spending on military programs—the figure is certainly higher than official pronouncements—but China's economic growth ensures that military programs are well-funded even in the context of a national development plan that prioritizes civil programs over martial. This fertile ground for continued, rapid modernization is made even more productive given that military and dual-use technologies are flowing into China with few constraints.[5] There is little reason to believe that this situation will change substantially over the next decade, although it would be a mistake to assume any particular intent for military employment beyond the general missions already mentioned.

Two strategic advantages accrue to Beijing in decisions regarding resource priorities for modernization efforts: (1) resources are plentiful enough to develop and maintain a "hybrid" force as long as no immediate threats demand national mobilization; and (2) the general levels of regional and global stability needed for Chinese national development are provided under the current international security architecture enforced by the United States and its treaty allies. It is for this latter reason that Chinese use of force in the current geo-strategic environment is inimical to overarching

Chinese national development priorities, and why even a growing imbalance of power in Beijing's favor over Taipei is unlikely to lead to war in the absence of severe provocation by Taiwan. Building a PLA that is increasingly more capable of inflicting damage on Taiwan (and perhaps U.S. forward forces), without presenting an imminent force projection threat, appears to be Beijing's approach to restraining Taiwan from just such a provocation.

CHINA'S NATIONAL SECURITY FRAMEWORK

Many analysts, both in the United States and abroad, take it as axiomatic that China will rise to become a superpower over the course of the next 2 to 3 decades, with growth across all components of national power. Some analysts take exception, believing that China may well take a turn into political chaos due to the many challenges inherent in managing a domestic economy marked by a fragile financial system, the environmental and social problems associated with rapid growth, and a very uneven distribution of wealth. Among analysts in China, more moderate assessments prevail—many Chinese theorists believe that China will achieve the status of a mid-level developed country by mid-century, but must first overcome a variety of challenges in political, social, economic, and military realms.

Regarding a reemerging China's position in the world, some theorists (primarily realists) in China and abroad believe that the international power structure dominated by the United States and its allies will not accommodate a more powerful and influential China. Others believe that as long as Beijing's intentions are largely of a status quo bent, then conflict can be avoided if key international actors do not reflexively seek to contain China's developmental goals. Still others hold that even though conflict is not inherent in China's rise, Beijing intends to drastically change the global power structure, thus rendering significant conflict likely.[6] The ultimate direction of China's reemergence should its economic and political influence continue to grow, however, is likely not dependent on the workings of a preordained theoretical framework but rather on a number of emerging, interacting variables. These variables include, but are not limited to:

- Beijing's perception of a possible U.S.-Japanese containment strategy;
- China's ability to remain the central cog in Asia's increasingly linked export "workshop";
- The success of prevailing market mechanisms in meeting China's growing resource and energy appetite;
- Reactions to perceived Japanese remilitarization;
- The fielding of theater ballistic missile defenses by other nations and the implications for China's strategic deterrent force;
- The scope and scale of Taiwan missile programs; and
- The ability of the central government in Beijing to maintain internal order as the market economy drives internal demographic and political ferment.

China's policy approach to this complicated geo-strategic environment is found in the successive issuance of comprehensive five-year plans. Beijing's 11th Five-Year Plan, in place as of 2006, paints the picture of a government balancing a number of domestic priorities in the socio-economic domain, while maintaining a pragmatic approach toward security issues that impinge on its ability to sustain economic growth. The transparent part of the plan does not, however, provide insight into some of the intentions behind Beijing's rapid military modernization effort and the appearance of a mercantilist bent in certain economic and diplomatic initiatives. To better understand how the puzzle pieces fit together, and especially to better construe Beijing's intent in defense modernization, gaining an understanding of China's approach to building comprehensive national power (CNP) would be helpful.

Most civil and military leaders in Beijing appear to view the "grand environment" as a competition among rivals for relative gains in CNP.[7] As one Chinese source puts it: "Comprehensive national power is the basis for the national strategy and national defense strategy, and it is also the basis for analyzing the ratio of the international strategy and strength."[8] Developing CNP is a quantitative endeavor for the Chinese that involves a wide variety of factors encompassing tangible and intangible strength in political, economic, scientific, technological, military, cultural, and educational spheres. National development strategists must consider all elements of power, and resolve fundamental contradictions, in order for balanced development to occur. CNP development focused on a "strategic objective" that represents the "basic national interest" will yield stability and growth.[9] The "basic national interest" for China appears to be sustained economic growth with secure control of sovereign territory (from both internal and external threats).

Because China's approach to building CNP assumes a competition for influence and resources, there is legitimate cause for concern in certain aspects of Beijing's foreign policy and military modernization efforts. There is, however, an active debate in Beijing regarding the "zero-sum" nature of great power relations. Moreover, there is no assumption among China's leadership that a violent reordering of the international security architecture must accompany CNP growth.[10] One source posits that Chinese CNP rose from the eighth position in the world in 1980 to the sixth position in 2000—a time frame in which China did not employ military force, but instead reaped the advantages associated with integration in a world economy underpinned by U.S. security guarantees.[1]

Political-Military Strategies

- Diplomatic:
  - Multilateral engagement—strategic pragmatism
  - Selective regional leadership
  - "Peaceful Rise" posture (for mid-term only?)
- Economic:
  - Regional Free Trade Agreements/areas
  - Leverage competitive advantages globally
  - Fuel pan-Asian export engine
- Military:
  - Increase deterrent/coercive pressure on Taiwan
  - Robust, pragmatic "military diplomacy" program
  - Secure market and resource access if required

China's approach to building comprehensive national power is evident in the evolution of Chinese national security theory and the concomitant direction of political-military strategies. Over the past 2 decades, China's national security construct has undergone a sea change, with interstate competition replacing Maoist ideological conflict as the driving force behind foreign policy decision making.[12] In terms of diplomacy, this means that Beijing has identified those multilateral forums in which participation, and in some cases even leadership, is essential for securing national development objectives. In the economic realm, it means that Beijing embraces the trend of market globalization, particularly in those areas where China's competitive advantages allow for consistently high growth rates. In the informational arena—some would perhaps say propaganda arena—the new security construct informs an effort to portray China as a responsible rising power whose goals are commensurate with general regional stability and equitable development. Finally, in the military realm, Beijing's approach to national security drives a comprehensive force modernization and "professionalization" effort. The PLA's doctrinal shift away from classic "People's War" to "local warfare under high-tech conditions" and beyond—including current initiatives to "informationize" the force—has dramatically changed Chinese views on military campaign planning and operations.

It is possible that Beijing's intentions are more malign than current "peaceful rise" rhetoric indicates— i.e., that even if there is no drastic downturn in China's growth or no concerted effort on the part of the larger international community to contain China, Beijing would still seek to subvert U.S. and allied influence and access in Asia, and to undermine U.S. leadership globally. Chinese theorists sometimes speak of the requirement to maintain good Sino-U.S. relations "for now," indicating that this path could be abandoned when China's comprehensive national power reaches a certain level. Most indicators, however, seem to point to an acceptance on Beijing's part of the status quo security environment—as long as the Taiwan issue is manageable and no outside power blocks Chinese access to the markets and resources necessary for sustained growth.

PLA STRATEGY AND DOCTRINE

Chinese force modernization and deployment programs follow from the overarching strategic framework according to which Beijing defines threats to "the basic national interest." Chinese strategists do not envision a need for global power projection capabilities in the first half of this century and believe that only the United States, or the United States allied with Japan, presents a viable military threat to strategic interests in the near to mid term. These interests primarily include resolution of the Taiwan issue in Beijing's favor, security of energy resources and economic lifelines, and increasing Chinese leadership in Asian economic and diplomatic decisionmaking forums. Territorial or resource disputes between Beijing and Japan, India, or a unified Korea could conceivably be added to the list in certain future scenarios, as could disputes arising from shifts in access to energy in Central Asia and Russia. In all cases, the Chinese view their periphery as the focus for military concern. As a result, Beijing's military modernization priorities are the maritime, air, and missile programs needed to conduct short-duration, high-intensity operations against a U.S.-Taiwan or U.S.-Japan foe in peripheral seas to the east and south.

China will not have the capacity to dramatically alter the Asian security architecture via military competition for at least the next decade. Beijing believes, however, that strategic objectives are in reach if the Party can maintain internal order while the PLA develops capabilities to control China's immediate periphery. While definitions of the periphery have expanded due to the importance of distant sea lanes for energy and market access, the Chinese know that they will not conduct combat operations, other than limited anti-access activities, beyond the Asian continent or adjacent seas.

In the last 3 decades, Chinese military thinking has undergone a radical change, resulting in the development over time of three doctrinal templates. The first of these was the framework of "People's War under modern conditions"—a moniker that gave a nod to the Maoist boilerplate but in substance recognized that protracted wars of attrition were no longer suited to China's evolving interests and geo-strategic environment. By the early 1990s, with the first Gulf War serving as a powerful driver, this doctrine metamorphosed into what is commonly labeled "local, limited war under high-tech conditions." Chinese military theorists are now grappling with a third template that focuses on the correct mix of "informationized" and mechanized forces and concepts to conduct short-duration, high-intensity combat in the information era. This newest template is not fully formed, and debates continue as to the relative importance of "informationization" versus mechanization, the appropriate level of effort and funding for one over the other, the appropriate mix in the force structure, and other related issues. CMC member and Director of the General Logistics Department of the PLA, General Liao Xilong, states that mechanization is the platform upon which "informationization" must be built. They are inextricably linked, although the degree of prioritization in the programmatic realm is murky.[13]

Informationization at the operational level appears focused on providing an integrated platform for joint war zone command, control, communications, computer, intelligence, surveillance, and reconnaissance (C4ISR) connectivity, and for peacetime command and control (C2) within the PLA's Military Regions (MR). According to official Chinese media, the 11th Five-Year Plan tasks the PLA Informationization Work Office to move the PLA toward a "perfect universal transmission . . . and processing platform."[14] Recent programs to establish integrated joint communications and data transfer capabilities attest to the priority placed on this effort, and China's information technology sector is certainly capable of providing an effective architecture commensurate with the high level of resource commitment. As one senior PLA general notes, success in informationized warfare hinges primarily on "national information strength"—both in terms of global perception management efforts and domestic capabilities in key information technologies.[15]

One of the primary tasks of conducting informationized warfare is to transform traditional modes of mobilization to fit the conditions of modern warfare—the concept of "people's war" in a new era. For this reason, the modernization and reorganization of militia and reserve forces (the vast preponderance of which are ground force units and personnel) is to a great extent focused on bringing in high-technology qualified reservists and militia members—both to form new high-tech units (such as information and electronic warfare detachments), and to leaven existing or transforming units with more capable engineers and computer technicians.[16] According to a recent *PLA Daily* article, "specialized technical detachments" comprise 41 percent of reserve units; and the PLA has introduced a number of new reserve units, including "information protection and prevention detachments, satellite communications [units], and electronic interference systems."[17] The urban militia is clearly evolving to provide the warfighting force with high-tech support, providing access to an increasingly tech-savvy workforce.[18]

MODERNIZING THE FORCE

China's force modernization is governed by "strategic planning and strategic management [via] implementation of the scientific development concept in national defense and army-building."[19] While the "scientific development concept" is often ambiguously defined in Chinese sources, it focuses to a great extent on "bring[ing] into play the superiorities of the socialist system in concentrating our forces on major tasks."[20] This involves building a force prepared to meet both internal and external exigencies in an informationized society—and leveraging the advantages of an authoritarian regime to tailor the force in accordance with overarching national development priorities.

PLA strategists understand that for the wars they expect to fight in the next few decades, their focus must be on using the niche capabilities they have to counter the moves of a technologically superior adversary. The Chinese are carefully studying how American forces approach dominant command and control, surveillance and reconnaissance, rapid resupply, and quickly overwhelming an opponent with multidimensional firepower. PLA campaign planning; modernization programs; and research, development, and acquisition initiatives aim to defeat an adversary who brings a "local war under high-technology conditions" to China's neighborhood. While this sounds like a defensive focus, it is anything but. The PLA is committed to an offensive capability—to limited power projection and preemptive, or at least rapid, strikes against an enemy's critical vulnerability. Viewing the matter in this light, it is easy to understand why Chinese strategists are fixated on information dominance in the early stages of a fight, on the interdiction of enemy supply lines, and on strikes against key adversary high-tech weapons systems.

Within this framework, Chinese military planners look to accomplish a "quick battle to force quick resolution" but with an emphasis on preemptive and unexpected strikes to remove an enemy's technological superiority—what Chinese strategists call "structural destruction operations."[21] The Chinese believe that creating local and momentary momentum (especially air and information superiority) in a regional clash will allow them to defeat a more advanced adversary's plan and bring conflict to a close under Beijing's terms. This is a defeat criterion more focused on an enemy's strategy than on its military force. The Chinese seek to deprive an adversary of the ability to use operational and technical superiority to control strategic outcomes.

For the ground forces, however, the focus of modernization is not solely on fighting China's next war. With changing domestic demographics, and a number of factors undermining Communist Party control, many ground force units must also become more effective in conducting traditional domestic control activities, which encompass disaster relief efforts, local civil development projects, and support for People's Armed Police (PAP) forces in quelling public disturbances.

DOMESTIC CONCERNS AND FORCE DEVELOPMENT

Numerous internal forces drive Chinese decisions regarding force structure and capabilities. Foremost among these is the concern for maintaining public order and Party control in the face of rising discontent over corruption, illegal land grabs, and the adverse side effects of economic reform initiatives. While exact figures for incidents of internal unrest and protests are difficult to parse, it is certain that such incidents have increased in both scope and frequency in the past few years—with an estimated 74,000 incidents in 2004, and estimates as high as 87,000 for 2005. Party leaders are also faced with managing anti-Japanese and, to a lesser extent, anti-American sentiments in the populace. The Party has also created, and must maintain, the public expectation that progress on the Taiwan issue will follow from military modernization expenditures, diplomatic efforts, and economic integration initiatives. The various requirements of domestic pressure—looking both inward to a restive populace and outward to Taiwan, Japan, and the United States—ensure that the CMC will promote a force structure and posture plan that is far from homogeneous.

Public opinion matters in the People's Republic—perhaps more so than at any time since its inception. Economic reform buys loyalty for the moment, and the Party has more or less successfully harnessed rising nationalism for its own purposes. Relying on continued high growth rates and public concurrence that "national interest" equates to the Communist Party, however, is fraught with risk, given the myriad national development challenges facing Beijing. The Party's fragile position has force posture and budgetary implications for the PLA. The need to reinforce Party credibility and maintain the growth essential for domestic stability means that military spending will not overtake civil programs for the foreseeable future. As mentioned earlier, China is to some extent spared the hard prioritization choices of many other nations due to its ready reserve of capital. However, should growth slow in the coming years, hard choices will follow. Leadership concerns about public opinion, nationalism, and domestic satisfaction also mean that the PLA ground force will not be able to cut personnel drastically in order to shore up funding shortfalls or focus all available funds on a warfighting mission should one arise. In a national crisis, such as a Taiwan declaration of independence, the Party will likely have sufficient backing to mobilize the nation for war—but public security concerns likely will rise commensurate with the level of economic pain.

China has reduced its ground force by about 500,000 personnel over the past decade without a major upheaval in the civil economy or significant reduction in the ability to control public disturbances.[22] Many of the unit and personnel cuts resulted in growth of the PAP, thereby shoring up domestic control capabilities. But the presence of the PLA throughout the country provides a concrete link between central power and local life that the Party is unlikely to relinquish. Further minor reductions could accrue from consolidation of headquarters, support, and training organizations, but major unit reductions are unlikely.

Rising nationalism has accompanied the growth of a Chinese middle class; and the ever-increasing nationalistic impulse and its concomitant anti-U.S. and anti-Japan flavor could drive a fragile Party to turn a malevolent eye toward Taiwan, the last major reminder of Chinese national weakness. China has been content with maintaining the status quo in the Taiwan Strait (applying of course Beijing's definition of a "one-China" status quo), but believes that indefinite support of the current arrangement abets an American containment strategy and growth of a nationalistic Taiwan identity, thus the argument constructed by Beijing in passing the Anti-Secession Law.[3]

> **The Status Quo . . . and Taiwan's Place In It.**
>
> Accompanying China's rise is a sense on the part of its elites that, despite enjoying China's best security posture since the Opium Wars, Beijing is under siege from the U.S. and Japan. This "siege mentality" can lead to what might seem irrational behavior. While China may indeed be a "*status quo* power," Taiwan represents an element of the *status quo* that must change in the current international order to accommodate China's rise. Look no further than the Anti-Secession Law for proof—while pundits can argue about timing, content, and motivation, the fact remains that an "irrational decision" was made in order to lay down a marker toward change of the *status quo* in the Strait. From Beijing's perspective, Taipei has been inexorably laying down anti-*status quo* rhetorical, if not legal, markers over the course of the past 10 years.

Following the 2006 National People's Congress (NPC), Chinese Premier Wen Jiabao clearly stated that the starting point for improved Sino-Japanese relations is cessation of Prime Minister Koizumi's Yasakuni shrine visits.[24] However, a plethora of security issues plagues the relationship beyond the lingering historical animosities—and there is little that the Chinese can do in terms of ground force structure and posture to place any real coercive or deterrent pressure on Tokyo. China's predominant view of Japan is captured in a recent article by Li Wen of the Chinese Academy of Social Sciences:

> Japan is resorting to a "curved attempt at national salvation"—using the great vessel that is the United States to . . . expand its sovereignty and initiative to the greatest extent under the framework of the Japan-U.S. alliance. By taking the initiative to assume more responsibilities on behalf of the United States in East Asia in order to acquire political and military might, it hopes to ultimately come to a stand-off and confrontation with China in the Asia-Pacific region, and set up a regional order dominated by the United States and Japan.[25]

THE PEOPLE'S LIBERATION ARMY GROUND FORCE: MISSIONS AND PRIORITIES

China's national defense tasks as described in the 2006 *White Paper on National Defense* are:[26]

- Uphold national security and unity, and ensure the interests of national development;
- Provide, via the PLA, the source of strength for consolidating the rule of the CCP . . . and a solid security guarantee for sustaining this period of strategic opportunity for national development;
- Guard against and resist aggression [and] defend against violation of China's territorial sea and air space, and borders;
- Oppose and contain the separatist forces for Taiwan independence and their activities; and
- Take precautions against and crack down on terrorism, separatism, and extremism in all forms.

Hu Jintao refines the mission focus, stressing four areas (three "services" and one "function") to which the PLA must attend: provide the "forceful guarantee" for consolidation of CCP rule, provide a secure environment for economic development, protect national interests, and contribute to the maintenance of world peace.[27] Particularly telling is that "forceful guarantee" for consolidation of Party rule occupies a position of primacy — the concept of the Party army is alive and well.

The 2006 session of the NPC also shed light on PLA direction and mission apportionment. When discussing budget mechanisms, the NPC focused on "homeland defense"; countering "power politics" in the international arena; and dealing with secessionism, terrorism, and extremism. Interestingly, the NPC called for a switch from "maintaining the status quo" to a "limited growth development mode" — suggesting that China's official military expenditure of 8 percent of total national spending is below the international average of 12–16 percent of expenditures.[28] While it is unlikely that the priority of military spending relative to civil programs will change, it is obvious that relative percentages will be adjusted. This may indicate that more of China's hidden defense expenditures will make their way into the light — thus answering some of the criticisms levied by the international community regarding budgetary transparency. But it is more likely that the NPC is laying the foundation for moderate defense increases as a percentage of gross domestic product (GDP) under the current model — increases that in real terms are significant if prevailing growth patterns continue.

In translating broad goals and tasks into operational mission requirements, the ground forces are required to provide:

- Forces capable of domestic control and service support operations throughout the country;
- Forces capable of conducting defensive operations along the vast land and sea border;
- Forces capable of conducting amphibious and airborne/airmobile operations against Taiwan;
- Forces capable of conducting heavy mobile operations into potentially unstable areas on China's periphery to protect economic interests; and[29]
- Units and personnel capable of supporting global military-to-military exchange programs, United Nations (UN) peacekeeping operations, arms sale support, and foreign military training programs—which in turn support economic and diplomatic initiatives in regions critical for meeting future energy and market access requirements.

CONSTRUCTING A "HYBRID" FORCE

In an official news media interview, Academy of Military Science strategist Chen Zhou summed up the various mission areas the PLA must prepare to accomplish.

> The Chinese military must develop the ability to deal with diverse security threats and accomplish diverse military missions, not only being able to deal with traditional security threats, but also being able to deal with non-traditional security threats. Not only being able to deal with external threats in a situation in which there is a proportional imbalance in strategic forces, but also being able to deal with internal threats brought about by unstable factors domestically.[30]

The organizational command and control structure of the PLA from the General Staff Department (GSD) through the MRs down to Guard Commands, although convoluted in chain of command and reporting channels, reflects the various missions assigned by the party. Homeland defense missions play a large role, occupying a sizable organizational and resource niche within this structure.[31] Preparations for amphibious, airmobile/airborne, and special operations against Taiwan figure prominently as well. Much less obvious in the C2 structure are the ground force units that play an increasingly critical role in Beijing's pragmatic "military diplomacy" strategies across the globe.

Even after significant reductions, the ground forces represent about 70 percent of total PLA strength. During the first 3 decades of the People's Republic, the PLA as a technologically backward but disciplined force expected the vast majority of its soldiers to farm, shoot, assist local construction and relief efforts, and die in vast numbers in the unlikely event of foreign invasion—or in Chinese force projection operations on the periphery. Today, about a third of the PLA soldiers, officers, and noncommissioned officers (NCOs) are expected to be professional practitioners of modern warfighting and are being trained and equipped to achieve that status.

The main ground combat force of the PLA consists of 18 Group Armies spread across the seven MRs.[32] Two of these Group Armies, the 39th in the Shenyang MR and the 38th in the Beijing MR, are "Rapid Reaction Units," expected to deploy on notice for combat from garrison without personnel or equipment augmentation. Another three Group Armies—one each in Shenyang, Lanzhou, and Jinan—are being modernized for modern mobile warfare missions. Four to five Group

Armies (Nanjing and Guangzhou MRs) are focused on amphibious operations. These nine or ten armies, along with several special operations "Dadui" (about 1,000 troops each), the Air Force's 15th Airborne Corps, and two Marine brigades in the South Sea Fleet, represent the true ground combat force of the PLA.[33]

Local ground force units, some affiliated with Group Armies other than those listed above and some not, are responsible for border defense operations, garrison and/or infrastructure defense, or providing disaster relief and other services in their districts. Among these units, mission-specific specialization is also a requirement, even if the mission equates to more traditional local defense objectives. With a few exceptions, these units are lower on the totem pole for equipment upgrades—but all reap benefits from improved training programs and increased spending on quality-of-life initiatives. Geographic considerations continue to play a major role in PLA force posture—determining whether units in various locations will focus on jungle, mountain, or other mission capabilities. Only a decade ago, both main and local forces were heavily involved in a broad range of financial enterprises. While the warfighting Group Armies and their subordinates are largely free of these diversions, the extent to which local forces continue with such activities is less clear.

The linchpin of a more capable, "mission-specialized" PLA will be its NCO corps. Many analysts focus on the degree to which the PLA NCO corps will grow to resemble its U.S. counterpart in terms of small unit leadership capabilities. This may be a red herring—the primary expectation of the PLA leadership is that the NCO corps will provide systems-level technical expertise and enhanced unit-level readiness. The burgeoning NCO corps will be a particularly critical driver for the variegated ground force modernization strategy. A skills appraisal regime for NCOs is the focus of requirements advanced in the PLA's "Opinions on Strengthening the Cultivation of Well-trained NCO Contingent"—indicating a more rigorous approach to NCO training and professional development in specialized skills areas.[34]

For officers, PLA leaders expect a growing awareness of and facility in joint and combined arms operations in addition to expertise in primary skill areas. In the Shenyang Military Region, 170 officers since 2000 have reportedly participated in a cross-service exchange program with naval, air, and Second Artillery units.[35] At the major unit level, however, officers, soldiers, and NCOs alike are focused, generally more so than their western counterparts, on the specific operational campaign or local mission for which the unit is task-organized.

Ground force reserve and militia units have traditionally supported main and local forces in their districts or regions, and most will continue to do so; but recent reorganization activity indicates that certain units and personnel, particularly in high-technology specialties, may have primary responsibilities of broader scope and scale.

DEFENDING HEARTH, HOME, AND PARTY

Chinese strategists have embraced the term "homeland defense," encompassing as it does a number of PLA mission areas. Garrison, Fortification (coastal defense), and Guard Commands perform missions that would correspond to our understanding of homeland defense—as would a number of reserve and militia units that would mobilize to conduct "people's air defense" and similar operations. Domestic stability operations, however, expand the homeland defense concept, incorporating both crowd control and counterterror equipment and training requirements. Casting an even broader net, Chinese leaders and strategists also frequently discuss Taiwan under the homeland defense rubric, thus bringing "active defense" operations into the mix.

While the 1.7 million-strong Ministry of Public Security police force and the 1.2 million-strong paramilitary PAP are primarily responsible for domestic civil control operations, the physical presence of the PLA throughout the country in large numbers remains the Party's final defense. Capital Garrison, Garrison, and Guard Commands are particularly trained and equipped not only to defend key locations from foreign (primarily air) attack but also to protect from the enemy within. Some active and reserve forces throughout the country, particularly those unattached to Group Armies, continue to exist primarily as providers of stability, protection, and services to Provincial Military Districts and Subdistricts. Many of them receive specialized training focused on the local geography—mountain, desert, or jungle—but they are not now, nor are they likely to become, recipients of modern equipment in the near to mid term. At the same time, they are unlikely to face large cuts in manpower or funding, due to their important role in handling border defense and local support missions.

It is difficult to identify fully which units outside of the PLA's warfighting core are responsible for which primary homeland defense mission or combination of missions—crowd control and counterterror operations, disaster relief, agrarian or service sector support activity, key infrastructure protection, and border defense operations. The existence of priorities seems very likely, however, based in some cases on unit designations (Guard, Fortification, and Garrison Commands), and in other cases on training regimens. The 2006 *National Defense White Paper* clearly indicates that engineering units bear the brunt of agrarian and service support activities, and that the PLA and PAP together are responsible for disaster relief operations on a large scale.[36]

THE WARFIGHTING CORE: A "DEFENSIVE" FORCE?

Even for those main ground force warfighting units that grab the attention of military analysts, defensive capabilities continue to predominate. The PLA's defensive flavor has shifted, however, from traditional concepts of absorbing and depleting an attacking force to active defense, one that requires offensive capabilities to keep attackers away from China's political and economic centers of gravity. As offensive capabilities develop, and numerous programs to field force projection systems come to fruition, China will continue to couch its modernization effort in terms of defense—while western and regional onlookers accuse Beijing of developing offensive capabilities that belie the rhetoric of "peaceful rise."

Beijing counters with arguments concerning the threat to China's sovereignty from Taiwan "separatists" and their American arms suppliers. While the cross-Strait environment is relatively stable for the moment, Beijing's perception that political and social developments on Taiwan threaten China's fundamental sovereignty continues to spur military modernization programs that provide a broad capability base for operations against Taiwan and any U.S. forces that join the fray. Statements by Premier Wen Jiabao following the 2006 NPC sessions clearly reveal Beijing's belief that the current government in Taipei is pursuing a "constitutional re-engineering project aimed at *de jure* independence."[37] While missile, air, and naval programs predominate in "active defense" preparations, ground force restructuring and training in the key Group Armies also illustrate a desire to build a regional power projection capability for those forces. The warfighting core of the PLA will not be a fully modernized force, however, until maneuver divisions and brigades reflect in training the capability to manage the complexities of joint offensive warfare.

Despite Beijing's belief that the regional environment is conducive to peaceful development, border security remains a fixation for the Party. Completing a 10-year program under the auspices of the National Frontier Defense Committee, a reinforced architecture consisting of road patrol, defensive obstacles, and automated alarms and monitoring systems is in place for local forces to better conduct border defense operations, and for public security forces to maintain social order along distant frontiers.[38] While primarily defensive in nature, the improved road infrastructure also lays the foundation for improved mobility in areas where future threats to energy and resource interests might induce Beijing to deploy heavy mobile ground forces.

OFFENSIVE ASPIRATIONS

Faced with the requirement to build an amphibious and air-transportable force capable of responding to a call to arms in the Taiwan Strait—and also to have a heavy mobile warfare force for possible use in Central

Asia, the Korean Peninsula, or the Russian Far East—PLA force planners have clearly begun to restructure, equip, and train units for specific offensive missions. The 2006 *National Defense White Paper* states that "the Army aims at moving from regional defense to transregional mobility, and improving its capabilities in air-ground integrated operations, long-distance maneuvers, rapid assaults, and special operations."[39] Over the course of the past decade, the PLA has built at least four major amphibious training bases, with about one quarter of the PLA's maneuver divisions and brigades focusing on training for amphibious operations.[40] The special operations and airmobile capabilities needed in support of missile and air strikes against Taiwan are also priorities in ground force development initiatives. Downsizing or retiring a number of old divisions in favor of modernized, task-organized brigades possibly improves the PLA's capability to respond to potential crises along the full length of China's northern border. It also produces units that can more effectively conduct amphibious operations against Taiwan or Taiwan-controlled islands in the Strait.

"Integrated joint operations" is the current buzz-phrase for training, equipping, and sustaining the PLA to conduct multiservice operations in an "informationized" environment.[41] While definitions of joint operations differ between Chinese strategists and their American counterparts, integrated joint operations specifically refer to multiservice campaigns controlled by a joint headquarters using an integrated C2 architecture. Analysts are unsure of the status of this architecture, but PLA and MR periodicals are running numerous articles referring to tests and experiments involving its components. An integrated architecture would overcome a major impediment to joint C2 and could potentially fuse data from intelligence-surveillance-reconnaissance (ISR) assets into a near-real time "sensor-to-shooter" targeting network.

The ability to conduct long-range precision targeting operations will increase greatly when China's growing constellation of space-based sensors and aerial surveillance platforms can pass near real-time data to a variety of PLA systems. While most PLA watchers rightly focus on the implications of ISR enhancement for cruise and ballistic missile systems in the Second Artillery, Air Force, and Navy arms, the ground force will also benefit from faster, more accurate targeting solutions for its short-range ballistic missiles and long-range, course-correcting artillery systems. While data link, data relay, and data fusion program details are obviously shrouded in secrecy, we can safely assume that systems linking and fusing data between space, air, and terrestrial systems likely will be available to combat commanders in 5 to 10 years.

Logistics is a key area of concern in integrated joint operations—legacy logistics support systems for the PLA are stove-piped, producing notoriously slow and inefficient service. However, an automated "tri-service logistic interaction platform" was reportedly introduced recently in a subdepartment of the Beijing MR (following a similar fielding in the Jinan region).[42] Of particular interest is that the platform was reportedly introduced to provide joint logistic support to the "Beijing Theater of Operation" rather than to the Beijing MR, thus stressing the wartime mission.

Recent developments in the helicopter force indicate that the General Staff is well aware of the need for air assault capabilities to address shortfalls in contingency mission areas, as would exist in a landing campaign against Taiwan or a mechanized campaign on the Korean border, in Siberia, or along China's Central Asian periphery. The *Chinese Naval Encyclopedia*, published in 1998, discussed in some detail the command and control requirements for helicopters in a vertical amphibious assault—clearly revealing aspirations to overcome some of the beach landing constraints presented by Taiwan landing sites.[43] It is a mystery to many analysts, this author included, as to why the PLA has not poured more resources into rapid development of air assault capabilities. The force remains small, with limited transport capabilities, but a recent *PLA Daily* report indicates that the PLA has a coherent, focused plan for changing the situation. An official from the Army Aviation Department of the General Staff states that the Army aviation force would concentrate on "rapid delivery of military strength, precision strike, effective air domination, and timely support [shifting] from a unitary arm to an integrated arm, from a force with only transport capability to a force with both transport and strike capabilities, and from playing a supplementary role to playing the main combat role in battle."[44]

China is involved in a joint project with France's Eurocopter to develop civil transport helicopters. While this would not necessarily translate to improved military capability, past history indicates that European design assistance in rotary aircraft provides a foundation for improvements in army aviation systems.[45] A sharp increase in transport and Army attack aircraft along with a production spike in amphibious lift capacity would be key indicators of a possible shift in Beijing's use-of-force intentions.

Changes to the structure of two mechanized Group Armies in Lanzhou and Shenyang have caught the attention of some analysts, particularly as two Chinese academics recently posited that China needs to prepare to go to war to secure oil if required.[46] It is unclear if these units are particularly focused on providing a limited force-projection capability in regions where border security and resource access are increasingly important challenges, but it is certainly conceivable that the PLA is posturing itself to have heavy mobile forces in these areas for future contingencies.[47]

While some of the PLA's force reorganization and reorientation efforts seem focused on the offensive use of heavy, task-organized maneuver brigades and divisions, it is important to keep this effort in perspective. A recent *PLA Daily* article discusses armament readiness in Xinjiang (Lanzhou MR) units. The Director of the Armaments Department for the Nanjiang Military District in Xinjiang proudly pointed to a new "three one-third" armament management system that served to keep two-thirds of combat systems combat-ready at all times.[48] For western armies, a 66 percent readiness rate for combat systems would hardly be a point of pride. Training for heavy mobile warfare power projection is also not much in evidence—certainly not to the extent of amphibious training. Significant sustainability and command and control challenges face the PLA as it considers expeditionary warfare requirements, and efforts to address shortfalls in these areas are still in the early stages.

INTERNATIONAL MILITARY SECURITY COOPERATION: "MILITARY DIPLOMACY" IN ACTION

Tao Shelan, noted PLA strategist, posits:

> We must view international security cooperation from the perspective of national strategy . . . having a high regard for military operations other than war, participating actively in UN peacekeeping operations,

international cooperation against terrorism, and disaster-relief operations, launching bilateral or multilateral joint military exercises, and making a greater contribution to upholding regional peace and stability.[49]

In an article on conducting "People's War under informationized conditions," the Hunan Military District Commander, Major General Zheng Zhidong, states that "establishing an extensive international unified battlefront" is a pressing task for the PLA.[50] Part of this effort has been, and likely will continue to be, bent on internationally isolating Taiwan, but there are several other objectives that drive security cooperation in the broader context.

China views military diplomacy as a key contributor to overall diplomatic efforts, including under this rubric UN peacekeeping missions, military exchanges, international disaster relief support, joint exercises, and bilateral and multilateral security dialogue. Weapons and materiel sales and support packages are doubtless managed as an element of this diplomatic effort as well. Official media characterized China's military diplomacy as "pragmatic and active" in 2005, encompassing exchanges with 138 countries and including 41 bilateral or multilateral cooperation projects. Highlighted events included disaster assistance following the Indian Ocean tsunami and Hurricane Katrina; U.S. Secretary of Defense Donald Rumsfeld's visit to China (especially showcasing his visit to the PLA Second Artillery); the Seventh Round of Defense Consultative Talks with the United States; the first-ever "joint drill" with Russian forces; "fruitful cooperation and dialogue" with the Shanghai Security Cooperation Organization (SCO) and Association of Southeast Asian Nations (ASEAN) nations; joint patrols with Vietnam in Beibu Bay; and "joint drills" with the navies of Pakistan, India, and Thailand.[51] Since 2002, Beijing claims to have held 16 joint military exercises with 11 countries.[52]

The PLA has increased military exchange and training activity across the globe, establishing itself since 2000 as a regular contributor to UN peacekeeping operations. According to the 2006 *National Defense White Paper*, China currently has 1,487 military peacekeepers serving in nine mission areas and in the UN Department of Peacekeeping Operations.[53] Much of China's focus has been on countries in Africa, the Middle East, Central Asia, Latin America, and Southeast Asia—where access to energy and markets will become increasingly important as China's economy grows. It is difficult to determine whether Beijing has identified specific units or personnel to focus on "military diplomacy" missions—and if so, to ascertain their training and modernization objectives. This would be an interesting area for further study as China's military exchange and security assistance programs increase in number.

Africa.

Beijing's *African Policy White Paper* of January 2006 clearly delineates a requirement to promote exchanges and training programs with African militaries, and to more closely cooperate in such areas as intelligence exchange, counterterrorism, and counterdrug and small arms proliferation activities. China's peacekeeping operations in Africa are indicative of the importance it places on the region. Beijing's ability to work with African governments whose policies constrain Western involvement allows China to intertwine financial, military, and commercial components into negotiations on energy access and developmental aid. The Sudan provides the best example. Khartoum provides about 5 percent of China's oil imports; Beijing provides a variety of arms.[54] In 2004, China reportedly deployed approximately 4,000 troops to Sudan, and currently has 435 troops there.[55] At the same time, Beijing has with relative success managed the challenge to its image as a responsible international actor — while delaying and watering down UN sanctions, China has provided troops for the Darfur peacekeeping mission.[56]

ASEAN.

A coherent East Asian community is at best nascent, but China's involvement with ASEAN provides a foundation for a significant bloc in which security issues will become more and more enmeshed with economic interaction.[57] China's large and variegated ground force potentially provides security assistance expertise for a variety of transborder challenges. Indonesia and China instituted a so-called "strategic partnership" in April 2005, signaling a "new era" that includes the sale of surface-to-surface missiles and potentially a wide range of other military ware. Although Jakarta has been careful to stipulate that this development is not an anti-American action, China certainly moved into the gap left by the United States when it imposed restrictions on military sales and contact following Indonesian Army abuses in the East Timor secession.

The Middle East.

In a recent state visit to Beijing, Saudi King Abdullah bin Abdulaziz inked an agreement that would increase annual oil exports to China by 39 percent — it is not implausible to assume that this high-level agreement will be followed by discussions on cooperation in other areas, to include military exchange and training activity. Beijing's close relations with Tehran provide the model for China's strategic pragmatism in the region — while maintaining a public counterproliferation stance regarding Iran's nuclear program, Beijing remains fully engaged on commercial and security assistance fronts.[58]

Latin America.

China's military assistance efforts in behalf of Venezuela also highlight the link between energy, international security issues, and China's military-to-military activities.[59] Of particular note was the provision in 2005 of several hundred troops, the majority of them special operations forces, to provide Venezuela with special operations training and President Chavez with additional personal security.[60]

Central Asia.

Central Asia represents an area of extreme importance to Beijing on a number of fronts: energy access, counterterrorism alliances on China's unstable northwestern border, and an opportunity for political influence in a region of concern to both Washington and Moscow. Beijing is positioning itself as a major player in what one pundit calls the "new Great Game."[61] In 2005, China and Russia (albeit with different emphases) upped the ante in terms of political jousting with the United States in Central Asia when they spearheaded issuance of a SCO statement calling on Washington to set a date for withdrawal of U.S. forces from Central

Asia. China's interests in the region will increase in tandem with Beijing's investment in Kazakhstani oil production. The PLA has conducted counterterrorism exercises with SCO member states for several years running and will strengthen military-to-military ties with participating states on this foundation. China's willingness to overlook unsavory internal issues in neighboring states has allowed Beijing the flexibility to weave a web of influence that furthers its political, security, and economic interests. China's support for Uzbekistan following recent unrest and human rights abuses there is yet another example of this "great game" being played out in the region.[62]

TOWARD 2020: DEPLOYING THE NEXT GROUND FORCE

Barring a catastrophic collapse of the Chinese economy, or massive destabilization caused by some other unforeseen factor, Chinese military modernization will continue apace for at least the next 2 decades. There is no sign in the current fourth generation leadership, and perhaps even less so in the fledgling fifth, that total, inflation-adjusted expenditures for defense hardware and software will decrease. A decrease in the number of missions on the PLA ground force's plate is also unlikely, unless a political *rapprochement* with Taiwan occurs. Even in that event, maintaining amphibious-capable forces on the eastern seaboard might remain a priority if the Japanese relationship sours further or Beijing perceives an increasing threat of war on the Korean Peninsula.

While "stratagem" remains paramount in the PLA lexicon—and Beijing continues to establish the economic and diplomatic framework for increased regional influence—China's leadership nonetheless firmly believes that hard military power must prop up a "peaceful rise." Missile, naval, and air programs will continue to receive the most funding and thus the most attention; but the PLA will also intently pursue special weapons programs; improved airborne, air assault, amphibious, and mobile warfare capabilities; and niche special operations capabilities. Wherever the potential battlefield in Asia—whether in the East China Sea (Taiwan, the Senkakus), the Korean Peninsula, Siberia, the South China Sea, India, Vietnam, or Central Asia—the Chinese seek the ability to project task-organized combined arms forces to the fight while conducting a limited range of advanced anti-access operations against a responding, technologically superior force.

The tenets of Beijing's doctrine in such a fight will likely include rapid application of the principles of mobile warfare and preemptive strikes against an enemy's C4ISR and logistics capabilities. The PLA will rely heavily on conventional missile strikes supported by special operations using small numbers of highly trained teams. Denying the enemy critical reconnaissance information during force flow will be a top priority. China will probably achieve by 2015 the key enablers in realizing this vision—greatly improved PLA over-the-horizon detection and targeting based on the deployment of new space-based sensors, long distance air reconnaissance drones, and airborne early warning (AEW) platforms. Integration of space-based sensors with aerial reconnaissance aircraft will represent a real and credible threat to U.S. forward bases, C4I nodes, logistics assets, and forward-deployed forces should a crisis in the Taiwan Strait escalate to armed conflict involving the United States.

To accomplish objectives on the Asian landmass or on adjacent islands against a regional foe (and potentially against responding or forward-deployed U.S. forces), China will continue to focus on an elite core within the larger PLA ground force. Recent reorganization within ground force aviation units indicates that Beijing is serious about building an air assault capability for the mobile warfare units making up this core. Trends in the airborne forces indicate improved heavy equipment para-drop capabilities. By roughly 2015, the Chinese likely will have a task-organized, airmobile capability, backed by heavier forces that can be inserted by air or para-dropped.

The Chinese have already shown the production capability to increase amphibious lift capacity in an extremely rapid fashion—whether they choose to spike production for a regional conflict in the next 2 decades depends on their perception of threats in the East and South China Seas, and their assessment of the cost in political terms as neighbors view this activity with alarm. The PLA has made great strides in developing fire support for mobile operations, and in 2015 they will have a formidable inventory of state-of-the-art, long-range, course-correcting rocket systems. Special operations forces (SOF) and capabilities are also seen by Beijing as keys to success in targeting pivotal enemy vulnerabilities and maintaining control of the pace of a campaign. SOF teams likely will be well-trained to conduct anti-reconnaissance and C2 disruption operations, involving deep-attack raids and sabotage. These teams are trained in para-drop operations, motorized airfoil parachuting, and seaborne delivery; and they likely will be equipped with portable communications jamming equipment.

In many ways, the idea of "People's War under informationized conditions" is more than mere rhetoric—it speaks to a genuine desire to develop over the next decade the national mobilization plans and niche operational capabilities to gain control of a local conflict rapidly. For the ground forces, the definition of "local" will not change over time as much as it will for the Navy and air forces, but it will include China's entire land border and islands within China's maritime territorial claims.

Because of the potential conflicts that the PLA might face and the continued need for a significant military presence throughout the country for purposes of Party reach and control, it is unlikely that the ground forces will undergo another large personnel cut in the next decade. A reduction on the order of up to 100,000 personnel in this timeframe is tenable, but larger cuts would be likely only in the event that the Party removed a major mission from the PLA's plate, for example, transfer of additional units to the PAP (which would spare the economy the burden of an influx of newly unemployed workers). Reaching some agreed "one-China" frame work with a future Taiwan leadership might be another example.

STRATEGIC IMPLICATIONS FOR THE REGION AND THE UNITED STATES

Over the course of the next 10–12 years, Beijing hopes to build a ground force with task-organized units capable of conducting mobile warfare, special operations, and amphibious and airmobile operations beyond its borders, while maintaining a large homeland defense cohort to protect political and economic centers of gravity. If the current unremarkable force development trends on Taiwan itself continue, by roughly 2012 the PLA will need only to ramp up troop transport production (or civil asset mobilization) to project a force that could overwhelm Taiwan's defenses rapidly enough to complicate severely any U.S. response.

Trends and developments in Central Asia and the South China Sea do not portend Chinese use of force in these regions, with the political costs of adventurism in these areas promising to be extremely high. Ground force modernization efforts, however, do not overlook the mobile warfare and amphibious operations capabilities that will strengthen Chinese credibility as Beijing wields soft power to accomplish political and economic objectives in ASEAN and the SCO. U.S. security planners and analysts should carefully watch specific force structure developments and training activity among the units responsible for these missions as clues to Beijing's intent in dealing with potential crises in these critical regions.

Analysts will also do well to deliberate upon the skills being developed by PLA ground force units charged with nontraditional roles. Many of these troops and units will be involved in increasingly active military diplomacy programs, and will expand contact via training and exchange programs with militaries across the globe. The PLA is certainly combat-inexperienced, but it is drawing lessons regarding both traditional and nontraditional operations from a large number of recent or presently active conflicts. PLA experience also grows with increasing participation in UN operations.

Continued incremental modernization of the ground forces is inevitable given Beijing's perception of the security environment along its periphery. China is not building a ground force poised for large-scale power projection, but it is developing a force capable of conducting integrated joint operations both within and several hundred kilometers beyond its borders. While Beijing frames the PLA ground force as "right-sized" for primarily defensive missions, the warfighting core of the force is very large by current international standards, and increasingly capable of conducting offensive operations.

The likelihood of U.S. troops facing PLA ground forces in combat is very low, but a number of the areas of overlapping interest for Washington and Beijing represent potential security dilemmas in which crisis escalation may be difficult to avoid. U.S. security planners and analysts should devote the time and resources to better understand PLA ground force task organization and scenario-specific training in order to make sound assessments of Beijing's preparations for handling developing crises. Beijing's perception of the competitive nature of international interaction does not necessarily foreshadow a reliance on military solutions to emerging challenges, but at the same time it does not allow for complacency on the part of potential competitors.

ENDNOTES - CHAPTER 6

1. "Preserving the State" is a chapter title taken from *T'ai Kung's Six Secret Teachings*, purportedly written to provide advice to Kings Wen and Wu of the Chou Dynasty in the 11th century B.C. Translated by Ralph Sawyer, *The Seven Military Classics*, Boulder, CO: Westview Press, 1993.

2. See Chapter Five, "Force Modernization Goals and Trends," in the *Office of the Secretary of Defense Annual Report to Congress on the Military Power of the People's Republic of China 2006*.

3. See Chapter Six, "PRC Force Modernization and Security in the Taiwan Strait," in the *Office of the Secretary of Defense Annual Report to Congress on the Military Power of the People's Republic of China 2006*.

4. Information Office of the State Council of the People's Republic of China, December 2006, Beijing. The full text of *China's National Defense in 2006* is available at *www.chinaview.cn*, December 29, 2006.

5. For a discussion of the acquisition of key technologies, see the U.S.-China Economic and Security Commission's 2005 Annual Report, available online at *www.uscc.gov/*.

6. For in-depth background on Beijing's approach to the international order and China's national development, see Alastair Iain Johnston, "Is China a Status Quo Power?" *International Security*, Vol. 27, No. 4, 2003, pp. 5-56; and for applications in the military realm vis-à-vis Taiwan and the United States, see Thomas J. Christensen, "Posing Problems without Catching Up: China's Rise and Challenges for U.S. Security Policy," *International Security*, Vol. 25, No. 4, 2001, pp. 5-40; and Ashley J. Tellis, "The Regional Perspective: Asian Attitudes Toward the Taiwan Conflict and Future Implications," a paper prepared for the conference "Preventing and Resolving Conflict Across the Taiwan Strait," co-sponsored by The Carnegie Endowment for International Peace and The China Reform Forum, Beijing, April 6, 2005.

7. For a clear statement from the PLA theorists, see Zhai Xiaomin and Yang Shouqing, *Military Intelligence Analysis and Forecast*, Beijing: National Defense University Press, 2000, Preface.

8. Yang Shouqing, "The Study of Comprehensive National Power," *Military Intelligence Analysis and Forecast*, Beijing: National Defense University Press, 2000, p. 174.

9. Liu Yichang, "Equation of Comprehensive National Power," *Chinese War Mobilization Encyclopedia*, Qian Shugen, chief ed., Beijing: Military Science and Technology Press, 2003, pp. 60-61.

10. For a number of excellent articles concerning the debate on the nature of shifting power in the context of China's reemergence, see Yong Deng and Fei-Ling Wang, eds., *In the Eyes of the Dragon: China Views the World*, Lanham: Rowman and Littlefield, 1999.

11. Yang Shouqing, p. 181.

12. Nan Li, *From Revolutionary Internationalism to Conservative Nationalism*, Washington, DC: United States Institute of Peace, 1999, p. 34.

13. Wang Peiyu and Fan Juwei, "When Watching the All-Army Exhibition on New Type Logistics Armaments, Liao Xilong Calls for Efforts to Upgrade the Technological Level and Construction

Quality of Logistics Armaments," *Beijing Jiefangjun Bao*, May 9, 2006, p. 1.

14. Shen Yongjun and Su Ruozhou, "PLA Sets to Push Forward Informationalization Drive from Three Aspects," *PLA Daily Online*, January 11, 2006.

15. For a comprehensive discussion of "informationization" objectives, see Zheng Zhidong, "Thoughts on Improving Preparations for People's War Under Informationized Conditions," *Beijing Guofang*, April 2005, pp. 19-20.

16. Zheng Zhidong, "Thoughts on Improving Preparations for People's War Under Informationized Conditions," *Beijing Guofang*, April 2005. See also *China's National Defense in 2006*, p. 24, for a discussion of specialized technical units as the new "backbone" of the militia, replacing infantry units.

17. "Specialists Recruited to Boost Military Capabilities," *PLA Daily Online*, December 30, 2005.

18. Wu Daxiang, "Clearly Recognize the New Mission of the Urban Militia," *Zhongguo Guofang Bao*, January 13, 2003, p. 3, reported in *Foreign Broadcast Information Service* (FBIS).

19. Paraphrasing key concepts of a lecture given by Zheng Shenxia, President of the Academy of Military Science. Reported in Lu Tianyu and Ou Shijin, "Deepen the Understanding of the Scientific Development Concept from Comparative Studies on Chinese and Foreign National Defense Modernization and Development: Cao Gangchuan, Li Jinai, and Chen Bingde Attend Theoretical Study Class for All-Army Senior Officials," *Beijing Jiefangjun Bao*, May 14, 2006, p. 1.

20. *Ibid.*

21. Zheng Zhidong, p. 19.

22. See *China's National Defense in 2006*, pp. 17-18, for a discussion of troop reductions and the "safety net" infrastructure in place to handle the large influx of former soldiers into the civilian realm.

23. The Anti-Secession Law was passed during the Third Session of China's Tenth National People's Congress on March 14, 2005. Article Eight of this law states:

> In the event that the "Taiwan independence" secessionist forces should act under any name or by any means to cause the fact of Taiwan's secession from China, or

that major incidents entailing Taiwan's secession from China should occur, or that possibilities for a peaceful reunification should be completely exhausted, the state shall employ non-peaceful means and other necessary measures to protect China's sovereignty and territorial integrity.

24. "We're Keeping a Close Eye on Secessionist Moves," *People's Daily Online,* March 15, 2006, at *english.peopledaily.com. cn/200603/15/eng20060315_250763.html.*

25. Li Wen, "PRC Scholar Cites Need of Good Relations with Japan for PRC's 'Peaceful Development'," *Beijing Dangdai Yatai,* July 15, 2005.

26. *China's National Defense in 2006,* p. 4.

27. *Xinhua,* September 29, 2005.

28. Peng Guangqian, "China's Defense Spending in Tune with Economic Growth," *People's Daily Online,* March 15, 2006.

29. Sun Xuefu, "Build a Military Force Commensurate with China's International Status," *PLA Daily Online,* April 28, 2006.

30. Tao Shelan, "Military Expert: Improving the Strategic Ability to Safeguard National Interests Is a Pressing Task for China's Military at Present," *Beijing Zhongguo Xinwen She,* May 9, 2006. Reported in FBIS, May 18, 2006.

31. *Xinhua Ziliao* web posting on "Peoples Liberation Army's Different Military Regions," reported in FBIS, May 18, 2006.

32. "Chinese Defence Today: Ground Forces Order of Battle," at *www.sinodefence.com/army/orbat/default.asp.*

33. Assessment of strength, structure, and mission for a "warfighting core" within the ground force is an imprecise science. I derived my judgments from unit data provided in the *Office of the Secretary of Defense Annual Report to Congress on the Military Power of the People's Republic of China 2006.*

34. Li Wuqin and Su Ruozhou, "Military Specialized Vocational and Technical Skills Appraisal under Trial Scheme Kicks Off," *PLA Daily,* May 9, 2006.

35. "Group Army Fosters New Type of Talented Military Elites," *PLA Daily Online,* December 28, 2005.

36. *China's National Defense in 2006,* p. 18, claims:

> The PLA has assisted in building new socialist villages . . . helped build over 48,000 small public projects . . . bringing immediate benefits to nearly 800,000 people. In addition, it helped build or enlarge 211 primary or secondary schools[, has] taken part in ecological engineering projects[, and has] planted 210 million trees and sown grass on more than 13 million square meters of land.

The paper also states that "the PLA and PAP have dispatched over 340,000 troops to take part in more than 2,800 emergency rescue and disaster-relief operations [evacuating] over 3.4 million people.

37. "We're Keeping a Close Eye on Secessionist Moves," *People's Daily Online,* March 15, 2006, at *english.peopledaily.com. cn/200603/15/eng20060315_250763.html.*

38. Li Yun and Zhou Meng, "A Historic Leap of China's Land Frontier Defense Infrastructure," *PLA Daily Online,* August 19, 2004.

39. *China's National Defense in 2006,* p. 5.

40. Dennis J. Blasko, "Chinese Army Modernization: An Overview," *Military Review,* September-October, 2005, p. 72.

41. A basic definition of integrated operations can be found at Li Chunli and Chen Yilai, eds., *Integrated Joint Operations Command,* Beijing: Military Science Publishing House, 2004.

42. Chen Hui and Zhang Xiaoqi, "First Tri-service Logistic Support Interaction Platform Established," *PLA Daily Online,* January 11, 2006.

43. Zhang Yun, "Amhibious Landing Command Control Unit," *Chinese Naval Encyclopedia,* Beijing: Haichao Press, 1998, pp. 200-201.

44. Li Chuanxin and Hu Junhua, "Army Aviation Force Seeks New Leapfrog Development in the New Year," *PLA Daily Online,* January 9, 2006.

45. "Helo Deal," *Aviation Week and Space Technology,* Vol. 164, Issue 5, January 30, 2006.

46. Wu Lei and Shen Qinu, "Will China Go to War over Oil?" *Far Eastern Economic Review,* Vol. 169, No. 3, April 2006, pp. 38-40.

47. Martin Andrew, "PLA Doctrine on Securing Energy Resources in Central Asia," *China Brief* (The Jamestown Foundation), Vol. 6, Issue 11, May 2006.

48. Wang Yanxiang and Ye Shi'an, "Xinjiang MAC creates a Dynamic Mangement System Method for Armaments," *Beijing Jiefangjun Bao*, May 15, 2006, reported in FBIS, May 18, 2006.

49. Tao Shelan, reported in FBIS, May 18, 2006.

50. Zheng Zhidong, p. 19.

51. "Military Diplomacy Helps China's Peaceful Development," *PLA Daily Online*, December 30, 2005, at *english.chinamil.com.cn/site2/columns/2005-12/30/content_374363.htm*.

52. *China's National Defense in 2006*, p. 33.

53. *Ibid*.

54. See General Administration of Customs of China data reported by *Reuters*, July 25, 2005; and "Human Rights Watch, *China's Involvement in Sudan: Arms and Oil*, November, 2003.

55. Josh Eisenman, R. Evan Ellis, Stephen Johnson, and Brett D. Schaefer, "Chinese Influence: Expanding in Both Africa and Latin America," Heritage Foundation Seminar, March 7, 2006, at *www.heritage.org/Press/Events/ev030706b.cfm*. See also *China's National Defense in 2006* for current figures regarding Sudan, Congo, and Liberia.

56. Ian Bremmer, "The Dragon Awakes," *National Interest*, Summer 2005, Issue 80, p. 135.

57. Jusuf Wanadi, "ASEAN and China Form Strategic Partnership," *Jakarta Post*, December 15, 2005, cited by *Taiwan Security Research*, at *taiwansecurity.org/News/2005/JP-151205.htm*.

58. Vivienne Walt, "Iran Looks East," *Fortune*, Vol. 151, Issue 4, February 21, 2005.

59. Bremmer, p. 136.

60. "China Trains Venezuelan Commandoes," *StrategyPage.com*, *www.strategypage.com/htmw/htsf/articles/20060120.aspx*.

61. Lutz Klevman, "Oil and the New 'Great Game'," *The Nation*, February 16, 2004, p. 13.

62. For an excellent overview of China's regional engagement strategy, see David Shambaugh, "China Engages Asia: Reshaping the Regional Order," *International Security*, Vol. 29, No. 3, 2004, pp. 64-99.

CHAPTER 7

PLA GROUND FORCE MODERNIZATION
AND MISSION DIVERSIFICATION:
UNDERWAY IN ALL MILITARY REGIONS[1]

Dennis J. Blasko

The Chinese *Defense White Paper* of 2004 acknowledged that "priority [has been] given to the Navy, Air Force, and Second Artillery Force" to strengthen the "comprehensive deterrence and warfighting capabilities"[2] of the People's Liberation Army (PLA) — a fact that has been apparent to foreign analysts for about a decade. Despite the precedence given to the development of the other services, the PLA ground forces (i.e., the Army) still comprise the vast majority of PLA forces. While the Chinese government has not provided an official accounting of the personnel distribution within the 2.3 million-strong PLA, most estimates usually credit the Army with about 1.6 million personnel (about 69 percent of the force, though this number and percentage are decreasing), the PLA Navy with about 255,000 (about 11 percent), the PLA Air Force with some 400,000 (about 17 percent), and the Second Artillery with around 100,000 (about 4 percent).[3] Moreover, ground force officers continue to retain the majority of senior leadership positions at headquarters from Beijing to local levels, though the PLA's leadership structure is changing gradually as officers from other services are assigned to senior command and staff positions. Although the Army is sometimes overlooked when outsiders concentrate on high-technology developments in the PLA Navy, Air Force, and Second Artillery, modernization of the ground force is an integral element of the overall, multifaceted, long-term transformation of the Chinese armed forces.

The 2006 *Defense White Paper* did not repeat the priority listing for the services, apparently because no major change in priorities had taken place since 2004. However, the 2006 document did disclose in general terms the PLA's "three-step development strategy" to build "informationized armed forces . . . capable of winning informationized wars" by the middle of 21st century (usually acknowledged to be 2049, or 100 years after the founding of the People's Republic of China [PRC]). Intermediate milestones were set at 2010 "to lay a solid foundation" and at 2020 "to make major progress" toward that final strategic goal "in accordance with the state's overall plan to realize modernization."[4]

Within this strategy, the Army seeks to move "from regional defense to transregional mobility, and improving its capabilities in air-ground integrated operations, long-distance maneuvers, rapid assaults, and special operations. . . . Priority is given to building Army aviation, light mechanized and information countermeasures units."[5] Many of these trends have been evident since 1999 as PLA ground forces, along with PLA reserve and militia units, have undergone significant modernization and increases in capabilities in *all* Military Regions (MRs) in preparation for a wide variety of potential missions. This chapter presents evidence affirming the trends in PLA ground force modernization identified by the 2006 *White Paper*. As the PLA explores new missions for the 21st century, the Army is building forces to maintain its relevance to

China's overall deterrence posture and, if deterrence fails, to PLA warfighting capabilities, even as the force shifts to a more maritime-oriented outlook.

THE FUNDAMENTAL ASSUMPTION: POLITICAL LOYALTY OF THE PLA

The fundamental underpinning of all aspects of PLA modernization is the political loyalty of the PLA to the Chinese Communist Party (CCP). Since taking over as Chairman of the Central Military Commission (CMC), Hu Jintao has added his mark to ensure the military's subordinate relationship to the CCP. A January 9, 2006, *Jiefangjun Bao* editorial set forth his "Three Provides and One Role" to define the duties of the PLA in supporting and defending the leadership of the party:

> ... in the new period of the new century, *our military should provide an important, powerful guarantee for strengthening the party's ruling position,* should provide a firm security guarantee for safeguarding China's opportunity for strategic development, should provide strong strategic support for safeguarding national interests, and should play an important role in safeguarding world peace and in accelerating the development of all.[6] [emphasis added]

The "Three Provides and One Role" are directly related to Hu's other theoretical contribution, the "scientific development concept." Despite its name, the scientific development concept is basically focused on the development of *people* as the key to modernization. Thus, the "Three Provides and One Role" are "an important reflection of the concept of scientific development in the military field," which is "a significant innovation in the party's guiding theory for the military, and points the direction for revolution[ization], modernization, and standardizing our military."[7] The 2006 *White Paper* repeats the content of the "Three Provides and One Role" in its section on "National Defense Policy" without mentioning Hu by name.

The January 2006 *Jiefangjun Bao* editorial also defined the mission of the party: "Our party is required to unite and lead the people around the country in completing three historic tasks: *accelerating modernization construction; reunifying the motherland; and maintaining world peace and accelerating common development*" [emphasis added].[8] A month earlier, another report had described Hu's application of those tasks to the military in "an important exposition on the historic mission of the armed forces in the new century." It will be useful to quote Hu at some length:

> This important exposition requires that our Army must not only pay attention to *national survival interests*, it must also pay attention to *national development interests*; it must not only *protect the security of the nation's land, territorial waters, and airspace*, it must also protect national security *in the electromagnetic sphere, in outer space, on the seas and oceans,* and in other respects; it must not only deal with traditional security threats to national sovereignty, unity, and territorial integrity, it must also deal with *nontraditional security threats* of which *terrorism is the focal point*; it must not only emphasize military and political security, it must also emphasize *economic and social security*; it must not only protect the overall situation of national reform, development, and stability, it must also *protect world peace and promote development* in common. ... It is theory which updates the place and role of the People's Army under the Party's long-term governance of the country. It is theory which updates the missions and functions of the People's Army in the midst of national development and the revival of the people. It is innovative theory for the People's Army's conduct

of peaceful missions internationally. It is theory which updates the objectives of the construction of our Army in the new century and the new era. It is innovative theory for the scientific development of our Army under the new circumstances.[9] [emphasis added]

In the mindset of the CCP, national survival is equivalent to survival of the party. The political officer system and party committee system are principally responsible for ensuring proper understanding of these requirements and to maintain ideological correctness and political obedience within the military. Commanders and political commissars at all levels work closely to maintain political reliability and are jointly responsible for all aspects of the performance of their units. When certain hardware and technological capabilities are lacking, the PLA sees its political system as adding ideological strength to the balance of power equation. One commenter put it this way:

> Experience shows that, given equality of weaponry and armaments, the fighting spirit decides the result of wars; given inferiority in weapons and armaments, a strong fighting spirit may make up for the inferiority in armaments, bring material strengths into full play, and create a miracle of using inferior weapons to defeat enemies armed with superior armaments under certain circumstances. Our military always pays great attention to cultivating the fighting spirit. Our military has been invincible and undefeatable throughout the long-term revolutionary wars because our military upholds a fearless spirit characterized by: "When fighting formidable enemies, we are braver than the enemies and we are stronger in the face of dangers and difficulties."[10]

From outside the PLA, it is often difficult to tell whether any of the above claims are merely rhetorical repetition of the party line or are truly believed as inherent strengths of the Chinese armed forces. Over time a more educated and sophisticated personnel force may question aspects of ideological training to a greater degree than would less educated peasants. If there were significant political disaffection in the ranks, however, it is unlikely that foreigners would read about it directly in the PLA press, though there may be hints in the context of some articles and internal PLA activities. The fear of chaos and the belief that the CCP and Chinese armed forces, imperfect though they may be, are the ultimate protectors against turmoil breaking out throughout the country, likely motivate the vast majority of PLA personnel to remain loyal to the party. There is little possibility the PLA, being a servant of both the party and the people, would take the lead in advocating political change within the party or the nation. However, it would likely encourage modifications to both government and party policies to assure that the support the Army needed for both its peacetime and wartime missions was forthcoming.

ARMY MISSIONS AND ROLE IN STRATEGIC DETERRENCE

Within the political framework outlined above, the armed forces of China have both external and internal missions. The PLA primarily is focused outward, the People's Armed Police (PAP) is focused domestically, and the militia is in general support.[11] The PLA and PAP also both have secondary missions that support each other's primary mission. The 2006 *White Paper* defines China's national defense policy as:
- Upholding national security and unity, and ensuring the interests of national development. This includes guarding against and resisting aggression, defending against violation of

China's territorial sea and air space, and borders; opposing and containing the separatist forces for "Taiwan independence" and their activities; taking precautions against and cracking down on terrorism, separatism, and extremism in all forms.

- Achieving the all-round, coordinated, and sustainable development of China's national defense and armed forces, with China pursuing a policy of coordinated development of national defense and economy.
- Enhancing the performance of the armed forces with informationization as the major measuring criterion.
- Implementing the military strategy of active defense, thus upgrading and developing the strategic concept of people's war, and work for close coordination between the military struggle and political, economic, diplomatic, cultural, and legal endeavors; use strategies and tactics in a comprehensive way, and take the initiative to prevent and defuse crises and deter conflicts and wars.
- Pursuing a self-defensive nuclear strategy.
- Fostering a security environment conducive to China's peaceful development.[12]

The 2006 *White Paper* elements listed above are slightly different from "China's basic goals and tasks in maintaining national security" as set forth in the 2004 *White Paper*, which specifically began by mandating deterrence of Taiwan independence ("to stop separation") and promoting reunification (which itself was a change from the priorities listed in the 2002

White Paper). The 2006 *White Paper* reference to China's nuclear strategy was also new, but other points from the 2004 *White Paper* are similar, as seen below:

- To stop separation and promote reunification, guard against and resist aggression, and defend national sovereignty, territorial integrity, and maritime rights and interests.
- To safeguard the interests of national development, promote economic and social development in an all-round, coordinated, and sustainable way, and steadily increase the overall national strength.
- To modernize China's national defense in line with both the national conditions of China and the trend of military development in the world by adhering to the policy of coordinating military and economic development, and improving the operational capabilities of self-defense under the conditions of informationalization.
- To safeguard the political, economic, and cultural rights and interests of the Chinese people, to crack down on criminal activities of all sorts and maintain public order and social stability.
- To pursue an independent foreign policy of peace and adhere to the new security concept featuring mutual trust, mutual benefit, equality, and coordination with a view to securing a long-term and favorable international and surrounding environment.[13]

The tone of the 2006 *White Paper* about Taiwan was not as intense as in previous years, but still firm. In 2004 the first mention of Taiwan was the following: "The situation in the relations between the two sides of the Taiwan Straits is grim." In 2006 the first mention of Taiwan had a different thrust: "The Chinese government has taken a number of significant measures to improve relations across the Taiwan Strait, thus promoting cross-Straits relations toward peace and stability." After these positive words, however, the following paragraph reminded readers about the gravity of the issue so far as Beijing was concerned:

> The struggle to oppose and contain the separatist forces for "Taiwan independence" and their activities remains a hard one. By pursuing a radical policy for "Taiwan independence," the Taiwan authorities aim at creating "de jure Taiwan independence" through "constitutional reform," thus still posing a grave threat to China's sovereignty and territorial integrity, as well as to peace and stability across the Taiwan Straits and in the Asia-Pacific region as a whole. The United States has reiterated many times that it will adhere to the "one China" policy and honor the three joint communiqués between China and the United States. But it continues to sell advanced weapons to Taiwan, and has strengthened its military ties with Taiwan.[14]

In the 2006 *White Paper*, this detailed discussion of Taiwan was positioned among those security challenges that "must not be neglected," which highlighted "growing interconnections between domestic and international factors and interconnected traditional and nontraditional factors."[15] The linkage here between a variety of internal and external factors is important to the PLA's allocation of time and resources. Certainly, military planning and training for potential Taiwan contingencies is high on the PLA's list. In the ground force, training for this mission is undertaken mostly in the Nanjing, Guangzhou, and Jinan MRs (see the section on *Amphibious Training* below). Meanwhile, the Army also trains for many additional missions in these and other MRs.

Among the other ground force training tasks is preparation for a variety of nontraditional security threats and challenges. These tasks have been mentioned frequently since 2002 and were included in the 2006 *White Paper*: "Security issues related to energy, resources, finance, information, and international shipping routes are mounting. International terrorist forces remain active, shocking terrorist acts keep occurring. Natural disasters, serious communicable diseases, environmental degradation, international crime, and other transnational problems are becoming more damaging in nature."[16]

Later, the *White Paper* reports that the PLA and PAP "dispatched over 340,000 troops to take part in more than 2,800 emergency rescue and disaster relief operations, involving more than 40,000 vehicles, [flew] more than 2,000 sorties (including the use of helicopters), [and] evacuated over 3.4 million people" in 2005 and 2006.[17] In addition, "PLA personnel have joined China's international rescue teams in international rescue operations after the Indian Ocean tsunami and the earthquakes in Pakistan and Indonesia."[18] The *White Paper* does not give specific numbers of troops deployed to these international rescue missions, and it is likely that actual numbers were small.

In another nontraditional security mission, for several years China has contributed significant numbers of military personnel and civilian police to United Nations (UN) peacekeeping operations. According to UN statistics, at the end of November 2006 China had deployed 1,659 military troops, military observers, and civilian police to 11 UN peacekeeping missions.[19]

In a major change to Chinese foreign and military policy prior to 2002, the PLA now engages in combined training exercises with many countries, mainly with nontraditional security types of scenarios. Consider the following synopsis broadcast on a Chinese military affairs discussion program:

> Since 2002, the relevant units of our Army have held 17 joint military exercises with 14 countries, including, in chronological order, Kyrgyzstan, Russia, Pakistan, Kazakhstan, France, India, and the United States. Among them, four were antiterror exercises on the ground. The first such exercise took place on October 10, 2002, namely, the joint antiterror exercise between China and Kyrgyzstan. It was held, together with the Kyrgyz armed forces, on Chinese soil along the border between China and Kyrgyzstan. The first outbound joint exercise for our Army, in which they went to another country with their own weapons and equipment, was with the armed forces of Tajikistan on September 22 this year. There have been [12] more joint exercises between China and other countries at sea. There has been just one large-scale comprehensive exercise, namely, the Sino-Russian joint military exercise from August 18-25, 2005.[20]

Taiwan, nontraditional security threats, and other factors threatening social stability were named together in the January 9, 2006, *Jiefangjun Bao* editorial referenced earlier as "factors of uncertainty surrounding China," and then continuing as follows:

> All social contradictions are influencing one another, and the number of factors threatening social stability has increased. Failure to appropriately prevent or handle problems in any field could impact or put a strain on periods of strategic opportunity that are important for national development. All this requires the military to make achievements within the framework of safeguarding national security and national development.[21]

Fragile domestic factors reemphasize the need to coordinate and balance military modernization carefully with other aspects of national economic development. In such an environment, the military recognizes it must share resources with other national requirements and support the national economy at the same time the nation supports it.

With these internal constraints in mind, the PLA focuses on deterring or preventing war. The current *White Paper* makes multiple references to the PLA's deterrent role, as does the January 2006 *Jiefangjun Bao* editorial, which uses some familiar terminology:

> The most important point in ensuring that the military satisfactorily safeguards periods of strategic opportunity *is to use the awe created by military strength to curb or postpone the outbreak of war.* In the past, we have often said, "Maintain an Army for one thousand days to use it for an hour." In fact, great changes have taken place in the connotation of the phrase "use of the military." Deploying military forces for the exchange of fire is one form of "using the military." *Utilizing the great awe created by military strength is another form of "using the military."*[22] [emphasis added]

The PLA has written openly about deterrence for several years, including a full chapter in *The Science of Military Strategy*, first published in 2001. An excerpt:

> Strategic deterrence is a major means for attaining the objective of military strategy, and its risks and costs are less than strategic operations. . . . Warfighting is generally used only when deterrence fails and there is no alternative. . . . Strategic deterrence is also a means for attaining the political objective. . . . Without resolute determination and firm volition, deterrence is feeble.[23]

In order to achieve effective deterrence and accomplish its political goals (such as reunification of Taiwan with the mainland), China must have, according to PLA doctrine, a capable fighting force: "Strategic deterrence is based on warfighting. . . . The more powerful the warfighting capability, the more effective the deterrence."[24] Although the PLA Navy, Air Force, and Second Artillery have leading roles in the Taiwan scenario, the Army also maintains an important role in deterrence with its potential to occupy the island physically by force. But the Army's role in deterrence extends to many other potential scenarios. Along China's 22,000 kilometers of land border with 14 nations, the Army principally shoulders the task of deterring foreign aggression or terrorist intrusions, despite the low probability of major land invasion in the contemporary international environment. Coping with the potential influx of North Korean refugees is probably a "nonconventional security threat" foremost on the minds of PLA planners in Beijing and northeast China. The Army is likely to take the lead in many antiterrorist scenarios on China's borders and shares that responsibility with the PAP and civilian police force throughout the country, especially in the run-up to the 2008 Summer Olympics. A strong, disciplined ground force also serves as a deterrent against domestic unrest even as the Ministry of Public Security police and the PAP are being equipped and trained to act as the primary forces to control internal stability.

No matter what the current likelihood of any specific scenario, the Chinese leadership understands that the capabilities to perform any of these missions cannot be built overnight. Therefore, the Chinese leadership believes it must invest, to some degree, in the entire spectrum of military capabilities required by a growing power, especially one that is not integrated into a web of military alliances. Land power remains an important component in the range of capabilities the PLA must acquire.

Military power, however, is only one element of comprehensive national power. China's strategic deterrence is built also upon its political, economic, scientific, technological, cultural, and diplomatic components. With the shadow of growing military capabilities constantly in the background, Beijing is currently using other elements of national power in an attempt to achieve its strategic objectives.

With political loyalty as its underlying foundation and building a credible military component for strategic deterrence as its goal, the PLA ground forces are engaged in the overlapping and interrelated processes of "Army building" and "preparation for military struggle." If, however, at any time during this long-term period of military modernization and transformation China's civilian leaders should call upon the armed forces to accomplish a mission, the PLA leadership will obey the orders of its chain of command. PLA planners will put together military force options that incorporate capabilities developed to date, using all assets available to them, to achieve their assigned missions. In the meantime, the PLA leadership appears to be keeping its civilian masters well informed of the progress the force has made, as well as the obstacles remaining.

ARMY BUILDING, OR THE REVOLUTION IN MILITARY AFFAIRS WITH CHINESE CHARACTERISTICS

The concept of Army building or national defense construction was associated with the "Revolution in Military Affairs (RMA) with Chinese Characteristics" in the 2004 *White Paper*, which included a complete chapter on the "Revolution in Military Affairs with Chinese Characteristics," with subsections on the following topics:

- Reducing the PLA by 200,000 personnel,
- Strengthening the Navy, Air Force and Second Artillery Force,
- Speeding Up Informationalization,
- Accelerating the Modernization of Weaponry and Equipment,
- Implementing the Strategic Project for Talented People,
- Intensifying Joint Training,
- Deepening Logistical Reforms,
- Innovating Political Work, and
- Governing the Armed Forces Strictly and According to Law.

Thus the "RMA with Chinese Characteristics" can be interpreted to cover the entire range of elements of PLA military modernization and transformation. In particular, mechanization and informationalization are essential elements. In 2006, the PLA was said to be "speeding up the revolution in military affairs with Chinese features and enhancing in an all-round way its capabilities of defensive operations under conditions of informationization."[25]

PLA publications often treat the "RMA with Chinese Characteristics" and "preparation for military struggle" as two different concepts but with significant overlap between them. For example, many elements of both the "RMA with Chinese Characteristics" and "preparation for military struggle" are contained in Hu Jintao's exposition on the PLA's historic mission:

> We must continue to arm our officers and soldiers with updated ideology, emancipate our minds further, update our point of view, firmly establish and implement in full the scientific development concept, break free of traditional views and unsuitable customs and methods which constrain the development of combat power, boldly explore new paths, new methods, and new measures in military training, political work, personnel development, and integrated support, and let the historic mission stimulate the enthusiasm of officers and enlisted personnel throughout the armed forces to promote *the revolution in military affairs with Chinese characteristics* and *prepare well for military struggle*.[26] [emphasis added]

In order to examine some of the various elements of the military modernization process, it may be helpful to categorize Army building and the "RMA with Chinese Characteristics" in terms primarily of personnel, force structure, equipment, and professional military education, while considering preparation for military struggle to focus on missions, doctrine, training, and logistics, accepting, of course, that there are many overlaps between these two broad categorizations. Taken together, they describe the process of modernization and transformation in the force. The following sections discuss recent developments in the two halves of Chinese military modernization—"RMA with Chinese Characteristics" and preparation for military struggle.

Personnel Policies.

Long before Hu's scientific development concept, the PLA was concentrating on improving the quality of its personnel. The counterintuitive decision to reduce the period of conscription to 2 years in 1999 was accompanied by a concerted effort to develop a professional noncommissioned officer (NCO) corps. Many observers might have predicted a longer period of conscription was necessary to properly train soldiers in a more technologically advanced PLA. Instead, the years of conscription appear now to be primarily a probation period for identifying those who might be qualified for and interested in extending their service as NCOs (and, to a lesser extent, as officers). Conscripts perform a variety of necessary tasks and provide manpower to make things run, but many technical tasks are assigned to an expanding NCO corps. NCOs now perform duties, such as squad leader, that previously were assigned to conscripts, albeit in their last years of service. They also are taking over duties previously performed by officers, e.g., company mess officer (an important responsibility), to the extent that nearly 70,000 officer positions have now been shifted to NCOs.[27] Other former officer duties now being assigned to NCOs include "aviation technician, captain of small-sized transportation boat, electric and mechanic chief, observation and communication chief, navigation chief, small warehouse chief, confidential archives keeper, club director, auto service unit head, driving instructor, head of soldiers training unit, nurse, and [logistics manager] in People's Armed Forces Departments."[28]

The PLA is in the midst of a continuing process to seek the right ratios of conscripts to NCOs to officers and uniformed civilian cadre (*wenzhi ganbu*). Officers and uniformed civilians felt the brunt of the latest 200,000-man reduction, with 170,000 such slots being eliminated.[29] Over 60,000 of the personnel eliminated were from MR and provincial Military District (MD) headquarters and administrative units.[30] Most, if not all, of these personnel wore ground force uniforms. At the same time, the absolute number of NCOs appears to be on the increase. The process of building an educated officer and NCO corps is enshrined in the "Strategic Project for Talented People," which began in August 2003 and passes milestones in 2010 and 2020.[31]

Perhaps as a lesson learned from other militaries, in the past year the PLA has implemented a new personnel policy permitting the hiring of "nonactive duty" contract personnel. (These workers are referred to by a number of names, including *feixianyi gongqin renyuan* and *wenzhi renyuan*.) This new category of personnel is composed of civilians who are hired for specific periods of time to perform service and support tasks in higher headquarters, noncombat units, hospitals, schools, etc. Though they perform many of the same functions as NCOs and PLA uniformed civilian cadre (*wenzhi ganbu*), nonactive duty contract personnel are managed separately and are not in career competition with the *wenzhi ganbu*. They were first reported being integrated into the force in the spring of 2006.[32] By the end of the year, over 20,000 NCO posts were filled by contract civilian personnel.[33] While these contract workers wear military uniforms (without insignia or rank) and may be required to go to the field to support combat operations, they are not counted among the PLA's active duty numbers.[34] There is some irony in this development in that it occurred exactly at the end of the reduction of 200,000 active PLA personnel.

Better pay and living standards for PLA personnel are necessary in order to attract and retain qualified soldiers. As living standards improve among the civilian populace, resolving quality-of-life issues is important in maintaining morale in the armed forces. Personnel pay raises have been acknowledged as a significant part of overall increases in the officially announced Chinese defense budget over the past decade. In July 2006 on average, pay for many personnel in the PLA doubled. Likewise, there has been a major effort to upgrade living facilities for all forces. In late 2005, General Logistics Department statistics reported that "30 percent of barrack areas have reached the standard of new-concept barracks, and 50 to 80 percent of Chinese military barracks will be transformed in the next 10 years."[35] Construction of new barracks is underway in all parts of the country, especially in border areas. One general said of the improvements: "The soldiers eat meat at every meal, can bathe every day, and can also get on the Internet to study; when some of them go home on leave, they can hardly wait to get back here." However, not all funding for new construction is new money from central funds. The 16th Group Army's land sales in downtown Changchun are an example of a policy allowing the sale of excess property to acquire money for essential projects:

> The organs and subordinate units of a group Army of Shenyang MR originally had their barrack areas in seven locations in three districts of Changchun City. They took advantage of the policy of making flexible use of real estate, and put on the market for auction scattered barrack areas in three locations, totaling over 20,000 square meters; these were in much sought-after locations in the city center, and *thus raised 129 million yuan for construction*; through exchange they obtained all the land needed for construction of a new barrack area and also gained 84 million yuan in compensation. Today, their new barrack area occupies 560,000 square meters in the Changchun City development zone, with a construction area of 80,000 square meters.[36] [emphasis added]

According to the report, "The whole Army now has several thousand idle barrack areas." This policy raises additional questions of how widespread the practice of land sales has become. Does anybody pay taxes on these transactions, and how, if at all, are the proceeds of these sales accounted for in China's overall military expenditures? The potential for graft and corruption in these sales is obvious, and central oversight over the practice is mandatory. In fact, in the summer 2006, a 5-year effort to combat military corruption was announced with the audit of "983 leading officers, including 26 at army level, 135 at divisional level, and 822 at regimental level," to determine "if there are any irregularities involving budgetary work, building projects, equipment procurement, investment, real estate projects, and profitable services."[37]

Force Structure.

The 2004 *White Paper* contained a revelation about provincial MDs, Military Subdistricts (MSDs), and People's Armed Forces Departments (PAFDs) that surprised at least this one PLA-watcher: these local garrison headquarters serve as elements of their respective local governments *and* are under *dual command* of the military from the MR headquarters *and* the local governments in which they serve:

> The provincial commands . . ., sub-commands, . . . and the PAFDs of counties, autonomous counties, cities and municipal districts *concurrently act as the military service organs of the people's governments* at corresponding levels and are responsible for the military service work in their respective areas *under the leadership of the military organs at higher levels and the people's governments* at corresponding levels.[38] [emphasis added]

These various headquarters interface directly with local government counterparts for conscription, mobilization, demobilization, and coordination of military needs with local economic development requirements (using the National Defense Mobilization Committee system), and in response to national and local emergencies such as war or natural disasters. A number of local governments have built joint military-civilian command centers. In keeping with the military's subordinate relationship to the CCP, these local relationships reflect a military subordinate to local government (and party) authority, not a separate power sitting on a mountaintop of its own.

MD headquarters command local PLA forces (including border and coastal defense units and some logistics units) and reserve units in the province. The 2006 *White Paper* changed its translation for MSD, using the words "prefectural military commands," not "subcommand" as in 2004. More importantly, however, it gave a more complete listing of MSD responsibilities: "The main tasks of a prefectural military command are to oversee the military training, political work, and equipment management of the militia and reserve force, organize and conduct wartime mobilization, and undertake military service registration and enlistment. Prefectural military commands in border areas are also in charge of the military, political, logistical, and equipment work of border defense troops as well as border defense duties, talks, and meetings, and border management, protection, and control."[39] In effect, this last sentence identifies the MSD headquarters as being in command of border defense forces.

The 2006 *White Paper* also included the most detailed description of the composition and responsibilities of PAFDs to date:

> PAFDs are organizations set up by the PLA *in counties* (banners, county-level cities, or municipal districts). . . . A people's armed forces department consists typically of a military affairs section, a political work section, and a logistics section. Its main tasks are to oversee reserve force buildup, combat readiness, military service [i.e., conscription], and mobilization, and command militia operations. The *grass-roots* PAFDs established by the state *at the level of township (town) or sub-district* are non-active-duty organizations. They are manned by full-time staff that are under the dual leadership of the local Party committees and governments at the same level and military organs at higher levels.[40] [emphasis added]

This description identifies two distinct levels of PAFDs: county-level and "grassroots-level" in townships. County-level PAFDs are manned by active duty PLA ground force personnel. Grassroots PAFDs, however, are not part of the active duty PLA and instead are local government organizations. While county-level PAFD personnel wear ground force uniforms like all other active duty members, grassroots PAFD cadre wear similar Army uniforms, but with different cap and collar insignia and rank epaulets. As local government workers, they are paid by the local governments despite their duties in direct support of the PLA, such as conscription, and their role in commanding militia units. The *White Paper* did not say how many grassroots PAFD personnel are found throughout China, but the number is likely to be significant.

Total active duty PLA ground force personnel are estimated to amount to somewhere around 1.5 to 1.6 million personnel, including main force units; local defense forces such as border and coastal defense units; most personnel assigned to MR, MD, MSD, and county-level PAFD headquarters (personnel from other services are increasingly being assigned to the

Joint Logistics Departments at MR headquarters and their subordinate units); most personnel assigned to logistics subdepartments including hospitals and supply/repair depots; and the staff, faculty, and students assigned to Army-related professional military education academies and schools.

Main force units are considered "mobile combat troops" and include group armies with their subordinate divisions, brigades, and regiments as well as some independent units under the command of MR or MD headquarters. PLA Army main force units continue to streamline their numbers and modify their composition. As of 2007, main ground force units are organized into 18 group armies along with a number of independent units (divisions, brigades, and regiments/groups).[41] Maneuver forces (infantry and armored units) consist of approximately 35 divisions and about 42 brigades. They are supported by roughly 40 more artillery, surface-to-surface missile, air defense, and anti-aircraft artillery (AAA) divisions and brigades, plus various functional entities such as Special Operations Force (SOF), reconnaissance, Army aviation (helicopter), engineer (combat and pontoon bridge), communications, chemical defense, electronic warfare, and "high technology" units.[42] Based on this estimated order of battle, if local forces, administrative headquarters, nondeployable logistics and maintenance/repair depots and bases, and personnel assigned to professional military education institutions are subtracted from the total number of ground forces, then the main force combat and combat support units currently may comprise only about half of the total ground force personnel number (somewhere around 700,000 to 800,000 personnel).[43]

In 2005, the PLA announced that the number of divisions would decrease while the number of brigades would increase. In particular, the number of group armies with an all-brigade structure will increase.[44] Appendices I to VII show that four of the 18 group armies (the 20th, 27th, 40th, and 47th) currently have this structure, with organizational levels extending from group army to brigade to battalion. Thus organized, all-brigade group armies are much smaller than their corps-level counterparts in the West, making all-brigade group armies closer in size to U.S. Army and Marine divisions.

As also can be seen from the order-of-battle listing in the appendices, a few other group armies have only one maneuver division assigned (the other units being brigades). These divisions would be likely candidates for future downsizing, if it has not occurred already. Currently, only the 38th Group Army has an all-maneuver division structure; previously the 39th Group Army also had this structure, but it appears that its 190th Mechanized Infantry Brigade was recently restructured from division size. Most group armies now appear to be structured with a combination of maneuver divisions and brigades, but if the 2005 announcement proves to be accurate, several of these group armies will lose or downsize their divisions in the future. Some or all of the estimated nine armored divisions in the ground force inventory may be high on the list for conversion to brigades.

As the number of divisions decrease, several infantry divisions have been reported as being restructured to consist of two infantry regiments and one armored regiment. Some of these divisions apparently were in a lower readiness category and did not have organic armored regiments in the past (a full-strength infantry division would have had three infantry regiments and one armored regiment, along with artillery, AAA, and other supporting units). This situation allowed for one of the three infantry regiments to be converted to an armored regiment, with new main battle tanks used to equip some of these new armored regiments. Restructured divisions would have fewer personnel than full-strength divisions, but more firepower and mobility than their predecessors. Restructured divisions have been identified in four MRs, including at least three divisions opposite Taiwan, suggesting that more such divisions may already exist or others may be restructured in the future. Currently, at least the six infantry divisions below are believed to have a three-regiment structure:

- 6th Independent Mechanized Division, Lanzhou MR;[45]
- 37th Infantry Division/13th Group Army, Chengdu MR;[46]
- 40th Infantry Division/14th Group Army, Chengdu MR;[47]
- 86th Infantry Division/31st Group Army, Nanjing MR;[48]
- 91st Infantry Division/31st Group Army, Nanjing MR;[49]
- 121st Infantry Division/41st Group Army, Guangzhou MR.[50]

The number of mechanized infantry units in the ground force continues to grow. Following the identification of the 149th Mechanized Infantry Division/13th Group Army in the Chengdu MR, mechanized units are now found in all MRs. Five mechanized infantry brigades have also been identified in the ground force. The 2006 *White Paper* notes the priority on building "light mechanized" units, without defining the term. At present, only the 127th Light Mechanized Infantry Division/54th Group Army is specifically identified as such, though part of the 149th Mechanized Infantry Division may also be a light unit. The two amphibious mechanized infantry divisions (the 1st and the 124th) in the Nanjing and Guangzhou MRs may also be considered as light mechanized units.[51]

Heavy units (all mechanized infantry and armored units) are approaching nearly half the number of total maneuver units in the Army. While this element of modernization improves ground mobility, personnel protection, and firepower, it also increases fuel consumption and maintenance requirements. Mechanized units are most efficiently moved over long distances on the ground by rail or on heavy equipment trailers, and they require special sealift capabilities for transport over large bodies of water. From the standpoint of unit morale, modern mechanized units possess a cachet that many PLA ground force units lacked in the past.

In the late 1990s, a Nanjing MR artillery brigade was transformed to become the first ground force short-range ballistic missile (SRBM) brigade. In 2006, *PLA Daily* carried a photograph of this unit, showing it to be equipped with the DF-11/M-11 SRBM.[52] Additionally, a second ground force SRBM brigade has been identified in the Guangzhou MR at Shantou, though it may not yet be operational. As of March 2006, a Guangzhou Library website contained an entry for "Guangzhou MR 75810 Unit, Guangdong Province Shantou Missile Base."[53] As in the Nanjing MR, the Guangzhou MR headquarters department has a "conventional missile department," indicating supervision of SRBM assets.[54]

In the mid-1990s, both the 38th and 31st Group Armies created mixed air defense brigades. This new type of unit upgraded traditional AAA units by integrating tactical surface-to-air missiles (SAM) along with AAA gun units. In the following decade, similar units were established throughout the country so that approximately half of all group armies now have air defense brigades. With the PLA's emphasis on integrated air defense operations throughout the country, this trend is likely to continue.

Despite having both SAMs and AAA gun units in the same organic formation, evidence suggests that actual integration of their combat operations has been slow to develop. As late as the summer 2006, the PLA newspaper announced that in peacetime "different types of AA artilleries would train separately while at wartime they each fought in their own way, thus it was hard for them to form an 'iron fist of air defense'."

However, new "digitalized equipment" has allowed units "to form a three-dimensional fire-net to fight simultaneously against the incoming high-altitude, mid-altitude, and low-altitude targets." Reportedly, this was the first time "different AA artilleries [were] organized into the same artillery group in a mixed way in training as well as in combat."[55] Similar challenges exist in integrating SAM units with AAA units, but progress in communications connectivity and training appears to be underway. This is an example of how force structure, new equipment, training, and doctrine all must be coordinated in order to turn the potentials of new equipment into actual operational capabilities — a process that may take years for a single unit to accomplish and decades to be achieved throughout the entire force.

Despite also being identified as a priority for development, the Army's helicopter contingent is still relatively small for such a large ground force. Army Aviation units are found in each MR (sometimes with more than one unit per region) and in a growing number of group armies. It is nearly certain that their numbers will continue to increase as new aircraft, both domestic and foreign, are added to the force. *The Military Balance 2006* reports something over 375 Army helicopters of all types in two training and eight operational regiments.[56] This estimate of both the number of airframes and units may be somewhat low, but in any case helicopters must be a high-demand, low-density item in the PLA.

In 2005, *The Directory of PLA Personalities* identified an Army Aviation regiment in both the 26th and 54th Group Armies in the Jinan MR, making them the third and fourth group armies with organic helicopter assets (others are found in the 38th and 39th Group Armies).[57] In October 2006, *Jiefangjun Bao* reported on a new Army Aviation regiment in a group army in South China, which was formed around 2004. This fifth Army Aviation regiment subordinated to a group army possibly is part of the 31st Group Army in the Nanjing MR.[58] The assignment of helicopter units to group armies is important because it makes training with the organic infantry units more routine and less encumbered by bureaucracy than having to go to MR headquarters to coordinate training. The identification of the Army Aviation unit in the 26th Group Army suggests that this formation deserves closer attention.[59]

The 2006 *White Paper* contains a unique chapter on "Border and Coastal Defense" which discusses the distribution of labor among PLA forces and Ministry of Public Security forces (including elements of the PAP) in maintaining security along China's land borders and coasts. This chapter acknowledged that "in 2003, the PLA border defense force took over the defense of the China-DPRK [North Korean] border and the Yunnan section of the China-Myanmar border from the border public security force." Furthermore, it described the organization of border defense units as "a three-level structure, namely, regiment, battalion and company," while coastal defense units have "a five-level structure, namely, division, brigade, regiment, battalion, and company."[60] Although the *White Paper* does not go into greater detail about the number of border and coastal defense units, this structure is consistent with what is known from other sources.

Based on analysis of information from a number of Chinese websites and official media sources, the order of battle for this portion of PLA local forces is estimated to be approximately 57 border defense regiments, 9 border defense battalions (not part of regiments), 4 patrol craft groups, 2 coastal defense divisions, 3 coastal defense brigades, 29 coastal defense regiments, and 4 coastal defense battalions. Border defense units are configured mostly as light infantry, with regiments estimated to have three battalions of three companies plus up to an additional four companies subordinate to regimental headquarters (for a total of up to 13 companies per regiment).[61]

The patrol craft groups are Army ship units tasked to monitor inland waterways along China's borders, such as the Yalu or Songhua Rivers. Coastal defense units may be infantry, artillery, or may perhaps be equipped with surface-to-surface anti-ship cruise missiles, such as the SILKWORM. Assuming that these units are slightly smaller than main force units of equivalent organization level (for example, a border or coastal defense regiment may have as few as 2,000 personnel), a total of approximately 200,000 Army troops assigned to permanent border and coastal defense responsibilities seems reasonable.

New Equipment.

Though much attention has been paid to new weapons and equipment in the PLA Navy, Air Force, and Second Artillery, the Army has also received large amounts of new gear since the late 1990s. Most of the ground forces' new equipment is produced by Chinese defense industries and is being distributed to units in all MRs. Unlike the Navy and Air Force, which have received significant numbers of several types of weapons systems from Russia, the ground force has mainly received only Russian Mi-17-series helicopters and some precision-guided artillery munitions and technology. On the other hand, Chinese defense industries have supplied everything from new main battle tanks, AAA guns and SAMs, unmanned aerial vehicles (UAVs), and small arms to logistics and repair vans and other support equipment (all part of the "mechanization" of the force). A major element of the new equipment entering the force is electronics, such as night vision devices, computers, and communications gear, produced by Ministry of Information Industry factories (part of the drive to "informationalize" the PLA). Many new types of electronics can be grafted onto existing systems provided adequate sources of electricity are available.

The Type 96 main battle tank is the most widely distributed new tank in the force and has been observed in units throughout the country. *The Military Balance 2006* estimates that there are some 1,200 Type 96 tanks (roughly equivalent to 12 armored regiments) in a force of around 8,580 total tanks.[62] The more advanced, more expensive Type 98/99 main battle tank has been deployed in smaller numbers, estimated at 80 for the entire Army. (This number is less than the complement for a full armored regiment, and, if accurate, suggests the tank may be issued in battalion or company-size units to a few divisions or brigades. The Type 98/99 has been seen in the 6th Armored Division/387th Group Army in Beijing MR and in the 3d Armored Division/39th Group Army in Shenyang MR). Both the Type 96 and Type 98/99 have appeared in photographs as being equipped with reactive armor to increase protection from anti-tank weapons. The new light amphibious tank, the Type 63A, has been issued to both ground force (about 400) and PLA Navy Marine units (150). HQ-7 and HQ-61 mobile SAMs are part of the new air defense brigades, along with new AAA guns. The Type 95 quad 25mm gun/SAM self-propelled system (PGZ 95) has also been observed in exercises, probably as the organic air defense system for armored or mechanized regiments. The PTL02 Assault Gun has been reported deployed to four MRs (likely in the 127th Light Mechanized Division and the 162nd Motorized Infantry Division in Jinan MR, the 6th Independent Mechanized Division in Lanzhou MR, the 121st Division in Guangzhou MR, and the 149th Mechanized Infantry Division in Chengdu MR).[63] It would also appear to be a good candidate for deployment to the two amphibious mechanized infantry divisions. Some long-range multiple rocket launchers, such as the A-100 (WC232) or WS-1/2, likely have been deployed to artillery brigades subordinate to some group armies, but exact numbers and locations are uncertain. A helicopter, the Z-10, designed solely for the attack role, is under development and will likely enter the inventory within a few years.

PLA units such as the 149th Mechanized Infantry Division and some SOF units are experimenting with small all-terrain vehicles, capable of being carried internally by Mi-17 helicopters, to increase the mobility of infantry forces in difficult terrain (for example, for reconnaissance units in mountains). The Type 95 5.8mm family of small arms is gradually replacing older models based on the AK-47 design. Of the weapons mentioned above, a decade ago only a few examples, such as the Type 96, PGZ 95, and HQ-7, were beginning to enter the inventory. Significant numbers of these and other weapons did not enter the force until around 2000 and later.

Currently, new weapons and support equipment are entering the ground forces in all corners of the country. However, often complete complements of weapons for all subordinate elements are not issued at the same time. Thus some units likely will have mixes of old and new weapons for some time to come. The number of various types and modifications of many major items of equipment (for example, *The Military Balance 2006* identifies five distinct types of main battle tanks and three types of light tanks), many of which are nearing the end of their operational life, would appear to complicate both repair and maintenance support. Some old weapons are probably kept in units for cannibalization to keep other weapons operational. Continuing to downsize some divisions to brigades (or to the three-regiment structure) will require fewer new weapons than if full-strength divisions were upgraded completely.

Reserve Units and Militia Forces.

Since about 1998, the reserve force (which is made up of both PLA reserve units and the militia) has undergone many of the same transformations and modernizations as the active PLA. Reserve units often receive hand-me-down equipment from active forces when the latter are upgraded to new gear, but some new equipment has also entered into the reserve force. Local governments are instrumental in funding operational and training requirements for the reserves and allocating space for training areas (which recently are being consolodated for efficiency purposes). Reserve units and militia are now more frequently integrated into larger exercise scenarios than in prior decades.

Army reserve units number approximately 40 divisions, 25 brigades, and several regiments.[64] Prior to 1998, few if any brigade-size units were found among Army reserve units. Since then a major development has been the creation of a reserve logistics support brigade in each MR. These units support both active and reserve forces in training. A number of other types of brigades have also been newly formed. In the spring of 2005, four new reserve brigades were reported by the Chinese media, three of which were AAA units:

- Zhangjiakou Army Reserve AAA Brigade,[65]
- Qinhuangdao Army Reserve Artillery Brigade,[66]
- Yichang Army Reserve AAA Brigade,[67]
- Shanxi Reserve AAA Brigade.[68]

New reserve AAA regiments have also been reported, but it is unclear whether these are independent units or subordinate to existing reserve divisions. Other new reserve Army units have been formed, such as the Tibet Army Reserve Mixed Brigade, and reserve and militia units to support the other services are also gradually being formed.

Reflecting China's preoccupation with being attacked by long-range aircraft and missiles, fully one-third of the number of PLA reserve divisions and brigades are AAA units. Urban air defense is an area of major emphasis for the militia, with many units reported to be training with shoulder-fired SAMs. Another important task for the militia is repair of civilian infrastructure—such as roads, railroads, and electricity grids—after China is struck by long-range weapons.

Professional Military Education.

As the PLA has downsized, its professional military education system has also restructured in size and course content. In the past few years, some 15 military academies and educational institutions have been closed, turned over to local control, reorganized, or reformed into comprehensive military training bases to fulfill new requirements. With the growth of the NCO corps, NCO education at six NCO academies and in NCO courses at officer academies has been emphasized. Academy course content has been modified to include instruction on the PLA's new doctrine. The 2006 *White Paper* says,

> the PLA has 67 military educational institutions [total for all services], which are divided into two types: those for academic credentials and those for pre-assignment education. The former offers undergraduate education for pre-commission officers and graduate education for officers. The latter consists of elementary, intermediate and advanced level institutions and NCO schools, and offers pre-assignment training and rotational training for active-duty officers and NCOs. Some pre-assignment educational institutions also offer graduate courses in military science.[69]

As fewer new officers are needed for the smaller active force, the yearly intake of officers, from various sources, has been reduced. In 2006 the number of high school graduates entering the officer academies was cut in half, down to 10,000 from 20,000 in previous years.[70] Also, the number of freshman national defense students in civilian universities on PLA scholarships was reduced to 10,000, down from 12,000 in 2005.[71] The number of college graduates who are not national defense students but who volunteer to join the PLA after graduation is not known. For both 2005 and 2006, the number of enlisted soldiers entering officer academies was 5,000.[72]

Since the beginning of 2005, three former Army academies (at Guilin, Dalian, and Jinan) have been transformed into regional comprehensive training bases for individual and unit training. Soldiers and units will rotate in and out of these bases for short periods of time to receive a variety of individual skill training or engage in unit drills. This change helps accommodate the need for additional training facilities as the ground force transforms and also reflects the need for fewer infantry lieutenants in the downsized force.

PREPARATION FOR MILITARY STRUGGLE

The chapter on National Defense Policy in the 2004 *White Paper* defined "preparation for military struggle" as follows:

> The PLA takes as its objective to win local wars under the conditions of informationalization and gives priority to developing weaponry and equipment, to building joint operational capabilities, and to making full preparations in the battlefields. Meanwhile, it adheres to the people's war concept and develops the strategies and tactics of

the people's war. To meet the requirements of integrated and joint operations, the PLA endeavors to establish a modern operational system capable of giving full play to the overall efficiency of the armed forces as well as the national war potentials. The PLA conducts more training and exercises with specific objectives in order to raise its capabilities in coping with various crises and contingencies.

"Preparation for military struggle" thus concerns itself primarily with how the PLA will fight in case it is called upon to perform any of a variety of military missions to win local wars. Force structure, personnel, and equipment all contribute to the preparation for military struggle, but its main emphasis is how to conduct integrated joint operations with the forces at hand. Though the Taiwan scenario currently may rank number one on the PLA planning agenda, it is the first among many missions the PLA believes it must be prepared to undertake. The guidelines for training are drawn from the universe of PLA missions: "The missions, tasks, and responsibilities for the armed forces are the sources of the driving force for military training. Missions determine tasks, and tasks bring along training."[73]

Training Guidance and Trends.

All components of the PLA are focused on improving their joint operational capabilities through realistic training. The developments described below fall under the rubric of "improving [Army] capabilities in air-ground integrated operations, long-distance maneuvers, rapid assaults, and special operations" as laid out in the 2006 *White Paper*.

PLA training consists of annual regional and service training tasks supported by a large body of guidance known as the "Military Training and Evaluation Program" (MTEP). In January 2006, the General Staff Department declared that the year's military training should focus on:

1. Strengthening realistic mission-oriented training;
2. Training according to the Military Training and Evaluation Program;
3. Exploring integrated training;
4. Giving prominence to joint training, including operational and tactical exercises and coordinated technical drills to increase integrated joint operations capabilities;
5. Enhancing the information quality of officers and men;
6. Standardized training management;
7. Regulating training evaluation and examinations; and
8. Intensifying personnel training, including transformation and reform of military schools and academies to prepare personnel for "military struggle."[74]

Many of these points concentrate on fixing shortfalls identified in the past. For example, currently commanders and staff at all levels are attempting to perfect the process of evaluating training and unit readiness. In prior years, honest assessments of training sometimes did not percolate up to higher headquarters. Standardized tests and evaluations, including live fire evaluations, are now conducted to judge quality of training. Units that perform poorly often are required to undergo remedial training before starting a new training cycle. Many of the training procedures the PLA now employs would be very familiar to foreign professional military observers.

Throughout the country, for training commanders and headquarters staff live exercises and computer simulations are a major emphasis. Much of this training concentrates on individual skills, but coordinating individual efforts as part of integrated staff planning is also a main concern. To better coordinate staff work, specific training for chiefs of staff has also been conducted. To help organize and coordinate staff efforts, the PLA has experimented in grouping battlefield functions together into systems such as command and control, intelligence, reconnaissance and early warning, fire support, and comprehensive logistics support so as to integrate the efforts of disparate units.

In 2004 the senior PLA leadership determined that all of the military's new equipment, capabilities, and various components of the armed forces were not being incorporated sufficiently in much of the training throughout the country. In response, the leadership developed a new phrase, "integrated joint operations," to refocus efforts on incorporating *all* existing and developing capabilities into joint training. At the same time, Beijing assigned the Chengdu MR to take the lead in the integrated joint training pilot project for experimenting in how to optimize interservice operations.[75] (At least one division in Lanzhou MR has also been assigned the duty of experimentation in integrated joint operations.[76]) Much of this training is now being conducted by the 13th Group Army and the 33rd Air Division, both headquartered in Chongqing. The two divisions of the 13th Group Army (37th Infantry Division and 149th Mechanized Infantry Division) are involved in this work. The 149th is also testing many new items of equipment, such as small, light all-terrain vehicles, and working closely with helicopter units in airmobile operations training.

An important result of this experimentation appears to be the initial steps in developing doctrine for PLA Air Force aircraft to provide close air support (CAS) to ground troops.[77] In the past, the PLA did not include CAS as an Air Force mission (focusing instead on preplanned, centrally controlled battlefield air interdiction missions to support ground forces), but now 13th Group Army and the 33rd Air Division appear to be exploring the tactics and techniques for this important joint mission, much of it a result of new communications equipment supplied to the force. Army Aviation helicopters are also integrated into fire support operations for ground units. Army Aviation units continue to work with both SOF and infantry units in all MRs to develop airmobile concepts suitable for the PLA. Helicopter operations include air landing of personnel, personnel descent by ropes from hovering aircraft ("fast rope" and rappelling), external sling-loading of equipment and fuel bladders, and internal transport of light equipment, as well as reconnaissance, fire support, command and control, electronic warfare, and resupply.

In recent years political officers have been called upon to increase their tactical proficiencies in military skills. The creation of the "Three Warfares," — psychological warfare, media war, and legal war — has given the political officer system tactical responsibilities on the battlefield it did not have in the past. The incorporation of the "Three Warfares" into training and campaign planning is an important element of "integrated joint operations." Small units, usually composed of soldiers from political departments in unit headquarters, have been reported to practice loudspeaker and leaflet delivery operations to demoralize the enemy or encourage his surrender. Political officers also perform the function of press or public affairs officers and lawyers in battlefield headquarters. Additionally, political officers are in the forefront of the psychological hardening training the PLA provides its troops to prepare them for the sights, sounds, and stresses of the modern battlefield. Many exercises and drills include fire, smoke, and noises to add realism to training.

Great emphasis has been placed on "new equipment" training and functional logistics and armament training. The Chinese defense industries that produce the PLA's weapons and equipment, in conjunction with PLA research institutes and PLA personnel serving as "Military Representatives" to factories, have developed a wide variety of simulators for use by units both before and after they receive new equipment. All types of simulators are utilized to help soldiers learn how to fire new weapons or to drive, operate, and maintain new equipment. Many units have built "high-tech" facilities for equipment simulators and computers. The use of simulators is less expensive than firing live ammunition, can be conducted more often by more individual soldiers, and subjects actual equipment to less wear and tear.

Effective units spend considerable time before new equipment arrives and after it is issued to ensure that unit personnel know how to operate and maintain the gear. This process is not as routine as might be expected, with reports often citing shortfalls such as the lack of training manuals, shortage of qualified instructors or technicians, and personnel being "scared" of using new equipment or refusing to take it to the field. Good commanders, however, find innovative ways to solve these problems with visits by experts or liaison with other units, schools, and factories. Similar to learning the basics of firing a weapon, logistics and armament units also spend a lot of time ensuring that personnel are technically trained to perform their functions in garrison before applying the same skills to austere field locations in support of actual troops. New means of communications available to support troops, such as video links and E-mail, allow soldiers in the field to get help from the rear in diagnosing field problems and expediting the flow of necessary supplies, spare parts, and medical care.

An important element of many training exercises is the deployment phase, a logistics-intensive period which may last for several days. Units moving to regional training areas may use their own organic transportation, but also increasingly incorporate multimode means, including rail movement for heavy equipment and fixed-wing airlift for headquarters and communications elements. In September 2006, *Jiefangjun Bao* carried a front page story about a Shenyang MR mechanized infantry brigade stationed in Liaodong (likely the 190th Mechanized Infantry Brigade/39th Group Army) conducting the "first" cross-regional mobility exercise of a full brigade to a training area in the Beijing MR. The exercise employed both road and rail transportation for an exercise against a Beijing MR armored brigade in Inner Mongolia.[78]

Though amphibious training receives the bulk of foreign attention, units in all MRs engage in training for many missions, such as border defense, defense *against* amphibious operations, high-altitude and desert operations, forest and urban operations, and especially anti-terrorist operations. Active duty units routinely incorporate reserve units, militia forces, and civilian support into training scenarios. In particular, local air defense, nuclear, chemical, and biological defense, and anti-terrorist drills frequently integrate active and reserve PLA units, PAP, militia, civilian police forces, and civilian support elements. These efforts are often coordinated in joint military-civilian command posts linked by modern communications and manned by military, party, and government officials. Though many foreigners believe the likelihood of mainland China being attacked is low, PLA ground forces continue the process of developing capabilities that cannot be created instantly or reconstituted quickly if allowed to atrophy.

Modern training with mechanized and informationalized forces is more expensive and harder on equipment than foot or motorized infantry training. New mechanized equipment and weapons with longer ranges require larger, more complex training areas than those for light infantry operations. Integrated joint operations training also requires large expanses of land, air, and sea spaces, to which the PLA often must coordinate with local governments for access.

Amphibious Training.

Amphibious operations are among the most complex of military operations. Such operations test individual training and skills, functional small unit training, and combined arms and joint operations abilities, and especially stress staff planning and logistics support. Much training in preparation for amphibious operations can be accomplished at locations away from the coast. Moreover, many military tasks practiced anywhere in the country can be applied to amphibious operations. However, the reverse is also true: many tasks undertaken during amphibious training are equally applicable to other scenarios.

Units from the Nanjing, Guangzhou, and Jinan MRs (and to a lesser degree Shenyang and Beijing MRs) practice amphibious operations annually. Over the past decade, the percentage of ground force units that have practiced amphibious operations *to some extent* might amount to between one-quarter and one-third of the current total ground combat force. Not all elements of these units have trained as extensively as others, with the two amphibious mechanized infantry divisions and a couple of armored units probably receiving priority for amphibious training. Some reserve and militia units also receive amphibious training.

Based on a review of articles in PLA newspapers from May to September 2005, it appears that unit-level amphibious training began in July in the Nanjing, Guangzhou, and Jinan MRs. In many cases, prior to moving to coastal amphibious training sites, units practiced related skills in training areas at or near their garrison locations. Training at the shore included basic skills, such as swimming, loading/unloading landing craft, beach assault, logistics support, etc., often starting with individual training and moving up to small unit and larger exercises.

Units often stayed at amphibious training areas for several weeks before redeploying to home stations. Many units deployed later in the season, which continued through September. No exercises involving multiple divisions or brigades controlled by group army headquarters were noted. Instead, most exercises appeared to be under division or brigade command. The following units could be identified in newspaper articles; however, it is possible that only elements of some of these units engaged in this training, and it is extremely probable that not all units which conducted amphibious training have been identified in the listing below:

- In the Nanjing MR, the 179th Motorized Infantry Brigade/12th Group Army and an armored division trained in July; the Anhui reserve infantry division and a Fujian reserve AAA regiment were also reported training that month. In August, a Jiangxi reserve regiment conducted basic loading and unloading exercises, and elements of the 1st Amphibious Mechanized Infantry Division/1st Group Army conducted amphibious training.

- In the Guangzhou MR, a divisional reconnaissance battalion conducted amphibious training in July, while several other units were reported training in their barracks areas. (Later, another reconnaissance unit or MR SOF unit may also have been involved in maritime training.) In August, the 163rd Division, the 124th Amphibious Mechanized Infantry Division, and the Air Defense Brigade from the 42nd Group Army, along with an armored brigade from the region, were reported training along the coast (though not necessarily in coordination with each other). In early September, the 121st Infantry Division/41st Group Army held an amphibious operations evaluation exercise.

- In the Jinan MR, a mechanized infantry regiment from the 127th Light Mechanized Division/54th Group Army conducted a landing exercise on the Bohai Bay in July. Additionally, both the 199th Motorized Infantry Brigade and the 8th Armored Division of the 26th Group Army conducted amphibious training in July. (The 8th Armored Division was reported to have conducted "several years of exploratory training.") By August, the 58th Mechanized Infantry Brigade/20th Group Army was engaged in training on the Yellow Sea and an artillery brigade moved to the Bohai.

From the unit identifications above, it seems likely that training in the Nanjing MR was under-represented in the media compared to the amphibious training in the Guangzhou and Jinan MRs. The number of reserve units involved in training in the Nanjing MR may account for the relatively small number of active units identified. The lack of reported larger exercises may have been the result of command attention focused on Peace Mission 2005 and "North Sword 2005," which would be conducted in August and September.

In the summer of 2006, a single armored division's amphibious training cycle in July was reported in multiple media sources (likely the armored division of the 12th Group Army in the Nanjing MR).[79] A close examination of the photos in these articles revealed that all the various reports focused on the training of a single division in that time frame. Amphibious training in the Jinan MR also began in July, while amphibious training in the Guangzhou MR appears to have begun by early August.

Peace Mission 2005.

The combined Chinese and Russian exercise, Peace Mission 2005, was the highlight of the PLA's 2005 training season and covered extensively by the news media in both countries.[80] Senior Chinese and Russian military leaders agreed to the concept of the exercise in the summer 2004. The scenario centered on an internationally/UN-approved intervention by Chinese and Russian forces to prevent an internal conflict escalating into a local war. After "opposition forces and armed formations" (with links to "terrorists on a global scale") of a third country seized cities and rural areas in the western part of the country, the government requested outside assistance to stabilize the situation. The mission of the combined force was to restore constitutional order in the state with minimal losses to the country. Both Chinese and Russian senior military officials stated the exercise was not targeted at any third party, country, or region.[81]

After planning and deployment phases, three main exercise events occurred from August 23 to 25, 2005, off the southern coast of Shandong peninsula, at Langyatai on the Longwan Bay southwest of Qingdao, and inland at a training area near Weifang. A total of about 10,000 Chinese and Russian Army, Navy, and Air Force personnel participated in the exercise, with approximately 1,800 Russians deploying to the exercise area.[82] All exercises were conducted within sight of reviewing stands and were, in effect, demonstrations of an hour or two in length, which highlighted tactical execution of combat and live fire drills, not interactive force-on-force maneuvers against live opponents.

The first day of the exercise focused on air and sea blockades to gain air and sea dominance. Fighters and long-range bombers struck at enemy air defense systems, command and control nodes, airfields, and armed forces. At sea, Russian forces included an anti-submarine warfare (ASW) ship, the *Marshal Shaposhnikov,* and the destroyer *Burnyy.* Among the

PLA Navy ships were the ASW destroyer Number 168 *Guangzhou* (Type 052B, *Luyang I* class), Number 136 *Hangzhou* (*Sovremenny*-class destroyer), Number 567 *Xiangfan* (*Jiangwei*-class frigate), and two diesel submarines.[83]

On August 24, after engineer units cleared obstacles on the beach at Langyatai, a PLA Navy marine armored battalion, using Type 63A light amphibious tanks, APCs, and small boats, and a Russian naval infantry unit equipped with BTR-80s, conducted amphibious landings in two waves of 30 to 40 vehicles each. The landings were supported by air and naval bombardments of shore defenses, as well as insertion of a SOF unit delivered by 18 helicopters and a personnel parachute drop from three transport aircraft.[84] The exercise was conducted in rainy weather, and the Russian news media reported that two Chinese Type 63A tanks and a Russian BTR-80 sank.[85]

The following day the exercise moved inland where long-range bombers and fighters delivered preparatory fires on and around a landing zone near Weifang. A Russian A-50 AWACS aircraft flew overhead (implying its control of these activities) and in-flight refueling was demonstrated. Chinese and Russian Il-76 transports in flights of four aircraft each dropped three airborne armored assault vehicles per aircraft. Two more IL-76s each dropped 86 paratroopers to link up with the vehicles on the ground.

The airdrops were supported by electronic jamming aircraft, fighters, and helicopter gunships. A flight of 18 helicopters followed by a second flight of 9 helicopters inserted infantry (perhaps including more SOF troops) into the battle area. Once on the ground, the airborne forces linked up with elements (at least a regiment) from the 127th Light Mechanized Infantry Division/

54th Group Army, and all forces conducted a live-fire demonstration against "enemy" forces that had retreated into defensive positions. This final ground assault was also supported by Air Force fighters.[86] The exercise, which included Type 96 main battle tanks, WZ-551 armored personnel carriers, and self-propelled artillery, was called a "forced isolation" drill to isolate, surround, and annihilate enemy forces.[87]

Peace Mission 2005 displayed many elements of modern joint operations over a limited period of time in front of senior officials from both countries. It demonstrated that the PLA and Russian military understand the complexity of modern campaigns and can plan, deploy, support, and control medium-sized military formations for a few days. The number (up to 18) and types (attack and transport) of PLA helicopters used in the mission reflected an increase over Army Aviation capabilities seen in previous years. Moreover, it is rare for PLA ground forces, SOF, airborne, and marines *all* to be integrated into a single exercise scenario in relatively close proximity to each other.[88]

To be sure, Peace Mission 2005 was a major step in the size, scope, and complexity for the two forces acting together. Nonetheless, the demonstrations were mainly military choreography to demonstrate the type of improvements in capabilities acquired by the PLA in recent years and the continuing relevance of the Russian military to international events. (Peace Mission 2005 was one of four military exercises conducted by the Russian armed forces beginning on August 16.[89]) The exercise was also a venue to exhibit the capabilities of Russian hardware for potential future sales. Forces actually involved in the exercise were too limited in number to be effective against a modern enemy larger than a small conventional force or guerrilla organization. Total ground forces involved equated to less than a division, including all Army, SOF, airborne, and marine units from both the Chinese and Russian sides.

Outside commentaries on Peace Mission 2005 varied in their conclusions. One saw, or thought it saw, "all components of a Taiwan invasion plan."[90] Another saw the exercise as "patently unrealistic against a terrorist organization, but quite suitable for operations against a regional naval power . . . aimed squarely at the governments in Pyongyang and Tokyo, to pressure North Korea to go back to the six party nuclear talks and Japan over its border claim to the Kurils."[91] Still another viewed the exercise as "hardly applicable to situations of ethnic conflict and even less so to modern counterguerrilla operations."[92] These observers appeared to see only what they wanted to see and were dismissive of the stated scenario as disingenuous on the part of the Chinese and Russians. While caution in judging exercise scenarios is always advised, sometimes announced scenarios may actually reflect true political objectives.

When the Chinese leadership wants to send a message to Taiwan, it has no qualms about doing it directly. An excellent example is found in media descriptions of amphibious exercises on and near Dongshan Island, which "closely resembles the west coast of Taiwan," in 2004:

> Dongshan Island exercises in the past have usually been conducted in three phases. Phase 1 was information war, with the focus on electronic countermeasures and paralyzing the enemy's communications and command systems. Phase 2 was a tri-service, Navy, Army, and Air Force sea crossing and landing operation. This mainly involved simulated missile attacks, air raids, a nighttime airborne landing, an Army-Navy island landing assault, special forces units attacking ports and airfields, and

> practice in street fighting. Phase 3 was counterattack against enemy reinforcements. Mainly this has been a simulated intervention by the U.S. Pacific Fleet....[93]

In the case of Peace Mission 2005, most obviously, the exercise area was held outside the Taiwan Strait (it seems unlikely that Moscow would have anything to gain by using this exercise in a blatant effort to intimidate Taiwan). Moreover, there was *no reported use* of ballistic missiles and little emphasis on information operations in Peace Mission 2005. Neither of the Army's two amphibious mechanized infantry divisions was involved in the exercise. Yet, all of these omitted elements are considered major factors in any expected campaign against Taiwan. But perhaps most importantly for Washington, there was no indication of a Phase 3 — prevention of intervention by the U.S. fleet (which would occur far away from shore) — something Moscow would certainly be reluctant to include as an intended message from this exercise. So while the actions on display in Peace Mission 2005 indeed *could* be used in a campaign against Taiwan, nearly *any* military exercise contains elements that *could* be a component of operations against Taiwan or any other potential enemy. (What military operation would *not* be considered a "component of a Taiwan invasion plan"? An anti-terrorist exercise perhaps?) But objectively speaking, significant aspects of an expected Taiwan scenario *were not* included in the actual exercise.

The criticism that Peace Mission 2005 operations were inappropriate for anti-terrorist, ethnic conflicts, or counterguerrilla operations seems to ignore the history of U.S. (and UN) operations in Somalia in 1992 and 1993 (which led off with an amphibious assault and eventually developed into fighting emphasizing SOF and helicopter operations); allied operations against the Taliban in Afghanistan in 2001 (which highlighted airborne operations and SOF supported by long-range bombers, fighters, and precision-guided munitions, as well as a naval blockade offshore); or, more recently, the 2006 Israeli incursion into south Lebanon against the guerrilla Hezbollah enemy (which was centered on tank, infantry, and artillery teams supported by helicopters and fighters, also supported by a naval blockade and precision bombing). Put into a larger perspective, it is at least conceivable that actions taken during Peace Mission 2005 were consistent with the exercise's stated scenario.

North Sword 2005.

A little more than a month after Peace Mission 2005, a much larger PLA joint operation was conducted in Inner Mongolia with several dozen foreign observers in attendance, including U.S. representatives. Though this exercise was considerably larger than Peace Mission 2005 and also included airborne operations, it was hardly noticed by the foreign press. As usual, however, the Chinese news media had several detailed descriptions of North Sword 2005.

In late September 2005, elements from two PLA armored divisions, Army Aviation units, and a PLA Air Force airborne division with supporting transportation and aviation units—numbering up to 16,000 personnel—conducted a four-stage exercise at Beijing MR's Zhurihe Combined Arms Tactical Training Base. The four phases included a change in unit alert level, long-range deployment and mobility, live-fire evaluation of armored and artillery units, and combat execution.[94] Foreign observers were present for only 1 day of the final phase, September 27, which consisted of a force-on-force exercise of Blue Force versus Red Force. The entire sequence of events would likely have taken approximately 2 weeks from alert to redeployment.

Although Chinese news reports usually spoke of two armored divisions in the exercise, the total number of troops involved (10,000 to 16,000) would indicate that both divisions did not participate at full strength. The Red Force was reportedly made up of elements of an armored division (said to have "the most advanced equipment and the greatest degree of digitization of our Army's armored units"), an air defense unit with SAM missiles, an electronic warfare unit, a helicopter unit, and other combat forces. The Blue Force was composed of elements of an armored division, an airborne element, and a repair battalion. Altogether over 2,800 tanks and other vehicles such as APCs, self-propelled artillery, command vehicles, trucks, and jeeps were said to be in the exercise, the "largest field maneuver" involving armored troops.[95] One report noted "several thousand airborne troops transported over 2,000 kilometers," which would indicate an airborne regiment from one of the two 15th Airborne Army divisions stationed near Wuhan. This drill was alleged to be both "the largest-scale long-range movement" and the first airborne exercise against an armored force in unfamiliar conditions.

The confrontational exercise began the day before the observers arrived with a Red Force preemptive attack on the Blue Force "stealthily" moving toward Red's assembly areas.[96] Initial operations included a day-long series of combined arms task force attacks and counterattacks, supported by helicopters and electronic warfare. At the end of the first day, Blue and Red Forces had reestablished their lines and were planning for the next day's operations.

Just prior to the Red Force's resumption of the offensive on the second day (when the foreign observers were present) at about 10 a.m., the Blue Force launched a ground counterattack supported by an airdrop of personnel, materiel, and heavy equipment behind Red lines. Blue Force helicopters, fighters, bombers, and ground attack aircraft—protected by airborne jamming of enemy radars—delivered preparatory fires throughout the battlefield. Blue Force SOF teams attacked Red SAM units to eliminate that threat to Blue air operations. UAVs circled above to provide intelligence for the Blue Force. Units employed fake buildings as camouflage to conceal armored vehicles, most likely self-propelled artillery, from Red observation in the desert.[97]

Faced with enemy to his front and rear, the Red Force commander ordered that "most of his force" remain in place to block the Blue advance while three combined arms teams counterattacked the paratroopers to the rear. In the process of this maneuver, the Blue Force dropped dummies as a deception and also delivered "a large number of antitank mines" to protect a flank from Red Force assault. Though the mines delayed the Red advance, mine clearance vehicles were brought forward, and the Red units were able to conduct an encircling maneuver around the Blue airborne force. Blue Force airborne assault vehicles were no match for the heavy Red Forces, resulting in the decision for the airborne force to scatter and rely on additional support from commandos and armed helicopters. The remainder of the day was a continuation of attack and counterattack by both forces.

The entire exercise was monitored in the training area's command center using various video and computer systems. Umpires controlled the maneuvers in the field and laser simulation devices were employed to assess firing accuracy.

If the Chinese reports are accurate, this exercise demonstrated many elements of integrated joint operations executed at division level that the PLA has emphasized since 2004. These include combined arms operations; airborne assault; battlefield air support; information operations including use of UAVs and deception activities; SOF or commando raids; and obstacle emplacement and clearing actions. It is noteworthy that the entire exercise took place during the day, and it is likely the airborne operations had been planned well in advance in order for the forces to arrive in the training area on time. A proper evaluation of unit readiness and exercise realism is difficult when based only on news media reports, but on-the-scene military observers had a unique opportunity to judge for themselves the success of many elements of the PLA modernization program.

However, some 9 months later, the 2006 report on the Chinese military by the U.S. Office of the Secretary of Defense (OSD) made no mention of this exercise or the participation of U.S. observers. By ignoring the exercise's existence, the Pentagon spared itself the obligation to comment on its content and importance. Instead, the report chose to emphasize China's lack of transparency, calling such activities "secondary areas of military activity," an undefined and apparently pejorative description.[98] Likewise, the Pentagon has never made any official comment about insights by U.S. observers who attended the first North Sword exercise in 2003. The presence of U.S. observers at both of these exercises and the amount of coverage by the Chinese press actually presents a rare opportunity to evaluate the completeness and accuracy of the Chinese reports against what the U.S. observers actually saw and heard while at the exercise. Such an assessment would be very useful to analysts, both inside and outside the U.S. Government, who use the Chinese media as a primary and supplementary source for information about the PLA. As for "transparency," the opportunity to observe the PLA in action at division level in a joint operation and compare that reality to the tenor of media coverage simply did not exist a decade ago.

PLA SELF-ASSESSMENTS

Outsiders with access only to Chinese news media reports (including both official and unofficial sources of information) must use great caution in coming to conclusions about PLA readiness and capabilities. To be sure, not all media reports are equal, and some certainly have greater authority than others. While over time many general trends and points of emphasis can be discerned, some specific capabilities and linkages to specific units can be manipulated through inaccurate reporting (intentional or not) or simple omission (e.g., not reporting important events). An argument can be made that the General Political Department propaganda ("publicity") department has reason to either exaggerate capabilities or understate progress depending on the target audience and topic. The General Political Department unquestionably has the ability to undertake concerted deception efforts in war or peace, which could be sophisticated enough to outwit attempts to double check sensitive information it seeks to protect. But the political system also has the mission to maintain morale in the force and educate the troops in their efforts at force modernization. So some degree of "truth in advertising" is required in the official news media.

With those caveats in mind, there is value to reading PLA self-criticism in its official new outlets. Though the Chinese military press acknowledges progress in developing PLA capabilities, authoritative essays repeatedly point out gaps that exist between PLA capabilities and requirements for modern combat. Shortcomings include personnel education and training levels, command and staff training, technological and equipment levels, training in joint operations, and funding available to the military. Examples of self-assessments from Beijing in 2006 include the excerpts below; regional and service newspapers provide many similar examples reinforcing and expanding upon these topics:

General Assessments.

> There is a gap between the current level of modernization in our military and the requirement that must be met in order to win regional informatized wars, there is still a gap between the current military power of our military and the requirement of fulfilling the historic mission of our military for the new period of the new century, and there are still some contradictions and problems in the military work that must be tackled as soon as possible.[99]

> At present, our military has yet to finish its tasks related to mechanization construction and is in the initial period of comprehensive development of informatization construction. The overall level of military skills is still low, informatized means used for military training have yet to be sufficient, and the model in which combat effectiveness is created has yet to completely shift to reliance on science and technology.[100]

> Now our Army has made tremendous achievements in building itself, but it is also faced with the problems

that its level of modernization does not meet the requirements of winning local war under informatized conditions and that its military capability does not meet the requirements of carrying out its historic missions at the new stage of the new century. Deep-seated problems need to be solved urgently, and some major relationships need to be grasped scientifically. . . . Ours is a large developing country, and the contradiction between the demand of Army modernization and the inadequate input will exist for a long time to come. So we should pay more attention to scientific management, optimize the allocation of resources, and increase the comprehensive efficiency of their utilization. . . . We should energetically carry forward the spirit of hard struggle, always implement the principle of building the Army through diligence and thrift, and do a good job in managing and using the limited military expenditure. This has provided an important method for our Army to follow the road of modern development with less input but higher efficiency and of bringing about faster and better development.[101]

Lack of Funds.

However, China is a large developing country. Money is needed in many aspects. The contradiction between the needs of military modernization construction and the short supply of funds will exist for the long run. Satisfactorily managing and using limited financial resources offered by the national government is a very practical issue before us.[102]

All-round national economic strength is continually rising, but in terms of China as a large developing country, our national economic strength is not yet powerful, and the scale of our national defense expenditure is very limited. We cannot compare our national defense input with developed countries; we must follow a road of national defense and Army modernization with Chinese characteristics in which expenditure is relatively small but returns are relatively high. In the management and use of national defense expenditure, we have always persisted in the guideline of building the Army with diligence and thrift, waged arduous struggle, done everything in diligent and thrift fashion, cherished and used well the money earned by the people's blood and sweat, used the limited national defense expenditure in maintaining normal operation of the world's most numerous Army, ensured the smooth promotion of preparations for military struggle, promoted the development of building all undertakings, and provided an effective security guarantee for safeguarding national security and stability and the period of strategic opportunity for economic and social development.[103]

Lack of Qualified Personnel.

At present, the quality of our officers and soldiers has comparatively improved. However, there is still a gap between the current level and the requirements that must be met in order to win wars. The low level of military quality and scientific and cultural quality has remained a noticeable problem with the structure of the quality of officers and soldiers. A serious shortage of professionals for commanders of joint operations and professional technicians is still apparent. All of these problems hold back and impede the development of military capability building.[104]

Some comrades think more about their personal interests than about their work and the cause. They rush to express their attitudes toward directives given by the higher-ups but do not make firm efforts to do solid work. Formalism and bureaucracy are rampant among some comrades. They are self-complacent, making appeals but not conducting profound investigations and studies. They pursue only superficially great momentum but ignore effectiveness in reality. Some comrades do not work responsibly in daily work and are afraid of shouldering responsibilities for problems. Some comrades are "yes men." They avoid contradictions, shy away from tough problems, and are loose in regulating their subordinates.[105]

Problems in Joint Training.

At present, our military is still comparatively weak in joint training. Some commanders have yet to strengthen their awareness of joint operations, the leadership and administrative framework and the operating mechanism for joint operations have yet to be completed, joint [operations with real troops] are still insufficient and there is still a comparatively wide gap between the current joint operation abilities and the requirements of actual battles.[106]

Technology Gap.

Over recent years, our military has made leaps-and-bounds progress in weaponry and armaments construction. However there is still a considerable gap between the current level and the requirement that must be met in order to effectively fulfill the historic mission of our military in the new period of the new century.[107]

However, there is a considerable gap between China and big military powers in the world. In light of the circumstance that we are weaker than others in "steel," we must be very strong in "morale." . . . Being satisfactorily prepared for military struggle is the most important, the most realistic, and the most urgent strategic task that is not only currently lying before our military, but also undoubtedly the best activity through which to cultivate the fighting spirit.[108]

At present, our military modernization construction is in the initial period in which mechanization, semi-mechanization, and informatization develop together. This determines that exploration in integrated training—which is related to mechanization, semi-mechanization, and informatization—is also in the initial period and our achievements in mechanization, semi-mechanization, and informatization are also in the initial [stage].[109]

Several of the assessments above were taken from a series by four "Contributing Commentators" in the *Jiefangjun Bao*, written from the perspective of each of the four General Departments from August 6 to 9, 2006. But as also can be seen, similar evaluations could be found in the months before and afterward. PLA leaders at all levels understand that the process of modernizing and transforming the entire force is a long-term effort and have set the target date of the year 2020 for completion of personnel and equipment efforts. Achieving the status of an upper tier, advanced military could take even longer. This is exactly the meaning of the "three-step development strategy" identified in the 2006 *White Paper*.

However, if so ordered by the government and party before it has completely achieved its modernization goals, the PLA will follow the command of China's civilian leadership and utilize its best units in the most appropriate way, supported by a large civilian effort, to achieve the political and military goals assigned.

The self-assessments included above were not unique to the spring and summer of 2006. Many similar articles could be found in previous years. Even so, the 2006 OSD *Annual Report to Congress* on the military power of China alleged that "misperception" could lead to miscalculation or crisis: "China's leaders may overestimate the proficiency of their forces by assuming new systems are fully operational, adeptly operated, adequately supplied and maintained, and well integrated with existing or other new capabilities."[110] While such a possibility "may" exist, even a cursory review of Chinese self-assessments available to the Pentagon reveals a much different degree of self-knowledge in the PLA than suggested by the OSD report. The self-evaluations listed above also do not support the conclusion of the U.S.-China Economic and Security Review Commission that "the Chinese military may be able to assimilate new weapons systems and technology at a more rapid pace than other nations."[111]

CONCLUSIONS

The evidence concerning personnel, force structure, equipment, and training developments outlined in this chapter reveals a ground force in all parts of the country modernizing to prepare to undertake a variety of military missions if called upon to do so. Moreover, based on authoritative articles in the military press, the senior leadership of the PLA appears to have a realistic understanding of both the strengths and weaknesses of the force.

Subtle indicators of widespread progress in the Army are Beijing's decisions to dispatch troops from Chengdu MR to participate in the UN peacekeeping mission in Lebanon, from the Jinan MR for the mission in Sudan, and from the Shenyang and Nanjing MRs in Liberia.[112] Assigning responsibility for high-profile missions to units outside of Beijing shows confidence in the abilities of these units and also gives other headquarters exposure to some of the complexities of planning for and executing overseas operations, especially when extended over several rotation periods. Though these deployments are conducted under administrative (i.e., noncombat) conditions and are relatively small in scale — about a battalion in strength per mission — they are not common operations conducted by the PLA. Successful completion of real-world missions also helps build morale in commands all over the country and increases the prestige of the PLA with the Chinese people.

Graft, corruption, and bribery are problems in the general Chinese society and spill into the armed forces. As the PLA modernizes, it will continue to fight the battle against corruption, particularly as more money becomes available to the force and increased interaction with local civilian logistics support companies becomes the norm. An insight into the scope of this challenge was provided by an article appearing in several Chinese news media outlets concerning the misuse of military vehicles, even in the capital city under the nose of the CCP and PLA headquarters: "Troops stationed in Beijing have made progress toward standardization and modernization, *but loopholes in management, idleness, and demoralization among a few military units are eroding the Army's reputation."* (emphasis added)[113] As the PLA seeks to be a model for society, such problems will likely result in rectification campaigns for many years to come. The full extent of many disciplinary and morale problems in the PLA is unlikely to be publicized widely in the Chinese news media, yet morale and discipline are important components of military readiness.

Some of these problems may be mitigated as the PLA becomes smaller and its personnel more highly educated and technically competent—two trends that are certain to continue. Although no specific plans for future personnel reductions have been announced, the active duty PLA ground force probably is still not optimized for the missions it faces and could probably be cut by at least another 100,000 without losing combat effectiveness. Ironically, a "right size" ground force will likely be a smaller but more capable force with more money spent on fewer units and personnel than at present.

In particular, the Army probably has more main force infantry units than it needs. Some infantry troops can be eliminated by downsizing more divisions to brigades or restructuring them with fewer regiments (as has been done in recent years). As the ground force becomes more mechanized and more mobile, perhaps additional units can be deactivated completely because of the creation of more flexible, powerful units. The PLA likely has too many local headquarters, administrative offices, and schools—most of which are manned by ground force personnel. If the new program of hiring "nonactive duty" contract personnel proves effective, additional active duty members could be shed, especially in the logistics fields. A concurrent growth in the number of reserve units is likely as the ground force modernizes and continues to reduce its size.

As the country's transportation infrastructure improves, especially its super highways and railroads, some units might be reduced because similar units from other parts of the country will be able to deploy across regions more rapidly than in the past. Land deployments, however, will require sufficient heavy equipment trailers and other support vehicles to transport tracked vehicles over long distances and continued training on rail deployments. Another key transportation factor will be increased air and sea lift provided by the Army's sister services and civilian support organizations to enable it to move long distances using joint capabilities. Ultimately, an appreciably smaller ground force transported by considerably larger air and sea forces will be necessary for China to project significant land power beyond its borders. Such a force will also require appropriate air and sea capabilities to defend it en route to its destination.

One component of the ground forces that may not be reduced in size proportionately is the PLA border and coastal defense force. Though technological improvements in communications, transportation, night vision, and intruder detection will enhance border surveillance efforts, the length of China's borders and the existence of terrorist, extremist, or separatist elements on its periphery, as well as uncertainty about stability in North Korea, argue for maintaining a credible force on the borders to deter overt violence and control illegal activities. But border and coastal defense units, along with most other local force units, would not likely be used for offensive force projection missions.

Though the PLA apparently has yet to make a major change to the ratio of ground to air and sea forces, a completely modernized PLA is likely to see growth in the relative sizes of the PLA Navy, Air Force, and Second Artillery at the expense of the Army.[114] The appropriate mix will take years, if not decades, to materialize (and will be much more costly than a motorized ground force). The composition of the future force, particularly if it entails a large increase in the size of the PLA Navy and Air Force, may also reveal something about China's intentions for its use. An indicator of a major move in the direction of greater "jointness" for the PLA, and especially for force projection, may be the assignment of a naval or Air Force officer to command a coastal MR. In the end, however, Chinese military planners and political leaders will likely seek to retain a relatively large ground force (somewhere around half of the total PLA[115]) to protect its borders, deter and repel potential invaders, provide options for land force projection in defense of Beijing's declared sovereign territory, and serve as a reminder to the Chinese population that the ground force is the government's and party's final line of defense to preserve domestic stability. This last mission is especially relevant if the size of the PAP is only 660,000 as reported in the 2006 *White Paper* and not up to 1.5 million as estimated by foreign analysts.

At this stage in the PLA's modernization process, judging from their official statements, China's senior military leaders do not appear to be overly eager to test their forces' capabilities in battle. They still see the need for up to 15 years before personnel improvement and equipment modernization programs play out. The longer the time frame the better, because it gives the forces more opportunity to practice their new doctrine using all elements and capabilities created in recent years. PLA officers understand the value of sweat on the training field in preparing the force for potential missions and realize that no single silver bullet will solve their military problems. At the same time, as it builds its strength, the PLA's new equipment and more complex exercises (both those that actually improve capabilities, as well as the firepower demonstrations put on for psychological impact) help in demonstrating China's determination to build its multidimensional strategic deterrence posture. Yet, as loyal servants of the CCP, the PLA leadership will obey its civilian leaders if ordered to use force against enemies threatening the party or state.

The PLA ground force of 2007 looks quite different from its predecessor in the mid-1990s. In 2020, it will be different still—likely significantly smaller and seeking to establish a role for itself in a more maritime-oriented overall force. However, the PLA's improvement in capabilities in absolute terms (as measured against itself) is only half of the equation for victory. Future PLA capabilities must also be measured relative to the capabilities of potential opponents and the likelihood that they, too, will continue to modernize and improve their own capabilities. No matter how much the PLA "mechanizes" and "informationalizes," future combat will not be any easier for it than in its light infantry days. More depends on the intellectual capabilities of the PLA's officers and NCOs to plan for and execute a new and unproven joint doctrine, the battlefield techniques and procedures they have developed through realistic training, and their ability to adapt to changing circumstances, than on the capabilities of any single weapon or weapons system the force may acquire in coming years.

Barring a major internal economic setback, it seems inevitable that Beijing will continue to increase the resources available to the military. But the force also faces constantly increasing costs for personnel, operations, and equipment. Unless official defense budgets are increased by even larger percentages than those of the last decade, it is likely the PLA will continue to stress economizing and the efficient use of centralized funds by its relatively large force, along with an unknown amount of support from local governments and an uncertain boost from other sources of income.

The PLA's missions may be modified as China grows and the international situation changes. New circumstances require close attention by China's neighbors and the United States. New evidence must be gathered and additional analysis undertaken to ensure that developments in the ground force are understood in the context of overall PLA modernization. Despite the important changes underway in the other services, it appears the Army will likely continue to be the single largest major component of the PLA for some time into the future. The ground force should not be overlooked in our zeal to understand the other dimensions of China's "non-transparent" military.

ENDNOTES - CHAPTER 7

1. This chapter updates information found in the author's book, *The Chinese Army Today: Tradition and Transformation for the 21st Century*, London: Routledge, 2006. Thanks are extended for research assistance provided by Luke Armerding, for comments by Colonel Frank Miller, and, as always, for the unique efforts of Ellis Melvin.

2. State Council Office of Information, "China's National Defense in 2004," Beijing, December 27, 2004.

3. International Institute for Strategic Studies, *The Military Balance 2006*, London: Routledge, 2006, p. 264. According to "China's National Defense in 2006," the Army "was the focus of force reduction (from 2003 to 2005), and its authorized number of personnel has been reduced by more than 130,000." Unfortunately, the Chinese government did not say how large the Army was before this reduction took place.

4. State Council Office of Information, "China's National Defense in 2006," Beijing, December 29, 2006. This same section on National Defense Policy further states, "China pursues a policy of coordinated development of national defense and economy. It keeps the modernization of China's national defense and armed forces as an integral part of its social and economic development, so as to ensure that the modernization of its national defense and armed forces advance in step with the national modernization drive." While defense modernization is clearly *an element* of the national modernization program, this statement is a slight modification to the previous formulation that "the Chinese government insists that economic development be taken as the center, while defense work be subordinate to and in the service of the nation's overall economic construction." Year 2000 China Defense *White Paper*.

5. "China's National Defense in 2006," chapters on National Defense Policy and The People's Liberation Army.

6. Article by JFJB Editorial Department Marks 50th Anniversary of Jiefangjun Bao *CPP20060215502001 Beijing Jiefangjun Bao (Internet Version-WWW)* in Chinese, Open Source Center (OSC), trans., January 9, 2006, p. 1. Hereafter JFJB Editorial Department Article.

7. *Ibid.*

8. Ibid.

9. Military Paper Hails Hu Jintao's Important Exposition on PLA's Historic Mission *CPP20051208510020 Beijing Jiefangjun Bao (Internet Version-WWW)* in Chinese, OSC, trans., December 8, 2005, p. 6. With the references to paying attention to "national development interests" and "on the oceans and seas," it is unclear whether these missions signify a major expansion of the PLA's outlook beyond Taiwan, going further to protect China's interests along sea lines of communication and potentially even to places where it has economic interests (e.g., such as the protection of access to natural resources necessary for China's development). If so, those military tasks would fall primarily on the other services, with the ground forces likely having only minimal input for some time into the future.

10. JFJB Commentator on Educational Campaign on Historic Mission of PLA (4) *CPP20060404515009 Beijing Jiefangjun Bao (Internet Version-WWW)* in Chinese, OSC, trans., April 3, 2006, p. 1.

11. According to China's National Defense Law, the Chinese *armed forces* are composed of the active and reserve units of the PLA, the People's Armed Police (PAP), and the militia. Like the militia, the PAP is a paramilitary organization which shares many commonalities with the PLA; nonetheless, by definition the PAP *is not* part of the PLA. The 2006 *White Paper*, for the first time, had an entire chapter on the PAP. Also for the first time, it gave a specific number for the total PAP force of 666,000. This figure is much lower than most foreign estimates of the size of the PAP, which ranged from around one million to about 1.5 million. The *White Paper* also identified the sub-elements of the PAP to consist of internal security forces; forces guarding gold mine, forest, water and electricity, and communications; and border security, firefighting, and security guard forces. The PAP is not funded out of the Chinese defense budget, but rather has "an independent budgetary status in the financial expenditure of the state." While the PAP has a secondary mission of supporting the PLA in local defense operations, as can be seen from the missions of its subordinate elements, its primary focus is on domestic security. The PAP has been compared to the Italian Carabinieri or French Gendarmerie; it has no direct analogue in the U.S. military.

12. "China's National Defense in 2006," National Defense Policy. Each bullet has additional information expanding upon the major theme.

13. "China's National Defense in 2004," National Defense Policy. The first task in the 2002 *White Paper* was "To consolidate national defense, prevent and resist aggression," followed by reunification.

14. "China's National Defense in 2006," The Security Environment.

15. *Ibid.*

16. *Ibid.* See also PRC Military's First Symposium on Unconventional Security Theory *CPP20061116710014 Beijing Jiefangjun Bao (Internet Version-WWW)* in Chinese, OSC, trans., November 16, 2006, p. 6.

17. "China's National Defense in 2006," The People's Liberation Army.

18. "China's National Defense in 2006," International Security Cooperation.

19. "UN Mission's Summary Detailed by Country," November 30, 2006, *www.un.org/Depts/dpko/dpko/contributors/2006/nov06_3.pdf*. These numbers are updated monthly.

20. CCTV-7 "Weekly Military Talk," Discussion on China-Foreign Joint Exercises *FEA20061218050546-OSC Feature-CCTV-7*, 1130 GMT November 19, 2006, OSC, trans. The 2006 *White Paper* says, "Since 2002, China has held 16 joint military exercises with 11 countries."

21. See Endnote 6. Despite impressive growth in the GDP, a multitude of problems threaten China's social stability and challenge Beijing's and local governments' leadership. These include graft and corruption, income and development disparities, nonperforming bank loans, the "floating population," demographic imbalances, water shortages, loss of arable land, pollution and environmental catastrophes, and natural disasters.

22. *Ibid.*

23. *The Science of Military Strategy*, Peng Guangqian and Yao Youzhi, eds., Beijing: Military Science Publishing House, 2005, p. 224. The concept of deterrence is also found in the 2002 *White Paper* in the tasks to "prevent and resist aggression . . . [and to] stop separation" and reiterated in several places in the 2004 *White Paper*, as seen in the first sentence of this chapter and the "basic goals and tasks" defined in this section.

24. *The Science of Military Strategy*, p. 228.

25. "China's National Defense in 2006," The People's Liberation Army. Note that in 2006, the term "Chinese features" was used in the English translation instead of "Chinese characteristics." The Chinese terminology did not change.

26. Military Paper Hails Hu Jintao's Important Exposition on PLA's Historic Mission.

27. "China's National Defense in 2006," The People's Liberation Army.

28. "Nearly 70,000 NCOs replace officers in 70-odd positions," in *PLA Daily* on-line, January 5, 2007, at *english.chinamil.com.cn/site2/news-channels/2007-01/05/content_697306.htm*.

29. "China Finishes Further Military Streamlining on Time," in *PLA Daily* on-line, January 9, 2006, at *english.chinamil.com.cn/site2/news-channels/2006-01/09/content_379998.htm*.

30. "China's National Defense in 2006," The People's Liberation Army.

31. "China's National Defense in 2004," Revolution in Military Affairs with Chinese Characteristics.

32. "Shenyang MAC Employs First Batch of Nonactive Duty Office Attendants," in PLA Daily on-line, 2006-05-26 at *http://english.chinamil.com.cn/site2/news-channels/2006-05/26/content_484628.htm*.

33. "China's National Defense in 2006," The People's Liberation Army.

34. A few photographs of the new category of contract workers are available. For example, see *mil.news.sohu.com/20060531/n243494839.shtml*. These were taken on the first day of their entry into service. It is possible that eventually contract workers will be issued some sort of distinctive insignia.

35. "Liaowang Article Discusses 'New-Concept' Military Barracks," *CPP20060105058002 Shanghai Liaowang Dongfang Zhoukan* in Chinese, No. 46, OSC, trans., November 17, 2005, pp. 35-39. Both quotes in this paragraph come from this source.

36. "Liaowang Article Discusses 'New-Concept' Military Barracks."

37. More on China Setting Up Military Auditing Body To Examine 1,000 Officers in 2006 *CPP20060720054039 Beijing Xinhua* in English, 1427 GMT, OSC, trans., July 20, 2006.

38. "China's National Defense in 2004," The Military Service System. The relationship between these levels of PLA headquarters and local government was repeated in the 2006 *White Paper*.

39. "China's National Defense in 2006," China's Leadership and Administration System for National Defense.

40. *Ibid*.

41. The 2006 *White Paper* for the first time acknowledges "18 combined corps," a fact that has been recognized for several years by foreign analysts, and defines them as "mobile combat troops." The *White Paper* does not, however, provide any count of units below that level.

42. This summary and the order-of-battle listing in Appendices 1 to 7 are based on *The Chinese Army Today*, Chapter 4, updated with information from the *Directory of PRC Military Personalities*, October 2006, *Directory of PLA Personalities*, October 2005, and information provided by Ellis Melvin and Chinese news media reports. This listing is as complete and up to date as possible, but recognizes its limitations and does not purport to be an authoritative compilation of all PLA Army units.

43. This gross approximation of 50 percent of personnel in the main combat force is derived by multiplying the number of group armies (divisions and brigades) and independent units by average personnel manning estimates.

44. "Chinese Military to be Restructured," in *People's Daily* on-line, July 13, 2005, at *english.people.com.cn/200507/13/eng20050713_195881.html*.

45. The *Directory of PLA Personalities*, October 2005, p. 241, alludes to an armored division in Nanjiang, Xinjiang. However, beginning in 2004, *Xinhua* reported on a mechanized infantry division in the region, with one regiment having converted to an armored regiment. See *news.xinhuanet.com/mil/2004-05/28/content_1495095.htm*. Based on a series of Chinese reports, it appears that there is no armored division in Nanjiang, but a mechanized infantry division with two infantry and one armored regiments, believed to be the 6th Mechanized Infantry Division. Chinese reports of the unit, some with photos of its distinctive patch, are found at *news.xinhuanet.com/mil/2005-09/19/content_3511680.htm, www.phoenixtv.com/phoenixtv/83888339152797696/20051028/673715.shtml*, and *www.tianshannet.com.cn/GB/channel6/57/200506/04/161602.html*. The 124th Amphibious Mechanized Infantry Division/42nd Group Army may also have a three-regiment structure.

46. E-mail from Ellis Melvin, October 1, 2003.

47. E-mail from Ellis Melvin, May 8, 2006.

48. E-mail from Ellis Melvin, November 2, 2006.

49. E-mail from Ellis Melvin, September 16, 2003.

50. E-mail from Ellis Melvin, August 17, 2006.

51. The 127th Light Mechanized Infantry Division was the first division transformed from its motorized infantry predecessor beginning in 1997. It was followed by the conversion of the two amphibious mechanized infantry divisions from motorized units in 2000 and 2001. The first mechanized infantry brigade was reported as being converted in 2002.

52. "Seeking New Growth Point of Fighting Power through Scientific Innovation," *PLA Daily* on-line, May 26, 2006, at *english. chinamil.com.cn/site2/news-channels/2006-05/26/content_484628. htm.*

53. Guangzhou Library web page at *www.gzlib.gov.cn/ serviceguide/service/08.asp*, accessed March 10, 2006. Currently, the web page no longer identifies the unit as a missile base. The *Directory of PLA Personalities*, October 2005, p. 223, alludes to an unidentified brigade, MUCD 75810, at Shantou (which previously I assumed to be an infantry brigade).

54. *Directory of PLA Personalities*, October 2005, pp. 122, 171.

55. "Mixed AA Artillery Group Goes Digital to Increase Prowess," in *PLA Daily* on-line, July 20, 2006, at *english.chinamil. com.cn/site2/news-channels/2006-07/20/content_532497.htm.*

56. *The Military Balance 2006*, p. 265. For a point of comparison, according to *The Military Balance 2006*, p. 31, the U.S. Army, which is a fraction of the size of the PLA ground force, has some 3,800 helicopters of all types.

57. *The Directory of PLA Personalities*, pp. 232, 234. The 26th and 54th Group Army units may actually have been in existence for several years, but not yet picked up by the *Directory*. Previously the helicopter unit in the 54th Group Army was considered subordinate to the MR.

58. "PRC: Training of Army Aviation Regiment with New Helicopter," *CPP20061031710005 Beijing Jiefangjun Bao (Internet Version-WWW)* in Chinese, OSC, trans., October 31, 2006, p. 11.

59. According to an E-mail from Ellis Melvin dated April 8, 2006, it appears likely that the 200th Motorized Infantry Brigade, formerly in the 26th Group Army, has been deactivated with appearance of a Unit Claims Office to handle affairs of a unit being deactivated. Previously there were rumors that the 200th was converted to a marine unit, but no evidence to support those rumors has been found. Perhaps the 8th Armored Division, currently believed to be in the 26th Group Army, is scheduled for downsizing to brigade status, thus giving the group army an all-brigade structure.

60. "China's National Defense in 2006," Border and Coastal Defense.

61. The border and coastal defense order of battle described here and found in detail in Appendices 1-7 is derived from E-mails from Ellis Melvin dated May 20, 2006; July 7, 15, and 21, 2006; and August 8, 2006; and a posting at *www.war-sky.com/forum/read/ content/tid-122571-fpage-1-toread – page-10.html* from June 17, 2005. Thanks to Dr. Taylor Fravel for providing this last web page and his interest in the topic.

62. All numbers of equipment in this paragraph are from *The Military Balance 2006*, pp. 265-267.

63. Qianlong website, March 27, 2006, at *mil.qianlong.com/4919 /2006/03/27/2420@3077214.htm*.

64. "China's National Defense in 2006," National Defense Mobilization and Reserve Force, says, "In recent years, while keeping its overall size unchanged, the reserve force has reduced the number of Army reserve units, while increasing the numbers of reserve units of the Navy, Air Force and Second Artillery Force ...." The Chinese government has never provided the number of reserve personnel or units, so knowing that the size of the force is "unchanged" or that Army reserve units have been reduced does not give any specific insight into the current size or structure of the force. *The Military Balance 2006*, p. 264, estimates "some 800,000" personnel in PLA reserve units; in 1997, the same source estimated over 1.2 million reserves and just 2 years later "some 500-600,000." Based on the number of reserve units in order of battle presented in Appendices 1-7, something around 550,000 Army reserve personnel would seem to be a reasonable, minimum ballpark estimate.

65. "Zhangjiakou Army Reserve Antiaircraft Artillery Brigade Formed April 27," *CPP20050511000012 Shijiazhuang Hebei Ribao (Internet Version-WWW)* in Chinese, OSC, trans., April 28, 2005.

66. "Qinhuangdao Army Reserve Artillery Brigade Formed April 28," *CPP20050511000033 Shijiazhuang Hebei Ribao (Internet Version-WWW)* in Chinese, OSC, trans., April 29, 2005.

67. "Hubei's Yichang Army Reserve AAA Brigade Activated April 29," *CPP20050511000085 Wuhan Hubei Ribao (Internet Version-WWW)* in Chinese, OSC, trans., May 1, 2005.

68. "Reserve-duty AA gun brigade trains multi-caliber gunners," *PLA Daily* on-line, May 23, 2006.

69. "China's National Defense in 2006," The People's Liberation Army.

70. "Military Colleges to Recruit 10,000 High School Graduates in 2006," *PLA Daily* on-line, May 11, 2006.

71. "112 Ordinary Universities to Take in Ten Thousand National Defense Students," *PLA Daily* on-line, April 30, 2006.

72. "Enrollment of Cadets from among Soldiers Wraps Up," *PLA Daily* on-line, August 3, 2006.

73. "Xinhua 'Sidelights' of All-PLA Military Training Conference," *CPP20060628005018 Beijing Xinhua Domestic Service* in Chinese, *1219 GMT*, OSC, trans., June 28, 2006.

74. "JFJB: GSH Makes Arrangements for Military Training in 2006," *CPP20060118502003 Beijing Jiefangjun Bao WWW-Text in English* January 18, 2006, OSC, trans.; and "General Staff Department Signs All-Army New Year Military Training Work," *Jiefangjunbao*, January 18, 2006, at www.chinamil.com.cn/site1/zbxl/2006-01/18/content_387044.htm.

75. "PLA Pictorial Profiles Integrated Training of Chengdu Theater 'Experimental Units'," *CPP20060531318002 Beijing Jiefangjun Huabao* in Chinese, OSC, trans., March 1, 2006, pp. 40-45.

76. PLA Pictorial, No. 6, 2005.

77. "PRC Military Magazine Carries Photos of Chengdu MR 'Integrated' Exercise," *CPP20060710318001 Beijing Xiandai Bingqi* in Chinese, OSC, trans., May 2, 2006.

78. *Jiefangjunbao*, September 6, 2006, at www.chinamil.com.cn/site1/xwpdxw/2006-09/06/content_576328.htm. Words like "first"

in headlines must be read with caution in order to understand exactly what the word applies to.

79. China News report, July 18, 2006, at *www.chinanews.com. cn/tupian/jsxw/news/2006/07-18/759800.shtml*; *People's Daily* on-line, July 19, 2006, at *english.people.com.cn/200607/19/eng20060719_ 284709.html*; and "A Coordinated Training," *PLA Daily* on-line, July 26, 2006, at *english.chinamil.com.cn/site2/news-channels/2006-07/26/content_537183.htm*.

80. A series of over 70 Chinese reports (including many photographs) on the exercise and forces involved can be found at *english.chinamil.com.cn/site2/special-reports/2005zelhjy/ exercises%20news.htm*.

81. Scenario for Russian-Chinese Military Exercise Outlined CEP20050824949010 *Moscow Krasnaya Zvezda*, in Russian, August 24, 2005; "China: 'Peace Mission 2005' Exercises Reflect Growing Sino-Russian 'Mutual Trust'," CPP20050827000080 *Beijing Jiefangjun Bao (Internet Version-WWW)* in Chinese, OSC, trans., August 27, 2005; and "China-Russia Drills Not Targeted at Other Countries: Russian FM," *PLA Daily*, August 27, 2005, at *english. chinamil.com.cn/site2/special-reports/2005-08/27/content_282257.htm*.

82. "Kommersant Says Russia, China Did Not Mention Peace Mission 2005 Exercise Losses," CEP20051013330001 *Moscow Kommersant* in English, OSC, trans., September 8, 2005.

83. "Scenario for Russian-Chinese Military Exercise Outlined"; and "China: 'Peace Mission 2005' Exercises Reflect Growing Sino-Russian 'Mutual Trust'."

84. "Chinese and Russian Troops Braved Bad Weather to Launch Amphibious Landing Exercise," *PLA Daily*, August 25, 2005, at *english.chinamil.com.cn/site2/news-channels/2005-08/25/content_ 281206.htm*.

85. "Kommersant Says Russia, China Did Not Mention Peace Mission 2005 Exercise Losses."

86. "Details on Combined Operations of Sino-Russian Forces in Joint Exercises, 25 Aug," CPP20050826000054 *Beijing Jiefangjun Bao (Internet Version-WWW)* in Chinese, OSC, trans., August 26, 2005, p. 4.

87. "China, Russia Launch Forced Isolation Drill," *PLA Daily*, August 26, 2005, at *english.chinamil.com.cn/site2/special-reports/ 2005-08/26/content_281920.htm*.

88. The majority of Chinese military newspaper reporting suggests that PLAAF airborne and PLAN marine training is conducted without direct integration into ground force exercises.

89. "Russia: Multiple Exercises a Test of 'Ivanov Doctrine' on War-Fighting Capacity," *CEP20050822949013 Moscow Nezavisimoye Voyennoye Obozreniye* in Russian, OSC, trans., August 19, 2005, p. 1.

90. Office of the Secretary of Defense, "Annual Report to Congress: Military Power of the People's Republic of China 2006," May 23, 2006, p. 2, at *www.defenselink.mil/pubs/pdfs/ China%20Report%202006.pdf*.

91. Martin Andrew, "Power Politics: China, Russia, and Peace Mission 2005," Jamestown Foundation, *China Brief*, Vol 5, Issue 20, September 27, 2005, at *www.jamestown.org/publications_details. php?volume_id=408&issue_id=3474&article_id=2370274*.

92. Mikhail Lukin, "Peace Mission 2005: A 1970s Template for Sino-Russian 'Peacekeeping'," Moscow Defense Brief, Vol. 2, 2005, at *mdb.cast.ru/mdb/2-2005/af/peacekeeping/*.

93. "China's Dongshan Island Military Exercises To Aim at Air Superiority Over Taiwan," *CPP20040703000045 Beijing Renmin Wang WWW-Text* in Chinese, OSC, trans., July 3, 2004.

94. "CCTV-7 Shows North Sword 2005 Exercise, PLA's Li Yu Meeting Foreign Observers," *CPP20060118502001 Beijing CCTV-7* in Mandarin, OSC, trans., 1130 GMT, September 28, 2005. These four phases have been reported as elements of other recent PLA exercises.

95. "Chinese Military Paper Details "North Sword 2005" PLA Exercise," *CPP20050929506005 Beijing Jiefangjun Bao (Internet Version-WWW)* in Chinese, OSC, trans., September 28, 2005; "PLA Airborne in '1st Live' Drill vs. 'Digitized' Armor Unit in 'North Sword'," *CPP20051108318001 Beijing Kongjun Bao* in Chinese, OSC, trans., September 29, 2005, p. 1; and "Xinhua Article Details PLA's 'North Sword 2005' Exercise Held at Beijing MR Base," *CPP20050927055066 Beijing Xinhua Domestic Service* in Chinese, OSC, trans., 1422 GMT, September 27, 2005. Based on the descriptions of the units involved, the Red Force might have been nearly at division strength with reinforcements as noted, while the Blue Force was likely to be a reinforced armored regiment with airborne augmentation under the command of a division headquarters.

96. Details of the daily events in this mock battle are found at "Chinese Military Paper Details 'North Sword 2005' PLA Exercise *CPP20050929506005 Beijing Jiefangjun Bao (Internet Version-WWW)* in Chinese, OSC, trans., September 28, 2005.

97. "China Launches Its Biggest-Ever War Exercises (1)," *People's Daily* on-line, at *english.peopledaily.com.cn/200509/27/ eng20050927_211190.html*. Give the PLA "A for Effort" here. But the camouflage designed as a brick building was inappropriate for the terrain, though the tent might cause enemy intelligence to question whether the emplacement was actually occupied.

98. Office of the Secretary of Defense, "Annual Report to Congress: Military Power of the People's Republic of China 2006," May 23, 2006, p. 14.

99. "JFJB: Promote Innovation in Military Work Using Scientific Development Concept," *CPP20060811720002 Beijing Jiefangjun Bao (Internet Version-WWW)* in Chinese, OSC, trans., August 6, 2006, p. 1.

100 . "JFJB Commentator on Promoting PLA's Informatized Military Training (5)," *CPP20060803720010 Beijing Jiefangjun Bao (Internet Version-WWW)* in Chinese, OSC, trans., August 3, 2006, p. 1.

101. "Qiushi Article by General Political Department on Scientific Development Concept," *CPP20060802710009 Beijing Qiushi (Internet Version-WWW)* in Chinese, No. 15, OSC, trans., August 1, 2006.

102. "JFJB: Scientific Development Concept as Guidance for Building Modern Logistics," *CPP20060814715022 Beijing Jiefangjun Bao (Internet Version-WWW)* in Chinese, OSC, trans., August 8, 2006, p. 1.

103. "PRC: General Liao Xilong Expounds on National Defense Spending," *CPP20060410510016 Shanghai Liaowang Dongfang Zhoukan* in Chinese, No. 11, OSC, trans., March 16, 2005, pp. 14-16.

104. "JFJB Commentator on Promoting PLA's Informatized Military Training (4)," CPP20060801720001 Beijing Jiefangjun Bao (Internet Version-WWW) in Chinese, OSC, trans., July 29, 2006, p. 1.

105. "JFJB Commentator: PLA Cadres' Dedication, Responsibility Key to Military Success," *CPP20060905715040*

Beijing Jiefangjun Bao (Internet Version-WWW) in Chinese, OSC, trans., September 4, 2006, p. 1.

106. "JFJB Commentator Urges Need for Improved Joint Operations Capability, Doctrine," CPP20060725720003 *Beijing Jiefangjun Bao (Internet Version-WWW)* in Chinese, OSC, trans., July 25, 2006, p. 1.

107. "JFJB Contributing Commentator on Need To Informatize Weaponry Development," CPP20060815715027 *Beijing Jiefangjun Bao (Internet Version-WWW)* in Chinese, OSC, trans., August 9, 2006, p. 1.

108. "JFJB Commentator on Educational Campaign on Historic Mission of PLA (4)," CPP20060404515009 *Beijing Jiefangjun Bao (Internet Version-WWW)* in Chinese, OSC, trans., April 3, 2006, p. 1.

109. "JFJB Commentator on Pushing for Exploration of Integrated Training," CPP20060327502003 *Beijing Jiefangjun Bao (Internet Version-WWW)* in Chinese, OSC, trans., March 24, 2006, p. 1.

110. "Annual Report to Congress: Military Power of the People's Republic of China 2006," pp. 24-25.

111. U.S.-China Economic and Security Review Commission, "2005 Report to Congress," November 2005, p. 126.

112. "Every Single Member of the Chinese Peacekeeping Troops Is Master Hand," *PLA Daily* on-line, March 29, 2006; "Chinese Peace-Keeping Force Formed for Sudan Mission," *People's Daily* on-line, September 9, 2005; "UN Awards Peace Medals to Chinese Peacekeeping Troops in Liberia," *PLA Daily* on-line, September 15, 2004.

113. "PLA Ordering Military Units to Monitor Vehicles; Army's Reputation Eroding," CPP20060903052009 *Beijing Xinhua* in English, OSC, trans., 1217 GMT, September 3, 2006.

114. The 2006 *White Paper* says, "The proportion of the Navy, Air Force and Second Artillery Force in the PLA has been raised by 3.8 percent while that of the Army has been lowered by 1.5 percent." These percentages have no meaning, however, because the Chinese government has never provided the previous breakdown of forces upon which to compare these numbers. Even if true, these percentages indicate only a minor adjustment in the overall ratio of forces, with the Army still being by far the largest service, at approximately two-thirds of the entire force.

115. The PLA will likely always consider itself primarily a land-based force and may for prestige purposes seek to maintain the size of its ground force as "the largest in the world." It could retain this title even with substantial future cuts in the size of the Army.

APPENDIX I

SHENYANG MILITARY REGION FORCES

16th Group Army, Changchun, Jilin.
46th Motorized Infantry Division, Changchun, Jilin
48th Motorized Infantry Brigade, Tonghua, Jilin
68th Motorized Infantry Brigade, Qiqihar, Heilongjiang
69th Motorized Infantry Division, Harbin, Heilongjiang
4th Armored Division, Meihekou, Jilin
Artillery Brigade, Yanbian, Jilin
AAA Brigade, Changchun, Jilin

39th Mechanized Group Army, Liaoyang, Liaoning.
115th Mechanized Infantry Division, Yingkou, Liaoning
116th Mechanized Infantry Division, Haicheng, Liaoning
190th Mechanized Infantry Brigade, Benxi, Liaoning
3rd Armored Division, Siping, Jilin
Artillery Brigade, Liaoyang, Liaoning
Air Defense Brigade, Liaoning
Chemical Defense Regiment
Army Aviation Regiment

40th Group Army, Jinzhou, Liaoning
Unidentified (UI) Motorized Infantry Brigade, Yixian, Liaoning
UI Motorized Infantry Brigade, Chifeng, Inner Mongolia
5th Armored Brigade, Fuxin, Liaoning
Artillery Brigade, Jinzhou, Liaoning
AAA Brigade, Jinzhou, Liaoning

Units Subordinate to MR or MD
191st Motorized Infantry Brigade, Dandong, Liaoning
Army Aviation Regiment
ECM Regiment
Special Operations Dadui, Huludao, Liaoning
High-Technology Reconnaissance Bureau, Shenyang

Border and Coastal Defense Units
Heilongjiang
1st Border Defense Regiment, Dongning
2nd Border Defense Regiment, Mishan
3rd Border Defense Regiment, Raohe
4th Border Defense Regiment, Fuyuan
5th Border Defense Regiment, Luobei

    6th Border Defense Regiment, Jiayin
    7th Border Defense Regiment, Heihe
    8th Border Defense Regiment, Tahe
    9th Border Defense Regiment, Mohe
    1st Patrol Craft Group, Jiamusi
    2nd Patrol Craft Group, Heihe
Jilin
    1st Border Defense Regiment, Linjiang
    2nd(?) Border Defense Regiment, Tumen
    3rd Border Defense Regiment, Hunchun
    Patrol Craft Group, Ji'an
Liaoning
    Border Defense Regiment, Dandong
    Coastal Defense Regiment, Haiyangdao
    Coastal Defense Regiment, Zhangzidao
    Coastal Defense Regiment, Shichengdao
    4th Coastal Defense Regiment, Guangludao
    Coastal Defense Regiment, Zhuanghe

Logistics Subdepartments and Units
    1st Subdepartment, Jilin, (merged with former 3rd Subdepartment)
    2nd Subdepartment, Shenyang, Liaoning
    4th Subdepartment, Jinzhou, Liaoning
    40th Subdepartment, Qiqihar, Heilongjiang
    Army Ship Transport Group

Reserve Units
    Liaoning Army Reserve 192nd Infantry Division, Shenyang
    Liaoning Army Reserve 1st AAA Division, Shenyang
    Liaoning Army Reserve 2nd AAA Division, Dalian
    Liaoning Jinzhou Reserve Logistics Support Brigade
    Liaoning Army Reserve Communications Regiment, Anshan
    Jilin Army Reserve 47th Infantry Division, Jilin city
    Jilin Reserve Artillery Division, Changchun
    Jilin Reserve Anti-Tank Artillery Brigade, Baicheng
    Jilin Reserve AAA Brigade, Changchun
    Jilin Army Reserve Communications Regiment, Tonghua
    Reserve Water Supply Engineer Regiment
    Heilongjiang Army Reserve Infantry Division
    Heilongjiang Army Reserve AAA Division, Daqing
    Heilongjiang Reserve Artillery Brigade
    Reserve AAA Brigade, Jiamusi, Heilongjiang
    Heilongjiang Army Reserve Chemical Defense Regiment, Harbin

APPENDIX II

BEIJING MILITARY REGION FORCES

27th Group Army, Shijiazhuang, Hebei
  235th Mechanized Infantry Brigade, Xingtai, Hebei
  80th Motorized Infantry Brigade, Luquan, Hebei
  188th Mechanized Infantry Brigade, Xinzhou, Shanxi
  UI Armored Brigade, Handan, Hebei
  7th Armored Brigade, Datong, Shanxi (Blue Force for the Combined Arms Training Base in Inner Mongolia)
  Artillery Brigade, Handan, Hebei
  AAA Brigade, Shijiazhuang, Hebei

38th Mechanized Group Army, Baoding, Hebei
  112th Mechanized Infantry Division, Baoding, Hebei
  113th Mechanized Infantry Division, Baoding, Hebei
  6th Armored Division, Nankou, Beijing
  6th Artillery Brigade, Pinggu, Beijing
  Mechanized Air Defense Brigade, Shijiazhuang, Hebei
  Chemical Defense Regiment
  Army Aviation Regiment, Baoding, Hebei

65th Group Army, Zhangjiakou, Hebei
  193rd Division ("Red 1st Division"), Xuanhua, Hebei
  194th Motorized Infantry Brigade, Huai'an County, Hebei
  UI Motorized Infantry Brigade, Chengde, Hebei
  UI Armored Brigade, Zhangjiakou, Hebei
  14th Artillery Brigade, Huailai, Hebei

Beijing Garrison
  1st Garrison Division, Beijing
  3rd Garrison Division, Shunyi

Tianjin Garrison
  196th Infantry Brigade, Yangcun, Tianjin municipality
  1st Armored Division, northern Tianjin municipality (the subordination of this unit to Tianjin Garrison is speculative)

Other Units Subordinate to MR or MD
  205th Motorized Infantry Brigade, Hohhot, Inner Mongolia

UI Motorized Infantry Brigade, Datong, Shanxi
AAA Brigade
Special Operations Dadui, Daxing, Beijing
Engineer Brigade, Nankou, Beijing (UN PKO force)
Engineer Water Supply Regiment

Border Defense Units
Inner Mongolia
1st Border Defense Regiment, Wulatezhongzhen
2nd Border Defense Regiment, Erlianhaote
3rd Border Defense Regiment, Dongwuzhumuqinzhen
4th Border Defense Regiment, Ejinazhen
5th Border Defense Regiment, Baotou Guyang
6th Border Defense Regiment, Haila'er
7th Border Defense Regiment, Manzhouli
8th Border Defense Regiment, Xinbaerhuzuozhen
9th Border Defense Regiment, A'ershan
Patrol Craft Group

Logistics Subdepartments and Units
5th Subdepartment, Taiyuan, Shanxi
6th Subdepartment, Fengtai, Beijing
7th Subdepartment, Shijiazhuang, Hebei
8th Subdepartment, Tianjin

Reserve Units
Beijing Garrison Reserve AAA Division
Beijing Garrison Reserve Chemical Defense Regiment
Tianjin Army 1st Reserve AAA Division
Hebei Army Reserve Artillery 72nd Division, Tangshan
Hebei Army Reserve Infantry Division, Shijiazhuang
Hebei Army Reserve Artillery Brigade, Qinhuangdao
Hebei Army Reserve AAA Brigade, Zhangjiakou
Hebei Army Reserve Brigade, Handan
Hebei Langfang Army Reserve Logistics Support Brigade
Shanxi Army Reserve Infantry 83rd Division, Xinzhou
Shanxi Army Reserve AAA Brigade, Taiyuan
Neimenggu Army Reserve 30th Infantry Division, Hohhot

APPENDIX III

LANZHOU MILITARY REGION FORCES

21st Group Army, Baoji, Shaanxi
 61st Division ("Red Army Division"), Tianshui, Gansu
 12th Armored Division, Jiuquan, Gansu
 Artillery Brigade, Zhongnig, Ningxia
 Air Defense Brigade, Linxia, Gansu

47th Group Army, Lintong, Shaanxi
 139th Mechanized Infantry Brigade, Weinan, Shaanxi
 55th Motorized Infantry Brigade, Zhangye, Gansu
 56th Motorized Infantry Brigade, Wuwei, Gansu
 UI Armored Brigade, Chengcheng, Shaanxi
 Artillery Brigade, Gansu
 AAA Brigade, Lintong, Shaanxi
 Engineer Regiment
 Communications Regiment

Units Subordinate to MR or MD
 4th Division ("Red Army Division"), Xinjiang MD
 6th Mechanized Infantry Division, Hetian, Xinjiang MD
 8th Motorized Infantry Division, Tacheng, Xinjiang MD
 11th Motorized Division, Urumqi, Xinjiang MD
 Artillery Brigade, Xinjiang MD
 AAA Brigade, Xinjiang MD
 Army Aviation Regiment, Xinjiang MD
 Special Operations Dadui, Qingtongxia, Ningxia
 ECM Regiment, Lanzhou, Gansu
 High-Technology Reconnaissance Bureau, Lanzhou

Border Defense Units
 Xinjiang
 1st Border Defense Regiment, Balikun
 2nd Border Defense Regiment, Qitai
 3rd Border Defense Regiment, Habahe
 4th Border Defense Regiment, Fuyun
 5th Border Defense Regiment, Emin
 6th Border Defense Regiment, Tuoli
 7th Border Defense Regiment, Bole
 8th Border Defense Regiment, Huochenghuiyuan

9th Border Defense Regiment, Zhaosu
10th Border Defense Regiment, Wushen
11th Border Defense Regiment, Wuqia
12th Border Defense Regiment, Tashenku'ergan
13th Border Defense Regiment, Zepu
Border Defense Battalion, Hami
Border Defense Battalion
Kashen Border Defense Battalion,
    plus eight Border Defense Companies,
    one independent Border Defense Battalion

Logistics Subdepartments and Units
    25th Subdepartment, Xining, Qinghai
    27th Subdepartment, Lanzhou, Gansu
    28th Subdepartment, Xi'an, Shaanxi
    29th Subdepartment, Xinjiang
    30th Subdepartment, Xinjiang
    31st Subdepartment
    32nd Subdepartment, Yecheng Xian, Xinjiang
    Motor Transport Regiment, Xinjiang MD

Reserve Units
    Shaanxi Army Reserve 141st Infantry Division
    Shaanxi Army Reserve AAA Division, Xi'an
    Hanzhong Reserve AAA Regiment (possibly subordinate to
        the Shaanxi AAA Division)
    Shaanxi Baoji Army Reserve Logistics Support Brigade
    Gansu Army Reserve Infantry Tianshui Brigade
    Gansu Army Reserve AAA Division, Lanzhou
    Xinjiang Army Reserve Infantry Division
    Urumqi Reserve AAA Regiment (possibly subordinate to the
        Xinjiang Army Reserve Infantry Division)
    Qinghai Army Reserve Infantry Division
    Qinghai Army Reserve Infantry Brigade
    Yinchuan Reserve AAA Regiment

APPENDIX IV

JINAN MILITARY REGION FORCES

20th Group Army, Kaifeng, Henan
  58th Mechanized Infantry Brigade, Xuchang, Henan
  60th Motorized Infantry Brigade, Minggang, Xinyang, Henan
  UI Armored Brigade, Nanyang, Henan
  Artillery Brigade, Queshan, Henan
  AAA Brigade, Shangqiu, Henan
  Engineer Regiment, Xinyang, Henan

26th Group Army, Weifang, Shandong
  138th Motorized Infantry Brigade, Laiyang, Shandong
  199th Motorized Infantry Brigade, Zibo, Shandong
  77th Motorized Infantry Brigade, Haiyang, Shandong
  8th Armored Division, Weifang, Shandong
  8th Artillery Brigade, Weifang, Shandong
  Air Defense Brigade, Jinan, Shandong
  Army Aviation Regiment, Liaocheng, Shandong

54th Group Army, Xinxiang, Henan
  127th Light Mechanized Infantry Division (*Tie Jun*, includes
    Ye Ting Independent Regiment) Luoyang, Henan
  162nd Motorized Infantry Division, Anyang, Henan
  11th Armored Division, Xinyang, Henan
  Artillery Brigade, Jiaozuo, Henan
  Air Defense Brigade, Xingyang, Zhengzhou, Henan
  Army Aviation Regiment, Xinxiang, Henan

Units Subordinate to MR or MD
  Special Operations Dadui, Laiwu, Shandong
  Electronic Warfare Regiment, Zhoucun, Shandong
  3rd Technical Reconnaissance Unit, Jinan, Shandong
  Pontoon Bridge Regiment, Pingyin, Shandong
  Pontoon Bridge Regiment, Mangshan, Luoyang, Henan

Coastal Defense Units
Shandong
  1st Coastal Defense Regiment, Chenghuangdao
  2nd Coastal Defense Regiment, Zhangshandao
  3rd Coastal Defense Regiment, Penglai

4th Coastal Defense Regiment
5th Coastal Defense Regiment, Rushan
Coastal Defense Regiment, Weihai
8th Coastal Defense Regiment, Laoshan
9th Coastal Defense Regiment, Rizhao

Logistics Subdepartments and Units
9th Subdepartment, Zaozhuang, Shandong
10th Subdepartment, Laiyang, Shandong
11th Subdepartment, Jinan, Shandong
33rd Subdepartment, Zhengzhou, Henan
34th Subdepartment, Xinyang, Henan
Army Ship Transport Group, Zhangdao
Army Ship Transport Group
Army Ship Transport Group

Reserve Units
Shandong Army Reserve 76th Infantry Division, Yantai
Shandong Reserve Artillery Division, Jining
Shandong Reserve AAA Division, Qingdao
Shandong Reserve Army Logistics Support Brigade
Henan Army Reserve 136th Infantry Division, Kaifeng
Henan Army Reserve AAA Division, Zhengzhou

APPENDIX V

NANJING MILITARY REGION FORCES

1st Group Army, Huzhou, Zhejiang
    1st Amphibious Mechanized Infantry Division, Hangzhou, Zhejiang
    3rd Motorized Infantry Brigade, Jinhua, Zhejiang
    10th Armored Division, Suzhou, Jiangsu
    9th Artillery Division, Wuxi, Jiangsu
    Air Defense Brigade, Zhenjiang, Jiangsu

12th Group Army, Xuzhou, Jiangsu
    34th Motorized Infantry Brigade, Chuzhou, Anhui
    36th Motorized Infantry Brigade, Xinyi, Jiangsu
    179th Motorized Infantry Brigade ("Linfen Brigade"), Nanjing, Jiangsu
    12th Armored Division, Xuzhou, Jiangsu
    Artillery Brigade, Xuzhou, Jiangsu
    Air Defense Brigade, Huai'an, Jiangsu

31st Group Army, Xiamen, Fujian
    86th Motorized Infantry Division, Fuzhou, Fujian
    91st Motorized Infantry Division, Zhangzhou, Fujian
    92nd Motorized Infantry Brigade, Quanzhou, Fujian
    UI Amphibious Armored Brigade, Zhangzhou, Fujian
    Artillery Brigade, Quanzhou, Fujian
    Air Defense Brigade, Xiamen, Fujian
    Army Aviation Regiment (possible)

Shanghai Garrison
    1st Coastal Defense Brigade
    2nd Coastal Defense Brigade

Other Units Subordinate to MR or MD
    Army Aviation Regiment
    31st Pontoon Bridge Brigade, Jiangsu MD
    Special Operations Dadui, Quanzhou, Jiangsu
    Chemical Defense Regiment, Nanjing, Jiangsu
    Surface-to-Surface Missile Brigade, Shangrao, Jiangxi

Coastal Defense Units
Jiangsu
    2nd Coastal Defense Regiment, Lianyungang
    3rd Coastal Defense Regiment, Nantonghaimen
    Coastal Defense Battalion, Sheyang

Zhejiang
    13th Coastal Defense Regiment, Shengsi
    15th Coastal Defense Regiment, Daishan
    17th Coastal Defense Regiment, Putuo
    18th Coastal Defense Regiment, Dinghai
    Shipu Coastal Defense Battalion, Xiangshan
    Dachen Coastal Defense Battalion, Taizhou
    Wenzhou Coastal Defense Battalion

Fujian
    11th Coastal Defense Brigade, Lianjiang
    12th Coastal Defense Division, Changle
    13th Coastal Defense Division, Jinjiang
    52nd Coastal Defense Regiment
    54th Coastal Defense Regiment, Dadeng
    56th Coastal Defense Regiment, Zhangzhou Dongshan

Logistics Subdepartments and Units
    13th Subdepartment, Wuxi, Jiangsu
    15th Subdepartment, Huai'an, Jiangsu
    16th Subdepartment, Nanjing, Jiangsu
    17th Subdepartment, Yingtan, Jiangxi
    18th Subdepartment, Fuzhou, Fujian
    Army Ship Transport Group

Reserve Units
    Shanghai Army Reserve AAA Division
    Jiangsu Army Reserve 2d AAA Division, Yangzhou
    Jiangsu Reserve AAA Division, Nanjing
    Anhui Army Reserve Infantry Division, Hefei
    Anhui Army Reserve AAA Brigade, Wuhu
    Fujian Army Reserve AAA Division, Fuzhou
    Fujian Zhangzhou Reserve Logistics Support Brigade
    Jiangxi Army Reserve Infantry Division, Nanchang
    Jiangxi Army Reserve Artillery Brigade
    Zhejiang Army Reserve Infantry Division
    Reserve Coastal Defense Regiment, Ningde

APPENDIX VI

GUANGZHOU MILITARY REGION FORCES

41st Group Army, Liuzhou, Guangxi
121st Infantry Division, Guilin, Guangxi
123rd Mechanized Infantry Division, Guangxi
UI Armored Brigade, Guilin, Guangxi
UI Artillery Brigade, Liuzhou, Guangxi
UI Air Defense Brigade, Hengyang, Hunan

42nd Group Army, Huizhou, Guangdong
124th Amphibious Mechanized Infantry Division, Boluo, Guangdong
163rd Division, Chaozhou, Guangdong
UI Armored Brigade, Guangzhou, Guangdong
UI Artillery Division, Qujiang, Guangdong
Air Defense Brigade, Chaozhou, Guangdong

Hong Kong Garrison
Infantry Brigade
Logistics Base, Shenzhen

Macao Garrison

Other Units Subordinate to MR or MD
132nd Infantry Brigade, Wuzhishan, Hainan
Surface-to-Surface Missile Brigade, Shantou, Guangdong (may not yet have reached operational status)
Army Aviation Regiment, Foshan, Guangdong
Special Operations Dadui, Guangzhou, Guangdong
32nd Pontoon Bridge Brigade, Hubei MD
Electronic Warfare Regiment
Technical Reconnaissance Bureau, Guangzhou, Guangdong

Coastal Defense Units
Guangdong
1st Coastal Defense Regiment, Nan'ao
2nd Coastal Defense Regiment, Guishan
4th Coastal Defense Regiment, Xuwen

Hainan
    10th Coastal Defense Regiment, Wenchang
    11th Coastal Defense Regiment, Danzhou

Guangxi
    1st Border Defense Regiment, Fancheng Huashishen
    3rd Border Defense Regiment, Pingxiang
    4th Border Defense Regiment, Longzhou
    5th Border Defense Regiment, Jingxi
    12th Coastal Defense Regiment, Weizhoudao

Logistics Subdepartments and Units
    19th Subdepartment, Hengyang, Hunan
    20th Subdepartment, Guilin, Guangxi
    21st Subdepartment, Guangzhou, Guangdong
    Army Ship Transport Group, Zhuhai
    Vehicle and Ship Transport Group, Qiongshan

Reserve Units
    Hunan Army Reserve Infantry Division, Changsha
    Hunan Hengyang Army Reserve Logistics Support Brigade
    Guangdong Reserve AAA Division
    Guangdong Zhanjiang Reserve AAA Brigade
    Guangxi Army Reserve Infantry Division, Nanning
    Hainan Army Reserve Division
    Hubei Army Reserve AAA Division, Wuhan
    Hubei Xiangfan Army Reserve Artillery Brigade
    Hubei Yichang Reserve AAA Brigade
    Shenzhen Reserve Chemical Defense Regiment

APPENDIX VII

CHENGDU MILITARY REGION FORCES

13th Group Army, Chongqing
    37th Division, Chongqing
    149th Light(?) Mechanized Infantry Division, Leshan, Sichuan
    UI Armored Brigade, Pengzhou, Sichuan
    UI Artillery Brigade, Chongzhou, Sichuan
    UI AAA Brigade, Mianyang, Sichuan

14th Group Army, Kunming, Yunnan
    31st Division, Dali, Yunnan
    40th Division, Kaiyuan, Yunnan
    UI Armored Brigade, Kunming, Yunnan
    UI Artillery Brigade, Yunnan
    UI AAA Brigade, Kunming, Yunnan
    Chemical Defense Regiment

Chongqing Garrison

Other Units Subordinate to MR or MD
    52nd Mountain Infantry Brigade, Nyingchi, Xizang
    53rd Mountain Infantry Brigade, Nyingchi, Xizang
    Army Aviation Regiment
    "Cheetah" Special Operations Group, Chengdu, Sichuan
    Special Reconnaissance Group, Chengdu, Sichuan (Technical Reconnaissance Bureau?)
    Electronic Warfare Regiment
    Technical Reconnaissance Bureau, Kunming, Yunnan
    Chemical Defense Technical Group

Border Defense Units
Yunnan
    1st Border Defense Regiment, Funing
    2nd Border Defense Regiment, Malipo
    3rd Border Defense Regiment, Hekou
    4th Border Defense Regiment, Pingbian
    5th Border Defense Regiment, Jinping
    6th Border Defense Regiment
    7th Border Defense Regiment, Jiangcheng

    8th Border Defense Regiment, Mengla
    9th Border Defense Regiment, Diqing
    10th Border Defense Regiment, Lancang
    11th Border Defense Regiment, Cangyuan
    12th Border Defense Regiment, Luxi

Xizang
    1st Border Defense Regiment, Shannanlongzi
    2nd Border Defense Regiment, Cuonei
    3rd Border Defense Regiment, Dingri
    4th Border Defense Regiment, Changdu
    5th Border Defense Regiment, Saga
    6th Border Defense Regiment, Yadong
    Jiangzi Battalion
    Gangba 2nd Battalion
    Luozha 5th Battalion
    Milin Battalion
    Motuo Battalion

Logistics Subdepartments and Units
    22nd Subdepartment, Kunming, Yunnan
    37th Subdepartment, Chongqing
    38th Subdepartment, Chengdu, Sichuan

Reserve Units
    Sichuan Army Reserve Infantry Division, Chengdu
    Sichuan Army Reserve AAA Division
    Sichuan Leshan Reserve Infantry Brigade
    Sichuan Dazhou Army Reserve Artillery Brigade
    Chongqing Reserve AAA Division
    Chongqing Army Reserve Logistics Support Brigade
    Guizhou Army Reserve Infantry Division
    Yunnan Army Reserve Infantry Division
    Tibet Army Reserve Mixed Brigade

PART IV:

THE PLA AIR FORCE

CHAPTER 8

FUTURE FORCE STRUCTURE OF THE CHINESE AIR FORCE[1]

Phillip C. Saunders and Erik Quam

INTRODUCTION

The People's Liberation Army Air Force (PLAAF) is in the midst of a major modernization campaign aimed at retiring and replacing obsolete aircraft designed in the 1950s and 1960s. While modernization has been underway in earnest for the past 15 years, China's Air Force is still in a transition phase, caught in the middle between the type of force the PLAAF fielded over its first 50 years and the development of a "new PLAAF" with modern equipment and capabilities.

This chapter seeks to illuminate the future force structure of the Chinese Air Force by examining current PLAAF modernization efforts and exploring the key decisions and tradeoffs likely to shape the aircraft and capabilities the PLAAF will pursue. Our focus is not on producing a quantitative estimate of the future PLAAF air order of battle or on analyzing the future Air Force organizational structure. These approaches have been performed by other knowledgeable analysts.[2] Instead, our emphasis is on the choices that will determine the future PLAAF force structure, with the goal of illustrating a range of possibilities and providing a guide to interpreting future developments.

The first section reviews PLAAF missions and describes five ways of thinking about the "right size" of the PLAAF. The second section describes the capabilities and limitations of the "old PLAAF" (circa 1995) as the Air Force acquired its first fourth generation fighters[3] and advanced surface-to-air missiles from Russia. It then examines the new systems China is acquiring and developing and the aspirations of the Chinese Air Force to build a "new PLAAF" capable of executing a broader range of missions. The second section concludes with an assessment of the progress the PLAAF has made in its transition to a modern Air Force. The third section of the chapter analyzes how decisions about the relative effort to be devoted to air defense vis-à-vis conventional strike missions and how the tradeoffs between foreign and domestic production and between high-technology and lower-cost systems, as well as the relative emphasis on support systems, will shape the future PLAAF. It argues that perceptions of the international threat environment (to include assessments of the likelihood of a crisis over Taiwan or a conflict with the United States) and budget concerns will have significant influence on the overall size of the future PLA and the speed of modernization.

The most likely path for PLAAF force modernization is to continue present efforts to build the Air Force using a variety of means, including continued procurement of advanced aircraft from Russia; continued domestic efforts to design and produce advanced aircraft; and incorporation of imported engines, avionics, and munitions into Chinese aircraft designs. However, the chapter sketches three alternative possibilities to illustrate a range of potential outcomes: (1) efforts to maximize capability quickly; (2) a high-technology Air Force; and (3) a domestically-produced Air Force. Although the choice of modernization pathways and decisions about tradeoffs will have a significant influence on future PLAAF force structure, it is already clear that the future PLAAF will be a significantly smaller, but more capable Air Force.

IS THERE A "RIGHT SIZE" FOR THE PLAAF?

This volume asks the question "What is the right size for the PLA?" For this chapter, the appropriate question is, "What is the 'right size' for the PLA Air Force?" To be useful, the concept of a right size must refer to capabilities as well as quantitative yardsticks such as number of personnel and aircraft, organizational units, and overall budget. Unfortunately for those seeking clear predictions, the right size depends heavily on which perspective is used to evaluate the future force. Moreover, some perspectives focus on relative capabilities, which imply taking the modernization efforts of the air forces of China's potential adversaries into consideration. The five perspectives on PLAAF modernization presented below highlight the reality that there is no single right size for the future Chinese Air Force, while illustrating some of the different considerations that will influence modernization efforts.

The first perspective focuses on China's external security environment, the military missions derived from potential threats, and the Air Force capabilities and force structure necessary to carry out these missions. Outside observers can analyze these factors, but it is China's subjective assessments — and the relative weight that China's internal assessment process places on different contingencies — that will determine how threats translate into requirements for Air Force capabilities. The poor performance of the Iraq military (which had more advanced weapons than the PLA) in the 1991 Gulf War highlighted how advanced U.S. military capabilities and operational concepts could make an enemy vulnerable, prompting intensified efforts to build a more advanced and capable PLA.

Most of the aircraft acquisitions and development programs shaping today's PLAAF were initiated prior to the Chinese leadership's intensified concern about the possibility of Taiwan independence that arose in the early 1990s. These included the Air Force's initial acquisition of Russian Su-27/*Flanker* fighters, the J-10 fighter development program, and efforts to acquire or build tankers and airborne early warning/airborne warning and control systems (AWACS) aircraft. These programs were all part of long-term efforts to create a modern Air Force that could respond to a range of contingencies.

The increased threat of Taiwan independence and the perceived need to be prepared to fight against the U.S. military if it intervened on Taiwan's behalf have accelerated Chinese military modernization and shaped it toward acquiring capabilities useful for a Taiwan contingency. China has emphasized building near-term combat capability through purchase and coproduction of Russian multirole fighters such as the Su-30, while placing less emphasis on some potential Air Force capabilities such as modern strategic bombers and air refueling, which are less critical given the relatively short distance between Taiwan and mainland China. PLA strategists are now beginning to look beyond the Taiwan issue and articulate the rationale for a Chinese military capable of longer-range operations in defense of sea lanes of communication (SLOCs) and China's expanding global interests, though it is unclear how persuasive this rationale will be to Chinese leaders.[4]

General assessments of the international security environment will influence overall Chinese defense budgets and the resources available for Army building, but specific contingencies might shape Air Force modernization more directly. One scenario would be a relatively benign security environment in which the Air Force concentrates on its air defense mission. This would imply greater emphasis on air bases and air defense assets along China's land and maritime borders, and a relative neglect of long-range strike capabilities. Another scenario would involve increased efforts to develop power projection capabilities to help protect China's SLOCs and to support Chinese claims to islands in the East China Sea and the South China Sea. This would imply increased emphasis on air refueling, antiship missiles, over-water flight training, long-duration maritime patrol and intelligence collection, and (perhaps) strategic bombing capabilities. This scenario would bring the PLAAF into conflict with PLA naval aviation over which service would have responsibility for these missions.

A third scenario would involve greater attention to potential threats from Japan and/or India. For the Air Force, this would involve greater attention to training for operations against well-equipped, technologically-sophisticated Air Forces. Geographically, the PLAAF might deploy its assets differently to improve its ability to operate or conduct air strikes against India or, to a lesser degree, Japan. A fourth scenario would involve preparations for potential conflict against the United States that ranges beyond Taiwan. Given U.S. Air Force capabilities, this would be the most demanding scenario for PLAAF force modernization. China's policy of not basing PLA forces overseas constrains the contributions tactical aviation assets (such as multirole fighters) can make to scenarios that require long-range operations. Air refueling can help extend the operational range of tactical aircraft, but is an imperfect substitute for overseas bases. Without overseas bases, the PLAAF might be at a disadvantage relative to the Navy and Second Artillery in fighting for budget resources for some scenarios.

A second means of assessing the right size for the PLAAF is to look at the potential military requirements associated with China's growing international interests. China's increasing integration into the world economy has created greater demand for resources (especially oil and gas) and access to international markets to support continued economic growth. This is stimulating a more activist Chinese foreign policy that might eventually require new military missions.[5] The extent to which China's expanding international interests translate into new military requirements for the PLAAF will depend on how Chinese leaders decide to pursue their interests and the relative value of military instruments (especially air power) in these efforts. To date, Chinese leaders have stressed China's peaceful development and downplayed the potential for using force to pursue Chinese interests. If this approach continues, the most likely new missions for the PLAAF would be strategic airlift to support Chinese contributions to international peacekeeping, disaster relief, and potential evacuation of Chinese nationals from conflict zones. A more aggressive Chinese approach to resource conflicts could generate requirements for an Air Force capable of expeditionary operations, but this appears unlikely.

A third approach for right-sizing the PLAAF would focus on the priorities of China's top civilian leaders, which encompass a range of strategic, developmental, and political objectives. From this perspective, the right size is a function of the leadership's estimate of the return on investments in Air Force capabilities relative to other uses of the resources. Chinese civilian leaders are clearly concerned with the need to keep defense expenditures in proper proportion to economic development efforts; the 2006 *Defense White Paper* calls for "coordinated development of national defense and the economy." However, defense and civilian industries can have positive synergies, so Chinese leaders might support some military expenditures (especially in research and development [R&D]) due to their benefits for the civilian economy. Chinese civilian leaders might also view defense spending increases as a means of helping to ensure the loyalty of the military to the Communist party. Significant portions of recent increases in military spending have been devoted to increased pay and improved living conditions for the military.[6] Investments in military capabilities give Chinese leaders increased international options, but acquiring certain capabilities (such as strategic bombers or an aircraft carrier) might also impose costs by stimulating adverse reactions from China's neighbors. Without more detailed knowledge of how Chinese civilian leaders think about the costs and benefits of various Air Force capabilities, it is difficult to derive a right size for the PLAAF from this perspective.

A fourth approach would focus on the relative return on investment in Air Force capabilities compared to other military capabilities. The right size for the PLAAF then depends on the relative contributions air power can make to the PLA's overall ability to perform its missions and execute its campaign plans. This requires a detailed examination and prioritization of PLA and PLAAF missions and responsibilities. The PLAAF's primary mission has long been air defense, with support for ground troops an important secondary mission. The air defense mission requires close coordination of both aircraft and ground-based air defenses such as surface-to-air missiles (SAMs) and anti-aircraft artillery (AAA).[7] Despite the long-standing secondary mission of supporting ground troops, the PLAAF has never been able to perform close-air support missions for ground forces and has had only a limited capability to perform bombing and interdiction missions in support of ground operations.

The 2004 *Defense White Paper* describes the PLAAF responsibility "for safeguarding China's airspace security and maintaining a stable air defense posture nationwide," noting that "the Air Force has gradually shifted from [a mission] of territorial air defense to one of both offensive and defensive operations." It highlights "the development of new fighters, air defense, and anti-missile weapons" and emphasizes training to "to improve the capabilities in operations like air strikes, air defense, information countermeasures, early warning and reconnaissance, strategic mobility, and integrated support."[8] The 2006 *Defense White Paper* stresses PLAAF efforts to speed up "its transition from territorial air defense to both offensive and defensive operations" and to increase "its capabilities in the areas of air strike, air and missile defense, early warning and reconnaissance, and strategic projection."[9] The white papers and other PLA doctrinal literature reveal that the PLAAF's air defense mission is now conceived of as a nationwide responsibility that incorporates both offensive and defensive actions. The emphasis on offensive operations, air strikes, and strategic mobility (coupled with the PLA-wide emphasis on joint operations and joint campaigns) imply a higher priority for operations that support ground forces.

These broad missions are translated into specific operational concepts and training requirements through campaign theory, which can be thought of as the PLA's operational doctrine. Throughout the 1990s, the PLA has been engaged in a major effort to revise and update its doctrine. This produced new PLAAF operational guidance in the form of a 1999 *"gangyao"* (operational regulation) titled "The Essentials of Campaigns of the Chinese People's Liberation Army Air Force."[10] In their contribution to the present volume, Kevin Lanzit and Kenneth Allen provide a fuller treatment of how doctrinal reforms and new operational concepts are influencing PLAAF modernization.[11] The PLAAF trains for three dedicated Air Force campaigns. The *offensive air campaign* employs air strikes on enemy territory to suppress or destroy enemy air defenses and to attack both strategic and campaign level targets. The *air defense campaign* seeks to establish air superiority over the war zone through several measures, including deterrence based on denial, resisting attack by targeting hostile intelligence and service platforms, and launching timely counterstrikes against enemy air bases and support assets. The *air blockade campaign* is designed to effect political coercion against the enemy via means such as air strikes that target ports and navigation routes. In addition, the PLAAF has major roles in two joint service campaigns: the *joint anti-air strike campaign* and the *airborne campaign*.[12]

The overall balance between offensive and defensive capabilities, the emphasis placed upon dedicated Air Force missions and campaigns, and the relative contributions the PLAAF can make to joint campaigns will all influence the right size for the PLAAF compared to other services. The 2004 *Defense White Paper* called for giving "priority to the building of the Navy, Air Force and Second Artillery Force," implying the need for greater investment in Air Force capabilities. However, ground force officers remain dominant within the

PLA, so that parochial service considerations are likely to continue to influence resource allocations.

A fifth approach would emphasize building the PLAAF into a modern Air Force capable of engaging and defeating other air forces.[13] Here the most ambitious benchmark would be the ability to engage and defeat the U.S. Air Force; a less ambitious goal would be to tackle advanced Asian air forces such as those of Japan and India. This approach implies an emphasis on the development of advanced fighter aircraft and force multipliers such as tankers and AWACS aircraft. In terms of force structure, such an approach would emphasize additional procurement of Russian aircraft, efforts to acquire advanced Western technology for Chinese platforms, and a reluctance to procure less-capable indigenous systems. Chinese air power advocates succeeded in persuading the Central Military Commission (CMC) to incorporate an Air Force component in China's Active Defense strategy in 2004.[14] But despite its appeal to Air Force officers, an air power-centric approach to warfighting is unlikely to be adopted by the PLA as a whole.

The five perspectives discussed above outline different ways of thinking about the right size of the PLAAF. Each suggests a different view about the role the Air Force might play in national security and what force structure would be appropriate. However, none provides a straightforward prediction as to what the future PLAAF will look like. In reality, future PLAAF force structure will be the product of a political process that incorporates some aspects of each of these perspectives.

THE PLAAF IN TRANSITION

The Old PLAAF.

The PLAAF was designed as a defensive force charged with the primary mission of air defense and a secondary mission of support for the ground forces. Air defense responsibilities included defending China's airfields, other critical infrastructure, political and economic centers, and ground forces.[15] The PLAAF was also charged with supporting ground troops via close air support and bombing operations, but has never really been able to perform this mission.

J-6 fighters and Q-5 attack planes, both variants of the 1950s vintage Soviet MiG-19 fighter, made up the numerical bulk of the PLAAF force through the mid 1990s. The J-6 is a second generation fighter designed primarily as an air defense interceptor; the Q-5 is a Chinese variant with ground attack capabilities.[16] At its peak, the Chinese Air Force deployed more than 3,000 J-6s in training and operational roles. The J-6 is a low-technology fighter, greatly inferior to the aircraft employed by modern air forces like those of Taiwan, India, and the United States. Although the PLAAF deployed vast numbers of J-6 fighters, their combat effectiveness was limited due to limited range, lack of on-board radar, and lack of all-weather capability.[17]

China made several attempts to produce more advanced fighters to replace or augment the J-6 in the 1970s and 1980s. The Chinese produced hundreds of J-7 fighters (a Chinese variant of the Soviet MiG-21 design) and several different variants of the J-8 fighter. Both the J-7 and J-8 were improvements over the J-6 in avionics and performance, but still lagged far behind the fourth generation fighters deployed in the Soviet,

United States, and Western European air forces in the 1980s. China attempted to upgrade the F-8-II using U.S.-built avionics under the "Peace Pearl" program, but this effort was aborted when the United States imposed sanctions on military exports to China after the 1989 Tiananmen incident. In the end, the PLAAF procured relatively small numbers of J-7 and J-8 aircraft, which upgraded Air Force capabilities somewhat but did not replace the J-6 as the principal aircraft. This may have reflected a decision to wait until more advanced aircraft were available from Chinese manufacturers before procuring new ones in large quantity. China did begin purchasing advanced fighter jets in the early 1990s with the acquisition of the fourth generation Su-27s from Russia as part of the effort to begin modernizing and updating the fighter force.[18] One objective in purchasing the Su-27 was to gain experience operating and maintaining an advanced fighter.

The ground-attack and bomber aircraft in the PLAAF inventory in the mid 1990s were also derived from late 1950s Soviet designs. The Q-5 attack fighter, a derivative of the J-6, is a close air support aircraft with ground attack and air-to-air combat capabilities. The Q-5's capabilities are limited by its relatively short range (about 800 km) and primitive avionics.[19] The Q-5 is capable of carrying a nuclear payload, and at one point several dozen Q-5s were designated for nuclear missions.[20] However, the Q-5's effectiveness as a nuclear delivery platform was compromised by its very short range. The H-6, the Chinese version of the Soviet Tu-16/*Badger* medium bomber, served as the PLAAF's primary dedicated bomber. Some H-6 bombers had a strategic nuclear role in the 1960s and 1970s, but it is unclear whether the PLAAF still has a nuclear mission.[21] The H-6's effectiveness in a traditional bombing role is limited by its range and slow speed, which make it highly vulnerable to modern air defense systems. At that time the PLAAF lacked air-launched cruise missiles that could have allowed the H-6 to then concentrate on a stand-off attack role.

Although this chapter concentrates primarily on aircraft, the PLAAF also has responsibility for long-range ground-based air defenses. (PLA ground force units also operate shorter-range SAMs and anti-aircraft artillery (AAA) to protect themselves.) The Chinese air defense strategy is centered on the belief that an air defense system needs to be "layered, redundant, robust, and operate as an integrated battle space air defense network."[22] Air defense systems are generally classified either as strategic, long-range systems that defend a wide area or as shorter-range, tactical systems used for defense of ground forces or point targets. The first Chinese SAM was the HQ-1, an SA-2 variant produced under license from the Soviets in the 1960s.[23]

Over the next several decades, China worked to develop and produce domestic SAMs, including Chinese versions of foreign systems such as the French short-range mobile Crotale system. Most of these systems were essentially obsolete by the early 1990s. In 1991, China bought equipment for four to six S-300PMU-1/SA-10 battalions from Russia, but did not take delivery until 1993. These missiles were deployed around Beijing and the Su-27 airbases at Wuhu and Suixi.[24] The S-300 gave China its first long-range air-defense system, with missiles capable of intercepting high and low altitude targets at ranges up to 150 km. The S300 also had a limited capability to intercept ballistic missiles.[25] Deployment of the S300 greatly enhanced the PLAAF's ability to control air space and conduct air defense missions.

The PLAAF of the 1980s and early 1990s was not equipped with aircraft capable of carrying out its missions of air defense and support for ground forces against a modern adversary. The limited capabilities of the Chinese attack and bomber force and the lack of communications with ground forces made them relatively ineffective in ground support missions.[26] PLA bombers were also extremely vulnerable to the modern air defense systems they would likely meet in nearly any regional conflict that might have required the aircraft to undertake bombing missions. However, the low-technology aircraft that China's defense industry could produce were also relatively inexpensive, so that the PLAAF compensated for technical limitations by procuring large quantities of aircraft. The PLAAF was one of the largest air forces in the world, but backward technology and obsolete aircraft constrained its ability to carry out its missions. Limited flight training time and unrealistic training aggravated the situation. In the late 1980s, PLAAF fighter pilots were flying only about 100 hours per year.[27] Most of those hours were conducted under good weather conditions, during the day, and with very little over-water training.

PLAAF limitations were evident when measured against the tasks it would confront during a Chinese invasion of Taiwan. PLAAF assignments would have included air defense of mainland China, achieving air superiority over the Taiwan Strait, and attacking airfields and strategic targets on Taiwan. The PLAAF would have had difficulty achieving these objectives against a Taiwan Air Force that historically enjoyed advantages in both technology and training. The PLAAF's acquisition of Russian Su-27 fighters in the early 1990s offered some challenge to Taiwan's technology lead, but the Taiwan Air Force began taking delivery of U.S. F-16 fighters and French *Mirage* 2000 fighters in 1997 to complement its existing F-5 and Indigenous Defense Fighter (IDF) aircraft. These systems restored Taiwan's unchallenged technological superiority and the ability to command the skies above the Taiwan Strait in the early stages of a conflict.

The potential involvement of U.S. forces in a Taiwan conflict scenario would have further increased the challenge for the PLAAF. The USAF had extraordinary advantages over the PLAAF in almost every respect except geography. In the mid-1990s, the United States was flying fourth generation fighters with much greater capabilities than China's most modern fighters. The United States also had AWACS, electronic warfare systems, and air refueling capabilities that China lacked. The PLAAF's operational limitations became more important as Chinese concerns about Taiwan independence began to grow in 1992-93. The U.S. deployment of two aircraft carriers in March 1996 following China's "missile tests" to intimidate Taiwan led the Chinese military to conclude that an attack on Taiwan would likely precipitate U.S. military intervention. As a result, China's planning for Taiwan contingencies began to take U.S. military capabilities into account. Although PLAAF modernization efforts were already underway, concerns about Taiwan independence gestures stimulated increased funding and efforts to build a modern Air Force capable of effective combat operations.

The New PLAAF.

The PLAAF is now in transition between the limited force consisting mainly of obsolete aircraft that it fielded in the 1980s, and the more advanced force it intends to field in the coming decades. The J-6 fighters that once made up most of the PLAAF fighter fleet have now been completely retired.[28] The remaining J-7 and J-8 fighters have been updated and remain in service, but these aircraft comprise only about 1,000 planes. The PLAAF's future aircraft are now beginning to enter the force, although the total numbers and precise mix of foreign and domestic aircraft remain an open question. The PLAAF now has 15 years experience operating the Su-27 fighter, as well as experience with the Su-30s and J-10s and modern surface-to-air missiles.

The Chinese vision for a new PLAAF embraces a highly trained modern Air Force equipped with high-tech aircraft, advanced precision-guided munitions, support aircraft that serve as force multipliers, and networked command, control, and intelligence capabilities that allow the PLAAF to fight and win a high-tech war under informationalized conditions. This force would not only be more capable of carrying out traditional missions such as air defense and support for ground forces against a modern adversary, but could also undertake offensive strikes against ground and naval targets further away from China's borders. The new PLAAF will be a smaller force, but composed primarily of more advanced third- and fourth-generation multirole fighters and fighter-bombers. It is uncertain whether China will decide to build or acquire new bombers, but the deployment of advanced cruise missiles should allow existing bombers to contribute more effectively to a variety of PLAAF missions, including antiship and ground-attack missions. The new PLAAF will also fully integrate support systems such as airborne early warning (AEW)/AWACS, aerial refueling tankers, intelligence collection, and signal jamming aircraft to increase the effectiveness of combat aircraft and enhance warfighting capability. Air force modernization will also include larger numbers of more capable air transports, which will enhance the effectiveness of PLAAF airborne forces for both internal security and external missions. The PLAAF will continue to update and modernize its ground-based air defenses, and will likely seek to develop more effective defenses against cruise missiles and ballistic missiles.

The PLAAF is building a more capable fleet of multirole fighters that will include both foreign and domestically produced aircraft. The foreign component will be comprised of the Su-27 and Su-30 fighters being procured from Russia. Those aircraft will be supplemented by the J-11, the Chinese-assembled version of the Su-27. Initial "coproduction" involved Chinese assembly of unassembled aircraft provided by the Russians, but the Shenyang Aircraft Corporation plans to increase gradually the proportion of domestically produced components for the J-11.[29] There were indications that the Chinese had concerns about the technological limitations of the Su-27s in the late 1990s and early 2000s. China complained that Russia was offering more advanced versions of the Su-27 to other customers. The Su-27SM system exhibited at the Zhuhai airshow was reported to have upgrades aimed at addressing China's concerns, including multifunction liquid crystal displays (LCDs) and a precision navigation system incorporating laser gyroscopes and a GLONASS/NAVSTAR receiver.[30] China has continued to purchase Russian-built Su-30s and to assemble J-11/Su-27 aircraft.

The J-10 is China's first domestically produced fourth-generation aircraft and will likely make up a large portion of the future Chinese force. The J-10 is a highly capable, multirole fighter strongly influenced by the Israeli *Lavi*, which was itself influenced by the F-16.[31] The J-10 is equipped with aerial refueling capabilities, which significantly improve its range and flexibility.[32] The J-10 has entered into serial production; some 60 aircraft (enough to equip about three Chinese aircraft regiments) are reportedly already deployed.[33]

The PLAAF may also field the *Xiaolong*/FC-1, an indigenously developed fighter that is the product of a Chinese-Pakistani joint venture.[34] Originally known as the Super-7, the project sought to upgrade the J-7 (MiG-21) fighter with a more advanced engine and upgraded Western avionics to provide a capable but less expensive fighter.[35] The PLAAF is reportedly not enthusiastic about acquiring the *Xiaolong*, but the aircraft's producer, the Chengdu Aircraft Industrial Group, is pushing for PLAAF purchases in order to validate the aircraft for foreign customers.[36] The *Xiaolong*/FC-1 would provide a less expensive alternative to the fourth generation aircraft the PLAAF is currently acquiring. The precise mix of the PLAAF fighter force will depend on decisions about the optimum ratio of high-tech to medium-tech fighters, itself a function of the imported-to-domestic ratio.[37]

A key limitation on China's ability to produce advanced aircraft has been its inability to produce suitably advanced engines. China's most advanced aircraft currently employ Russian engines. Chinese defense industries have made considerable efforts to develop the capability to produce modern jet engines. The Shenyang Liming Engine Manufacturing Corporation has now developed the *Taihang* aero-engine, also known as the WS-10, the first high-thrust turbofan engine to be domestically researched and developed in China. The WS-10 may eventually be installed in China's J-10 fighters and possibly also its Su-27 variants.[38] The ability to produce advanced engines in China would give the PLAAF increased flexibility in choosing between domestic and foreign fighters in the future. However, the WS-10 has reportedly not yet been installed in operationally deployed aircraft, and Russian manufacturers hope to continue to supply engines for Chinese fighters in the future.[39]

Along with fighters, the PLAAF will continue to modernize its ground-attack and bomber forces. The centerpiece of China's efforts to improve its ground-attack capabilities is the JH-7/FB-7 *Flying Leopard*. JH-7 development began in 1975, but the aircraft did not fly for the first time until 1988 and was not revealed publicly until 1998. Although the JH-7 is a multirole aircraft, its limited capabilities against modern fighters suggest that it will be used mainly for ground attack and antiship missions. The JH-7 is capable of carrying C-801/802 antiship missiles, and was initially deployed with PLA Navy (PLAN) Aviation units.[40] About 20 JH-7s are currently deployed with the PLAAF 28th Air Division in Hangzhou.[41] The PLAAF is reportedly unenthusiastic about the JH-7 and would probably prefer to acquire more advanced multirole fighters instead.

The direction of Chinese efforts to modernize its bomber force is less certain than those in behalf of ground-attack aircraft. Production of the H-6/*Badger* bomber has resumed, with an emphasis on production of a new variant possessing the ability to employ antiship cruise missiles and land-attack cruise missiles.[42] Chinese military websites show pictures of the H-6 and the modified H-6D with cruise missiles on them, as well as pictures of the H-6 firing cruise missiles from the air.[43] The H-6's vulnerability to modern air defenses suggests the aircraft will be employed primarily as a standoff platform to deliver cruise missiles from outside the reach of enemy air defense systems. The capability to locate enemy ships and pass target information to bombers and fighter-bombers would be critical for successful antiship operations.

It is unclear whether the PLAAF will procure a new bomber capable of penetrating air defense systems. The "Peace Mission 2005" joint military exercises conducted by Russia and China in August 2005 included Tu-22 and Tu-95 bombers.[44] Russian officials have openly discussed their interest in selling these aircraft to China.[45] The Chinese defense press has extensively discussed the pros and cons of the Russian Tu-22 and the Tu-95 bombers, but thus far there has been no decision to purchase either aircraft. Some critics note that both craft were designed in the 1960s, and that even with upgrades these aircraft would not mark a great technological leap forward.[46] Others argue that it is important to get experience flying supersonic bombers, and that the Tu-22 and Tu-95, being significantly more capable than the H-6, would therefore improve the capabilities of the PLAAF bomber force. Some argue that it is as important to procure the Tu-22 and Tu-95 now as it was to purchase the Tu-16 more than 40 years ago.[47] Chinese sources have stated that the only reason China would buy new strategic bombers would be to prevent the United States from entering any Taiwan scenario. Such a purchase would signal that China was moving towards a bomber fleet capable of long-range operations.[48] Some Chinese analysts believe that procurement of strategic bombers would cause a major shift in the balance of power in Asia.[49] However, these arguments highlight a potential downside to PLAAF acquisition and operation of advanced strategic bombers: the United States and China's neighbors are likely to view these systems as highly threatening advances in Chinese military capabilities.

Along with fighters and bombers, the PLAAF will devote significant efforts to develop and deploy force multipliers that will enhance the capabilities of its combat aircraft. These systems include tankers, AEW aircraft, electronic warfare and intelligence collection aircraft, as well as transports that will support a rapid-response capability for internal and external contingencies. The PLAAF already has a number of aircraft capable of aerial refueling. Though the Su-27 and J-11 fighters are not capable of air refueling, the Su-30 can be refueled by Il-78/*Midas* tankers.[50] China has ordered four Il-78 tankers from Russia, but delivery of the aircraft has reportedly been delayed due to production problems.[51] PLAAF J-8 and J-10 fighters can be refueled by HY-6 tankers, based on a modified H-6 platform. The PLAAF has a regiment of HY-6 tankers based at Leiyang in the Guangzhou MR to support its J-8 and J-10 fighters.[52] The HY-6 is capable of refueling two J-8II aircraft at the same time from a refueling pod extended from each wing. Expansion of the tanker force and delivery of Il-78 tankers from Russia will extend the range and endurance of the PLAAF's refuelable combat aircraft. However the mix of different tanker platforms and limitations on which aircraft each tanker can support will impose some limits on operational flexibility.[53]

China has made several efforts to acquire or develop AEW and AWACS capability, but current information suggests that only limited progress has been made. Some Chinese sources take the position that an AEW capability, which would relay aerial radar information to ground-based air controllers, would be better suited to the PLAAF than AWACS since it would require fewer changes in current operational practices.[54] China reportedly signed a deal in 1996 to acquire the A-501 *Phalcon* AWACS from Israel, but the purchase was cancelled in July 2000 after the Israeli government came under pressure from the Clinton administration.[55] China's initial effort to develop a domestic AEW capability used the Il-76 as a platform for the KJ-2000, equipped with an indigenously designed phased-array radar.[56] R&D on this system has reportedly made significant progress, but the program was set back by the crash of a prototype in June 2006 that killed some 40 technicians involved in the R&D effort.[57] A second domestic AEW program, the KJ-2, is being developed based on the Chinese Y-8X transport aircraft.[58] Both the KJ-2 and the KJ-2000 are to be equipped with data links compatible with the J-7, J-8, J-10, J-11, JH-7, and H-6. Both of the AEW aircraft carry an indigenously-developed phased-array radar.[59] If the KJ-2 and KJ-2000 are compatible with Chinese-built J-11 fighters, China might also be able to modify its Russian-built Su-27 and Su-30 aircraft to work with them. The 2004 Department of Defense report on Chinese military power projected that the PLAAF will have several AWACS or AEW aircraft by 2010.[60]

China is also making efforts to modernize its transport fleet. China currently operates about 13 Russian-built Il-76/*Candid* transports, and reportedly has ordered 38 more.[61] It is continuing production of the Y-8 and preparing for the introduction of the Y-9 transport.[62] The Y-8 is a medium-lift turboprop transport based on the Soviet Antonov An-12. This platform has also been adopted for various other missions including maritime patrol, AEW, electronic intelligence, unmanned aerial vehicle (UAV) carrier, and airborne radar test bed.[63] The Y-9 military transport was first shown in public at the 2005 Beijing Aviation Expo. The Y-9 is a medium-size tactical support aircraft that is an upgrade of the Y-8. It is capable of carrying 98 armed soldiers or paratroopers, or 72 seriously wounded patients plus three medics.[64] In addition to these dedicated military transports, Chinese airlines fly large numbers of commercial aircraft (including European and U.S. airliners) that could be pressed into service in a crisis.[65]

China's efforts to update its air defense capabilities are focused on building a modern integrated air defense system capable of both offensive counterair and defensive counterair operations.[66] For ground-based air defenses, this will involve continued procurement of advanced Russian SAMs and efforts to develop more capable domestic SAMs. Since 1999, China's air defense efforts have emphasized the "three attacks and the three defenses." The three attacks refers to efforts to develop air defense systems capable of attacking stealth aircraft, cruise missiles, and armed helicopters, while the three defenses refer to efforts to protect against precision strikes, electronic jamming, and electronic reconnaissance and surveillance.[67] Chinese analysts have been particularly impressed by U.S. cruise missile capabilities; defense against cruise missiles is likely to be an increasingly important element of Chinese air defense efforts.

China's ground-based air defenses have been significantly upgraded with wider deployment of Russian S-300 SAMs, the acquisition and deployment of new short-range mobile SAMS such as the *Tor* M1/SA-15 from Russia, and the development of new indigenously produced models.[68] The PLAAF has built on its initial deployment of Russian S-300/SA-10 SAMs by procuring and deploying the longer-range follow-on S-400/SA-20 system to extend air defense coverage over the Taiwan Strait. China is expected to deploy the S-300/PMU2 soon and is also working to reverse-engineer the S-300 to allow indigenous production.[69] China has developed a modern indigenous SAM, the KS-1A, which incorporates a phased-array radar and associated ECM systems for use against high-altitude reconnaissance aircraft, UAVs, helicopters, air-to-surface missiles, and other targets.[70] China is also developing the FT-2000 and FT-2000A antiradiation SAMs, which could be used against early warning and jamming aircraft. The extent to which these indigenously developed systems are actually deployed with PLAAF ground-based air defense units is unclear.[71] The net result is a major upgrade of Chinese air defense capabilities. However, the extent to which Chinese radars, surface-to-air missiles, and antiaircraft artillery are effectively tied together into a genuinely integrated air defense system also remains murky. The distribution of air defense responsibilities and assets among Army, Air Force, and Navy units greatly complicates efforts at effective integration.

The new PLAAF will be a high-technology force able to engage most modern air forces. It will be increasingly capable of conducting joint operations with other services and combined operations with multiple branches of the PLAAF. The Chinese have already made joint operations a priority in doctrinal revisions and training and will continue to do so in the future, although joint operations capabilities are currently still at an early stage.[72] Along with better equipment, PLAAF pilots will be better trained than the pilots of the old PLAAF. The PLAAF has begun to intensify all-weather training, increase the number of flight hours, and conduct more over-water flight and attack training.[73] The PLA is making significant efforts to improve the training and professionalism of its officer corps, which may improve its capability to command and execute more complex operations. Joint operations, increased force-multipliers, better training, and a more highly technological, more capable force will combine to give the new PLAAF greater operational capabilities, allowing China to project power away from the mainland and into Asia.

TRADE-OFFS IN PLAAF MODERNIZATION EFFORTS

The previous section described the PLAAF as an Air Force in transition between the "old PLAAF," composed primarily of obsolete aircraft and limited support systems, and the envisioned "new PLAAF," equipped with advanced aircraft and support systems capable of carrying out more ambitious missions and military campaigns. While the outlines of this new PLAAF are apparent, the precise mix of types and numbers of aircraft that will make up the force structure of the future PLAAF remains undetermined.

This section will first highlight and discuss key decisions that Chinese leaders will confront in the next decade, and then speculate on how their choices will influence PLAAF force structure. The decisions China reaches on allocating roles and missions among services and branches, and on the proportions for domestic versus foreign procurement, low-technology systems versus high-technology systems, and combat aircraft versus support aircraft, will be key shapers of PLAAF modernization efforts and future force structure. Chinese assessments of the international security environment and the resulting resource allocations for military and Air Force modernization will also influence the pace of modernization and the size of the Air Force.

Division of Labor.

One important choice is how key missions are divided among services, branches, and weapon systems. Missions such as air defense and conventional strike can be performed by several different types of weapon systems and assigned to different services or branches. How China's military leaders decide to allocate missions will have a significant impact on PLAAF force structure.

The air defense mission now belongs primarily to the Air Force, which operates most of China's fighters and most of its long-range ground-based air defenses. However the Chinese Navy also operates fighters with an air defense mission, while PLAN ships are armed with increasingly capable surface-to-air missile systems. The question is further complicated when the broad range of potential threats that Chinese air defenses must defend against is considered. These not only include aircraft, but also existing and emerging cruise and ballistic missile threats. China's "three attacks and three defenses" concept calls for efforts to develop systems capable of attacking stealth aircraft, cruise missiles, and armed helicopters and of defending against precision strikes, electronic jamming, and electronic reconnaissance and surveillance. Successful air defense will likely require a mix of active and passive defenses and a range of air defense capabilities.

Three considerations are especially relevant. The first is whether the existing division of air defense responsibilities between the PLAAF and the PLAN will be maintained.[74] Currently the PLAAF has responsibility for long-range ground-based air defenses and responsibility for air defense along China's land borders, while the PLAAF and PLAN aviation forces divide air defense responsibilities for China's sea borders on a geographical basis.[75] This division of labor creates potential gaps in coordination among ground-based air defenses, ground-controllers run by the Air Force, and the fighters operated by Navy aviation units. Increased joint training and efforts to build a truly integrated air defense system could ease these coordination problems if properly carried out. The extent to which the PLAN retains some air defense responsibilities will affect the number of PLAAF fighters assigned to air defense missions. In doctrinal terms, this question is evident in the potential overlap between the PLAAF's responsibility to execute an air defense campaign and the multiservice responsibilities in a joint antiair strike campaign.

A second consideration is the relative emphasis on ground-based air defenses versus fighters. China has made significant investments in surface-to-air missiles in recent years that have significantly improved its air defense capabilities. China's S-300 and S-400 SAMs have effective ranges long enough to reach most of the way across the Taiwan Strait, posing a serious threat to Taiwan fighters on air defense or potential strike missions. PLAAF SAM radars routinely "paint" Taiwan fighters while they are flying on the Taiwan side of the center line; PLAAF SAM systems are now assessed to be effective enough to make Taiwan Air Force attack missions against mainland targets very dangerous.[76]

A third consideration is China's interest in defending against threats posed by cruise missiles and ballistic missiles. These threats are generally best dealt with by ground- and sea-based defenses rather than aircraft. China's advanced Russian SAMs such as the S-300 have fairly good capabilities against cruise missiles, but only a limited capability against short-range ballistic missiles. However, China has limited long-range radar capability and no early warning radars or overhead IR sensors to provide warning of ballistic missile launches. Filling these gaps would require a significant investment in sensors and communications equipment. Moreover developing and deploying a significant ballistic missile defense capability would require a major commitment of resources. The threat posed by U.S. cruise missiles and by Taiwan's emerging cruise and ballistic missile capabilities may force China to devote resources to this task. A decision to pursue more comprehensive cruise and ballistic missile defenses, with the consequent additional drain on resources, would represent a significant national commitment on a vast scale.

While decisions about which active defenses to pursue will have the biggest impact on PLAAF force structure, the PLA also puts great emphasis on passive defenses as a means of surviving enemy air and missile attacks. Mobility, camouflage, decoys, underground shelters, and a logistics system that can support dispersed operations are all important parts of efforts to protect PLA forces from attack. The cost of developing and employing passive defenses is borne by all the PLA services, but Air Force investments in mobility and passive defenses will draw resources from other aspects of the modernization program.

Conventional strike, against both land and naval targets, is a second major area where decisions about the division of labor among PLA services and branches may have a significant impact on PLAAF force structure. China is currently pursuing a range of conventional strike capabilities, including ballistic missiles controlled by the Second Artillery, cruise missiles that can be fired from air, ground, and naval platforms, and strike aircraft (including both multirole fighters and bombers). Chinese military officers see value in having a range of conventional strike options, which will permit multidimensional attacks against targets such as aircraft carriers and provide for more flexible attack options in a Taiwan contingency. Nevertheless, decisions about whether ballistic and cruise missiles can substitute for a new strategic bomber and about how the PLAN and PLAAF will divide maritime strike responsibilities will play a large role in determining the forces in the future PLAAF. With the Air Force increasingly training over water and procuring advanced aircraft with aerial refueling capabilities, there will be the potential for the PLAAF to perform some missions heretofore assigned to PLAN.

It is unclear whether the PLAAF will develop or procure a new strategic bomber capable of penetrating modern air defense systems. China may ultimately decide that more accurate and longer range ballistic missiles and cruise missiles delivered from a variety of land, air, and naval platforms can substitute for strategic bombers. A related question is whether the PLAAF still has a nuclear delivery mission, and, if so, whether this mission will be a driver of future PLAAF force modernization. The 2004 *Defense White Paper* refers to nuclear missions for the Second Artillery and the Navy, but not the Air Force.[77] Recent studies of PLA strategic modernization have not found PLAAF interest in building or acquiring a new strategic bomber to perform nuclear strike missions.[78] At this point, it does not appear that the PLAAF has a nuclear role. If it re-acquires such a role in the future, China would most likely develop nuclear-armed cruise missiles that can be delivered by an H-6 bomber rather than develop a strategic bomber fleet.

Domestic vs. Foreign Production.

A second issue we have earlier noted is the balance between domestic production and foreign purchases of aircraft. China has historically sought to be self-reliant in military production, but this goal has so far eluded it. PLAAF weapon systems have mostly been reverse-engineered from Soviet/Russian systems, been purchased outright, or incorporated significant amounts of foreign technology.[79] China's defense industry reform and modernization efforts are intended to improve China's capability to develop and produce high-technology weapons, including advanced aircraft. China's defense industries are currently unable to develop and produce state-of-the-art weapons, with the aviation industry struggling to produce military aircraft at the mid-1980s level of technology. Chinese leaders must therefore strike a balance between domestic production of less capable aircraft and munitions on one hand, and acquisition of more capable weapons from foreign suppliers (with attendant uncertainties about excessive dependence and future support), on the other.

The degree to which the domestic-foreign trade-off actually constrains China's choices depends heavily on the quality of the weapons produced by China's domestic aviation industry. Chinese leaders are devoting significant attention to improving the country's overall science and technology base, including its military-industrial base. This includes significant resources devoted to R&D and the goal of building the capacity for "indigenous innovation." The effort to reform and improve China's defense production also included a major restructuring of the military procurement system in 1998. The reforms established (1) the PLA's General Armaments Department for the purpose of identifying military needs and managing procurement for the military; and (2) a civilianized Committee on Science, Technology, and Industry for the National Defense (COSTIND) to supervise the defense industry and coordinate procurement.[80] Additional reforms in July 1999 reorganized China's core defense industries into ten major enterprise groups. This reform created the China Aviation Industries Corporation I and II as the two holding companies that manage more than 100 industrial enterprises related to the aviation industry.[81] The goal is to raise quality and reduce costs by introducing increased competition into the defense industry, but the extent to which the reforms have succeeded is unclear.

These reforms hold the potential to improve significantly the capabilities of Chinese domestically-produced aircraft, but the pace and extent of the improvement is difficult to assess. At the moment, China still has only a limited ability to produce modern combat aircraft. China's most advanced indigenously-built fighter, the J-10, relies heavily on Israeli technology and design assistance to attain performance roughly equal to that of the F-16, a fighter introduced into the USAF in 1976. The JH-7, the most advanced aircraft China's defense industry has produced on its own, is equivalent to attack aircraft that entered service in the United States, Russia, and Europe in the 1960s, 1970s, and 1980s.[82] Advanced fighters often have extended development cycles (the U.S. F-22 took 20 years to develop[83]), but China's 20-30 year development cycle is producing aircraft that are well behind state-of-the-art fighters. The defense industry reforms described above, combined with efforts to provide increased and uninterrupted funding for aircraft R&D programs, will likely help speed up this lengthy development cycle. However, greater reliance on indigenous aircraft development programs rather than acquisitions of foreign aircraft would greatly slow the pace of PLAAF modernization.

PLAAF efforts to acquire modern combat aircraft have relied heavily on purchases of Russian aircraft, technical assistance from Russia and Israel in developing Chinese fighters, and imported components from Russia and European countries. Chinese leaders worry about excessive dependence on foreign sources for several reasons. First, changes in political relations could cut off the availability of technologies, components, spare parts, or maintenance support. China experienced this in 1960, when the Sino-Soviet split led the Soviet Union to withdraw technical assistance to Chinese military industries. A less severe version occurred in 1989, when the United States and Western European countries imposed an arms embargo after the Tiananmen massacre. This cut China off from U.S. assistance in modernizing its F-8-II fighter and limited China's access to spare parts for other weapons systems, such as the *Black Hawk* helicopters it had purchased from the United States. More recently, the United States has pressured Israel to cancel a contract to build an AWACS system for China and urged the European Union (EU) to retain its arms embargo against China.[84] Even if technology and maintenance support remain available, the need to send weapons systems to other countries for maintenance may limit their availability for military missions. China's Su-27 and Il-76 aircraft still depend heavily on Russia for spare parts and periodic overhauls.

A second concern is that the aircraft, avionics, ordnance, and technologies foreign countries are willing to make available to China may lag behind state-of-the-art systems. Russia has been willing to transfer a great deal of aviation technology to China, but other suppliers have been more cautious in what they are willing to sell. Since the available technologies were more advanced than what China could produce, China has pursued these deals. But as China's military technology base improves, limitations on the availability of advanced technologies will likely be a greater concern. Moreover, because Russia is making only limited investments in advanced military R&D, at some point Russian technology will be of less value in helping China close the gap between its aircraft and avionics and those of the United States and Western Europe.[85]

A related concern is that excessive dependence on imported components and technologies may permanently lock China into an inferior position. Without independent R&D capabilities, Chinese manufacturers will be unable to make advances beyond the technologies they are able to import. Indigenous innovation is the ultimate goal of China's R&D effort. While imported technologies are important in allowing China to catch up, continued progress will ultimately require the ability to conduct independent design and development work. China's leaders are committed to self-reliance in arms production, even though this is not practical at the present time. This point is evident in recent Chinese press reports about the J-10 and JH-7 aircraft stressing that China owns the intellectual property rights for these aircraft as a key achievement.[86]

Foreign procurement offers the ability to build an advanced, highly capable Air Force more quickly than waiting for Chinese manufacturers to move up the learning curve and produce more advanced aircraft themselves. Although the PLAAF's initial purchase of the Su-27 was intended to gain experience operating and maintaining a modern fighter, the subsequent purchase of additional Su-27s, the co-production of the F-11, and the purchase of Su-30s reflect an outright effort to build a significant combat capability based on Russian fighters.

How the foreign-domestic trade-off affects future PLAAF force structure will depend on continued availability of foreign aircraft that are more advanced than Chinese designs, the degree to which Chinese manufacturers are able to close the quality gap, the perceived urgency to build combat capability, and the availability of foreign engines, avionics, and ordnance that can be incorporated into Chinese aircraft. At present, China is primarily pursuing both foreign purchases and Chinese platforms incorporating foreign components, but the leadership's intention is for Chinese companies to master military aviation technologies and reduce China's dependence on foreign suppliers. The extent to which this goal can be attained remains to be seen.

High-technology vs. Cost.

A third issue is how the PLAAF will manage the balance between expensive, high-technology aircraft and less expensive, less capable aircraft that can be procured in greater numbers. This dichotomy overlaps with the foreign-domestic tradeoff discussed above, since foreign aircraft are generally both more expensive and more capable than aircraft produced in China. (However, domestic systems may actually be less economical if all the relevant R&D costs are included.) A senior Chinese Air Force officer indicated that PLAAF procurement costs for "foreign aircraft" were 10 times higher than for "domestic aircraft."[87]

All air forces confront the challenge of optimizing the mix of expensive high-performance aircraft and cheaper less-capable aircraft in their inventory. This challenge is compounded by continuing improvements in the technology employed in advanced military aircraft. Marginal improvements in capabilities are typically accompanied by major increases in costs. More advanced aircraft also usually have higher maintenance and operating costs. The result is a general pattern wherein the quality and capabilities of the aircraft in an air force's inventory increase, but the total number of combat aircraft declines. This dilemma is especially acute for countries such as China and India with large territories to defend and the consequent need for large air forces. As older aircraft that were produced in large quantities (such as the Chinese J-6 fighters and Indian MiG-21s) are retired, they are usually replaced with more capable aircraft but in smaller quantities. The net result is a significant decrease in the size of the inventory.

Table 1 compares changes in the number of combat aircraft in the PLAAF, the Indian Air Force (IAF), and the U.S. Air Force (USAF) from 1995-96 to 2006. The Chinese and U.S. Air Forces both shrank significantly. The number of IAF combat aircraft shrank only slightly (partly because of previous down-sizing in the early 1990s).

	China		India		United States	
	1995-1996	2006	1995-1996	2006	1995-1996	2006
Bombers	420	222	0	0	201	177
Fighters/ Ground Attack	4,426	2,421	815	763	2,501	1,565
Total Combat Aircraft	4,846	2,643	815	763	2,702	1,742

Source: *The Military Balance*, 1995-96 and 2006 editions, International Institute for Strategic Studies.

Table 1: Comparison of PLAAF, IAF, and USAF Combat Aircraft Quantities, 1995-96 to 2006.

Most air forces have a bias for more capable, high-technology aircraft, but this preference is usually tempered by budget realities. Although the general trend is for air forces to get smaller as they upgrade their fleet to more capable (and more expensive) aircraft, the amount of the size reduction depends on the costs of the replacement aircraft, which will in turn depend on the particular mix of high/low capability aircraft. The USAF has historically sought a mix of limited quantities of high-performance fighters and larger quantities of a less expensive fighter. This amounts to a compromise between having a few maximum performance aircraft for air superiority against an adversary's best fighters while also providing enough aircraft to carry out a range of Air Force missions. In the 1970s, the USAF implemented this high/low mix by procuring the F-15 *Eagle*, a more expensive fighter interceptor, and the F-16 *Falcon*, a medium performance fighter.[88] The USAF is currently deciding on the proper mix of F-22 *Raptor* interceptors and less-expensive F-35 joint strike fighters.[89]

Like other air forces, the PLAAF faces the dilemma of optimizing the mix of expensive high performance aircraft and less-expensive, less-capable planes. Historically, the Chinese military has sought to compensate for the inferior quality of its weapons systems by procuring large quantities. A broad effort is now underway across the Chinese defense industry and the PLA to improve the quality of Chinese weapons. In 1995, the CMC called for a shift "from an army based on quantity to an army based on quality."[90] For the Air Force, this requirement will be met by procurement of aircraft from two sources: purchases of advanced Russian fighters, transports, tankers, and (possibly) bombers and AEW/AWACS systems; and procurement of aircraft developed by the Chinese defense industry, such as the J-10, JH-7, and FC-1.

Within the fighter force, there will be two key decisions. The most important decision affecting PLAAF force structure is whether the PLAAF will procure less capable aircraft such as the FC-1 fighter and the JH-7 attack fighter in large numbers. The PLAAF has been reluctant to purchase either aircraft due to concerns about limited performance such as low speed and limited maneuverability. The JH-7 is currently in service in PLAN aviation units and one PLAAF regiment, however.[91] If the PLAAF decides (or is forced by civilian leaders) to procure the FC-1 or JH-7 in quantity, this would push the PLAAF in the direction of a larger, less capable fleet of aircraft.

The second decision will involve the mix of advanced Russian aircraft and less-capable Chinese aircraft. Will the future PLAAF consist largely of Su-27s and Su-30s, or will the J-10 become the PLAAF's mainstream advanced multirole fighter? At this point, it is difficult to know the exact cost and performance differentials between the Su-27/Su-30 and the J-10, but the J-10 will likely be less expensive and somewhat less capable. This decision will shape the composition of the high-end of PLAAF fighter capabilities. If rumors of a Chinese effort to develop an advanced fighter with stealth capabilities are correct, this would add another wrinkle to decisions about the proper mix of fighters. The PLAAF will confront similar procurement decisions with regard to transports (Russian Il-76 versus Chinese Y-8 and Y-9 transports), tankers (Il-78/*Midas* versus modified H-6 tankers), and AEW/AWACS (Russian A-50/*Mainstay* versus domestic AEW/AWACS aircraft). The tanker and AEW/AWACS decisions will likely be influenced by decisions concerning the fighter fleet.

Support Systems vs. Combat Aircraft.

A fourth issue is the balance between investments in support systems that serve as force multipliers, on one hand, and procurement of combat aircraft, on the other. The 2004 *Defense White Paper* declares the PLA's objective as winning "local wars under the conditions of informationalization." The *Defense White Paper* highlights the importance of informationalization as the correct orientation and strategic focus of PLA modernization. It notes that modernization will strengthen military information systems and speed up informationalization of main battle weapon systems, making full use of various information resources and focusing on increasing system interoperability and information-sharing capability. The *Defense White Paper* also highlights the need to accelerate the modernization of weaponry and equipment and to improve joint operational capabilities.[92]

Informationalization involves improving collection of intelligence about the position of enemy forces from a variety of sources, sharing that information widely among command elements and operational units, and controllilng forces in a networked manner to make optimal use of information for tactical purposes. As applied to the Air Force, this implies a greater investment in intelligence collection, command and control, and support systems that can help ensure that ground-based and air assets function in an integrated and more effective manner. This suggests that the PLAAF will increase its investments in support systems such as tankers, transports, AEW/AWACS, intelligence collection platforms, and electronic warfare systems. This investment, plus the communications capabilities needed to tie the sensors, command and control system, and weapons systems together, would compete with procurement of additional combat aircraft.

The PLA desire to proceed in the direction of informationalization is clear, having been reinforced by the 2006 *Defense White Paper*. The principal question is whether hardware capabilities, operational concepts, and training will support this shift in emphasis, especially under combat conditions. With respect to the Air Force, questions about informationalization center upon China's ability to build effective AEW/AWACS, intelligence collection, and electronic warfare systems and to integrate these systems effectively with aircraft and ground-based air defenses. Integration of avionics, information display, communications, and weapons systems is likely to be a particularly challenging task, especially given the PLAAF's mix of Russian and Chinese aircraft (and the fact that some Chinese aircraft incorporate foreign avionics). The degree of difficulty will be further heightened if the goal of improving joint operations capability requires integration across the services. The crash of a prototype during flight testing in June 2006 highlights the obstacles China will face in fielding an indigenous AEW/AWACS capability.[93]

It is clear that the PLAAF intends to develop and procure a range of support aircraft and capabilities to improve the combat effectiveness of its aircraft and air defenses. The chief uncertainties concern China's ability to procure foreign platforms and technology, the capabilities of indigenously-developed systems, the capacity to integrate sensors, weapons, and C4I systems using secure communications, and whether the PLAAF will develop the operational concepts and training necessary to use these new capabilities successfully in multiplying the effectiveness of combat forces.

Assessments of International Environment and Budget Constraints.

The decisions of China's military leaders regarding the four trade-offs discussed above will have a significant impact on the future force structure of the PLAAF. How China's top civilian leaders assess the overall international security environment and the resources they are willing to devote to military modernization will shape the overall budget and policy environment in which military modernization takes place.

A leadership reassessment of China's security environment might change the relative priority and resources devoted to military modernization. For example, the need to prepare for a possible conflict over Taiwan independence has been a key justification for increased military spending in recent years. If the Taiwan situation appears more stable and the international environment relatively benign, the need for increased military spending may be less persuasive to civilian leaders focused on domestic priorities such as promoting development and reducing inequality between urban and rural China. There are some indications that PLA strategists are beginning to look beyond the Taiwan issue and articulate the rationale for building a Chinese military capable of global operations in defense of China's sea lines of communication and expanding global interests. It is unclear how persuasive this rationale will be to Chinese leaders. Conversely, a downturn in Sino-U.S. relations could reinforce Chinese concerns about potential U.S. threats, thus bringing increased emphasis on military modernization. Concerns about Japan's emergence as a "normal country" with fewer restraints on its military could also heighten Chinese threat perceptions.

The civilian leadership's assessment of the security environment will have a major impact on the resources available for military modernization, but other factors will also influence military budgets. These include China's ability to continue rapid economic growth, whether China avoids a serious economic or financial crisis, the relative weight placed on military expenditures compared to other leadership priorities, and additional costs for social spending as China's population ages.[94] Barring an economic collapse, Air Force budgets are likely to increase even if China's recent pace of double-digit increases in real defense spending slows. Nevertheless, budgets will still be a constraint on Air Force modernization, forcing China's leaders to grapple with the trade-offs described above.

ALTERNATIVE MODERNIZATION PATHS

The most likely path for PLAAF force modernization is to continue present efforts to build the Air Force using a variety of means, including continued procurement of advanced aircraft from Russia; continued domestic efforts to design and produce advanced aircraft; and incorporation of imported engines, avionics, and munitions into Chinese aircraft designs. The Chinese preference is to shift gradually away from foreign procurement and use of foreign components as the Chinese aviation industry's capabilities to produce advanced aircraft and components improves. Three variations on this force modernization path illustrate alternative possibilities.

Efforts to maximize capability quickly. This path would likely flow from a leadership assessment that China's security environment was deteriorating and that more resources needed to be devoted to accelerate military modernization. The likely consequences would be increased procurement of foreign aircraft; redoubled efforts to acquire foreign AWACS, tanker, and transport aircraft; and accelerated production of both high and medium-quality indigenous aircraft. Efforts to replace imported components with Chinese-produced equivalents would be deemphasized in favor of buying increased stocks of critical foreign components. Given procurement and production lead times, this scenario would require at least 2 to 3 years to produce substantial gains in capability. The PLAAF's ability to absorb and employ additional aircraft would be constrained by its capacity to train pilots and maintenance personnel and the time needed to upgrade units to operate more advanced aircraft.

A high-technology Air Force. This path would emphasize advanced aircraft, support systems, and the C4I capabilities to integrate aircraft into informationalized operations. The PLAAF would focus procurement on Russian fighters and possibly the J-10 fighter, while procuring few if any FC-1 or JH-

7 aircraft. China might also explore the possibility of codevelopment of new advanced aircraft with Russian partners as a means of upgrading its aircraft inventory and improving the R&D capability of its defense industry. The PLAAF would retire older aircraft as more capable replacements entered the force in order to focus its resources on advanced aircraft. Investment in support aircraft such as AEW/AWACS and tankers would be a priority, with renewed efforts to procure foreign platforms and technology combined with intensified indigenous development efforts. The PLAAF would resist efforts to replace foreign engines and avionics with Chinese-produced equivalents that did not deliver the same performance or reliability.

A domestically-produced Air Force. This path would emphasize indigenous efforts to produce advanced weapons and seek to avoid reliance on foreign suppliers. It implies less emphasis on procurement of Russian aircraft, increased purchases of J-10 fighters (and possibly FC-1 and JH-7 aircraft), and intensified efforts to replace foreign engines and avionics with indigenous equivalents. Development of force multipliers such as AEW/AWACS, tankers, and transports would depend on how quickly the Chinese defense industry's R&D efforts progressed. (A "spiral" development model where initial capabilities were deployed and then improved over time would be a possibility.) This approach implies a more relaxed pace of modernization, but would lay a firmer foundation for future Chinese efforts to develop advanced aircraft. This path would likely result from leadership confidence that China's security environment was improving and that a military conflict was unlikely in the medium term.

CONCLUSION

The PLAAF hopes to build an Air Force consisting primarily of advanced aircraft integrated with effective support systems, with the capability of conducting offensive strike missions against ground and naval targets and effective air defense against advanced militaries. This Air Force would be capable of conducting and supporting joint operations, and would rely heavily on networking and informationalization to employ air power effectively. These aspirations will likely be constrained by the current technological limitations of the Chinese aviation industry and by the resources made available to support defense modernization. One of the biggest uncertainties is whether the PLAAF will choose (or be forced) to procure large quantities of less capable aircraft as a means of developing and strengthening the Chinese aircraft industry and/or supporting the leadership's goal of indigenous innovation and self-reliance. PLAAF decisions about how many J-10, FC-1, and JH-7 fighters to procure will be a key indicator. In theory, the defense reorganization of 1998 that established the General Armaments Department should give Air Force requirements greater weight in procurement decisions, but this may not be true in practice.

Chinese leadership perceptions of the international threat environment (to include assessments of the likelihood of a crisis over Taiwan or a conflict with the United States) and PLAAF budget allocations will have a significant influence on the overall size of the future PLA and the speed with which modernization takes place. PLAAF force structure will also be shaped by decisions about the division of labor on air defense and conventional strike missions and about the proper proportions between foreign and domestic production, high-technology and lower-cost systems, and combat aircraft and support aircraft. Regardless of the specific decisions on these issues, it is already clear that the future PLAAF will be a significantly smaller, but more capable Air Force.

APPENDIX I

PLAAF AIRCRAFT

FIGHTERS AND GROUND ATTACK

J-6

The J-6 is now in full retirement and no longer deployed.

J-7

The J-7, the Chinese version of the Mig-21, is a fighter interceptor for air defense and attack. The J-7 is tasked with providing local air defense and tactical air superiority. The active J-7 fleet includes upgraded J-7II/J-7IIA, J-7IIM and J-7C/D models. The upgrades feature advanced avionics, better radar, and better engines than the originals, as well as helmet-sighted Air-to-Air Missiles (AAM).[95] There are roughly 750 J-7s in active service, deployed in all seven Chinese military regions.[96]

J-8

The J-8 is a single-seat twin-engine second generation air superiority fighter with ground attack capabilities. The PLAAF flies upgraded J-8I, J-8II versions of the J-8. The J-8 is armed with a 23 mm Type 23-3 twin-barrel cannon. It carries the PL-2B and PL-7 AAM. It is also armed with 57 mm Type 57-2 unguided air-to-air rockets, launchers for 90 mm air-to-surface rockets and bombs.[97] The upgraded J-8E has been equipped with new radar and defensive electronics while some J-8Ds have been equipped with aerial refueling capabilities. It is possible that some J-8s have been armed with helmet-guided AAM, either the PL-8 or the PL-9.[98] Currently there are 245 J-8s deployed by the PLAAF.[99] The J-8 is likely deployed in the Nanjing, Beijing, Lanzhou, and Guangzhou MRs.[100]

J-10

The J-10 is a multirole, light-weight air superiority fighter and attack aircraft. There is also the J-10B, a two seat trainer with potential attack missions, as well as the Super-10, an advanced J-10 model possibly deployed in 2006. The J-10 is powered by the Al-31 FN turbofan engine, but may eventually employ a domestically developed turbofan engine. The J-10 is armed with PL-8 AAMs, with capabilities to carry the more advanced PL-11 and PL-12.

The J-10 also has the potential to carry the Vympel R-73 and R-77 AAMs, as well as C-801 or C-802 ASMs, YJ-8K (antiship) or YJ-9 (antiradiation) missiles. The J-10 can reach Mach 2.0, with a maximum ceiling of 18,000 meters and a combat radius of 1,100 kilometers, making it a viable aircraft for most potential conflicts China may face. The J-10 also has aerial refueling capabilities.[101] The J-10 entered large-scale production in 2006; 62 aircraft are already deployed.[102]

J-11/SU-27SK

The Su-27SK is a single-seat fighter, while the Su-27 UBK is a twin-seat trainer. The J-11 is the Chinese co-production model of the Su-27. Su-27SK is powered by the Russian Twin Al-31 FP thrust-vectoring aero-engine. Upgraded Su-27s are suspected to be armed with AA-10 and AA-11 AAMs.[103] The KnAAPO website also claims that the Su-27SK is capable of carrying six R-27R1 medium range AAMs, six RW-AE medium-range AAM, six R-73E short-range AAMs, and a 30mm automatic single-barrel cannon.[104] There are roughly 116 Su-27/J-11s deployed in the PLAAF. The J-11 and the Su-27s are now deployed in the Nanjing, Guangzhou, Chengdu, Shenyang, Jinan, and Beijing MRs.[105]

Su-30MKK

The Su-3MKK is a multirole, twin-seat fighter. It is powered by two Al-31F aero-engines and is equipped with in-flight refueling capabilities that extend its range.[106] The weapons package on the Su-30MKK consists of 30-mm cannon, missiles, and bombs mounted on 12 hard-points under the wing and fuselage. Air-to-air missiles include the R-27R1 (R-27ER1) guided medium-range missiles with semiactive seekers, 2 R-27T1 medium-range missiles with heat seekers, the RVV-AE medium range missiles with active radar seekers, the R-73E dogfight missiles with heat seekers. The air-to-surface weapons are comprised of up to 6 Kh-31P anti-radiation missiles with passive radar seekers, up to 6 Kh-31A antiship missiles with active radar seekers, up to 6Kh-29T (Kh-29TE) short-range missiles with TV-seekers or Kh-29L with semi-active laser seekers, two Kh59ME medium-range missiles with TV-commanded seekers, and up to three KAB-500Kr or one KAB-1500Kr guided bombs with TV-correlated seekers.[107] The PLAAF has roughly 73 Su-30s in their force, deployed in the Nanjing and Guangzhou MRs.[108]

Q-5

The Q-5 is a single seat, supersonic attack aircraft.[109] Recently, the upgraded Q-5D was fitted to carry and deliver Laser Guided bombs.[110] It is also armed with one 23mm canon with 100 rounds and can carry a 250 kg bomb in the fuselage as well as Norinco 90-1 rockets or four 130-1 rockets. It is also capable of carrying the C-801 antiship missile, the PL-2, PL-2B, PL-7, AIM-9 Sidewinder, or the R-550 Magic missile systems.[111] New trainer versions of the Q-5 are being powered by the Al-222-25F turbofan, suggesting that the PLAAF may be upgrading the Q-5 program. There are more than 400 Q-5s still deployed, mainly in the Shenyang, Jinan, Beijing, and Nanjing MRs.[112]

Xiaolong/FC-1

The Xiaolong/FC-1 (previously known as the Super-7; sometimes referred to as the FC-4) is a multirole fighter bomber and the most advanced fully indigenous fighter in domestic development in China. The FC-1 flew its first test flight in Chengdu on April 28, 2006.[113] It is the most advanced aircraft the PLAAF is domestically producing and is the product of a joint Chinese-Pakistani cooperative. It is powered by the WP-13F turbojet engine.[114] The Xiaolong has intercept and ground attack capabilities. It is not yet deployed into the PLAAF force.

JH-7

Initial aircraft have gone primarily to the PLANAF, although the PLAAF had received models by 2004. The JH-7 has an advertised maximum payload of 14,330 pounds, carrying up to 4 C-801/802ASMs, freefall bombs, and AAMs.[115] The PLAAF operates one regiment of JH-7 aircraft.

BOMBERS

Hong-5

The H-5 bomber is a light bomber that has been in production since the 1950s. It has a relatively short range for a bomber, roughly 2,400 km. There has been a slight resurgence in production over the past several years, despite the H-5 being rather obsolete. The PLAAF has roughly 94 H-5s deployed.[116]

Hong-6

The H-6 is a medium bomber based on the Tu-16, with land attack and sea attack missions. The H-6 has a maximum range of

2,672 miles and five plus hours of flying time. The H-6 has been converted to carry the Y-63, a 200-300km range cruise missile that was derived from the C-601 family of cruise missiles. Some H-6s have been converted to HU-6 refueling platforms.[117] There are currently about 128 H-6 bombers in service throughout China.[118]

TANKERS

H-6/Il-78

China has a number of H-6 bombers converted into tankers, which are deployed in the Guangzhou MR. China signed a contract with Russia to purchase 8 Il-78M tankers in 2005, although these aircraft have not yet been delivered.[119] The Il-78 has a cruising speed of 466 mph, with a refueling speed between 267 and 366 mph. The Il-78s can refuel China's Su-30s, while the H-6s serve the J-8Ds and J-10 aircraft.[120]

TRANSPORTS

Y-8

The Y-8 is based on the Antonov An-12 and serves as China's medium-weight military transport. The Y-8 is armed with twin 23 mm cannon mounted on the tail turret. The Y-8 has a reported payload of 20,000 kg, enough for 96 troops with vehicles and weapons, or 130 passengers.[121] It is unclear how many Y-8s are currently deployed in the PLAAF.

Y-9

The Y-9 was unveiled at the 2005 Beijing Aviation Expo. According to reports it is so similar to the Y-8 that it was hardly noticeable as a new aircraft. The Y-9, like the Y-8, is a medium transport but is slightly smaller than some variants of the Y-8. It is capable of carrying 98 armed soldiers or paratroopers, as well as 72 littered patients plus 3 medical staff members, or 98 lightly wounded patients at one time.[122] A relatively new aircraft, it is not yet fully deployed with the PLAAF.

Il-76

The Il-76 is a 1970s vintage Russian military transport with a 40-ton payload.[123] The Institute of International Strategic Studies estimates that there are 13 Il-76s in service in the PLAAF, with another 38 on order. The PLAAF has various other transports deployed, with a total force of about 295 transport aircraft.[124]

AEW/AWACS

The PLAAF AEW/AWACS aircraft in development are the KJ-2000 and the KJ-2, based on the Il-76 and the Y-8X, respectively.[125] China has been making a concerted effort to develop AEW/AWACS aircraft domestically; however, there have been complications as highlighted by the crash of a prototype in June 2006.[126] It is unclear if any AEW/AWACS aircraft are currently operationally deployed in the PLAAF.

ENDNOTES - CHAPTER 8

1. The authors thank Kenneth Allen of the CNA Corporation for his many suggestions and corrections, which significantly improved the quality of this chapter. The views expressed in this chapter are those of the authors and do not reflect the official policy or position of the National Defense University, the Department of Defense, or the U.S. Government.

2. Kenneth Allen, Glen Krumel, and Jonathan Pollack, *China's Air Force Enters the 21st Century*, Santa Monica, CA; RAND, 1995, p. xiv; Mark A. Stokes, "The Chinese Joint Aerospace Campaign: Strategy, Doctrine, and Force Modernization," in James Mulvenon and David Finkelstein, eds., *China's Revolution in Doctrinal Affairs: Emerging Trends in the Operational Art of the Chinese People's Liberation Army*, Alexandria, VA: Center for Naval Analysis, 2004, *www.cna.org/documents/doctrinebook.pdf*, pp. 221-306; and Ken Allen, "PLA Air Force Organization," in James C. Mulvenon and Andrew Yang, eds., *PLA as Organization: Reference Vol. 1.0*, Santa Monica, CA: RAND, 2002, pp. 346-457.

3. While Western and Russian analysts regard the Su-27/ *Flanker* as a fourth generation fighter, the PLAAF refers to it as a third generation fighter. This chapter will adopt the Western usage, which regards the MiG-19/J-6 as a second generation fighter and the MiG-21/J-7 as a third generation fighter.

4. See interview with Rear Admiral Yang Yi in Tao Shelan, "Military Expert: In China's Peaceful Development It Is Necessary To Uphold the Dialectical Strategic Thinking of Making the Country Rich and Building Up Its Military Strength," *Zhongguo Xinwen She*, May 15, 2006, Foreign Information Broadcast Service (FBIS)-CPP20060515072002; Sun Xuefu, "Forge a Military Force Commensurate with China's International Status," *Jiefangjun Bao*, April 28, 2006, FBIS-CPP20060502502001.

5. See Phillip C. Saunders, *China's Global Activism: Strategy, Drivers, and Tools*, Washington, DC: National Defense University Press, 2006, *www.ndu.edu/inss/Occasional_Papers/OCP4.pdf*.

6. The 2006 PLA pay increase doubled pay for most officers. Author's interview, Beijing, 2006.

7. Allen, Krumel, and Pollack, p. 114.

8. China's National Defense in 2004, State Council Information Office, December 27, 2004, *www.china.org.cn/english/2004/Dec/116032.htm*.

9. China's National Defense in 2006, State Council Information Office, December 29, 2006, *http://www.china.org.cn/e-white*.

10. For details and analysis, see David M. Finkelstein, "Thinking About the PLA's 'Revolution in Doctrinal Affairs'," in James Mulvenon and David Finkelstein, eds., *China's Revolution in Doctrinal Affairs: Emerging Trends in the Operational Art of the Chinese People's Liberation Army*, Alexandria, VA: Center for Naval Analysis, 2004, *www.cna.org/documents/doctrinebook.pdf*.

11. Kevin Lanzit and Kenneth Allen, "Right-Sizing the PLA Air Force: New Operational Concepts Define a Smaller, More Capable Force," in Roy Kamphausen and Andrew Scobell, *Exploring the 'Right Size' for China's Military*.

12. A principal source for information on Air Force and joint campaigns is Wang Houqing and Zhang Xingye, eds., *Zhanyi Xue [Science of Campaigns]*, Beijing: National Defense University Publishers, May 2000, pp. 346-66, 441-458, 474-484. In addition to Lanzit and Allen, "Right-Sizing the PLA Air Force," other useful sources include Ken Allen, "PLA Air Force Mobile Offensive Operations," *Taiwan Defense Affairs*, Vol. 3, No. 2, Winter 2002/2003, p. 2.; Lieutenant Colonel Thomas R. McCabe, USAFR, "The Chinese Air Force and Air and Space Program," *Air and Space Journal*, Fall 2003, at *www.airpower.maxwell.af.mil/airchronicles/apj/apj03/fal03/mccabe.html*; and Stokes, "The Chinese Joint Aerospace Campaign."

13. This air power-centric approach might be labeled *dakongzhunzhuyi*, "Big Air Force Chauvinism."

14. Lanzit and Allen, "Right-Sizing the PLA Air Force."

15. Allen *et al.*, *China's Air Force Enters the 21st Century*, p. xiv.

16. "Wen Wei Po Quotes 'Authoritative' Sources on Decommissioning of Jian-6 Fighters," FBIS CPP20051203506007, December 3, 2005; Thomas R. McCabe, "The Chinese Air Force and Air and Space Power," *Air and Space Power Journal*, Fall 2003, p. 77.

17. Allen *et al.*, *China's Air Force Enters the 21st Century*, p. 222.

18. "Su-27s Restart Sino/Soviet Arms Trade," *Flight International*, April 3, 1991, accessed via Lexis-Nexis, *www.lexisnexis.com*.

19. "Air Force, China," *Jane's Sentinel Security Assessment—China and Northeast Asia*, November 7, 2005.

20. Bates Gill and James Mulvenon, "China and Weapons of Mass Destruction: Implications for the United States," A Conference Report, National Intelligence Council, November 5, 1999.

21. The 2004 Chinese *Defense White Paper* mentions nuclear roles for the Second Artillery and the Navy, but not for the Air Force. State Council Information Office, "China's National Defense in 2004," December 27, 2004, *www.china.org.cn/english/2004/Dec/116032.htm*.

22. K. Sengupta Prasun, "China's New Air Defence Systems Unveiled," *Chinese Military Update*, Vol. 2, No. 3, September/October 2004, p. 7.

23. "Missile Systems," *China's Aerospace and Defence Industry*, *Jane's*, December 5, 2000.

24. *Ibid.*, pp. 33-34.

25. Shirley Kan, "China's Foreign Conventional Arms Acquisitions: Background and Analysis," Congressional Research Services, Library of Congress, November 6, 2001, p. 16.

26. Stokes, "The Chinese Joint Aerospace Campaign."

27. *Dangdai Zhongquo Kongjun* [*China Today: Air Force*], Beijing, China Social Sciences Press, 1989, cited in Allen *et al.*, *China's Air Force Enters the 21st Century*, p. 130.

28. "Wen Wei Po Quotes 'Authoritative' Sources on Decommissioning of Jian-6 Fighters," Hong Kong Wen Wei Po, FBIS CPP20051203506007, December 3, 2005; "Zaijian Lao Liu," *World Outlook*, Ocotober, 2005, accessed March 11, 2006, at *www.donggang.gov.cn/public/st.asp?id=6201&va=1&s=1*.

29. Rick Kamer, "Flankers of the People's Liberation Army," *Chinese Military Update*, Vol. 1, No. 7, January 2004, pp. 5-8.

30. Vladimir Karnozov, "Sukhoy Concerned Over Sales to China, Declining Orders," Moscow Voyenno-Promyshlenny Kuryer, FBIS, November 24, 2004.

31. "Air Force, China," *Jane's Sentinel Security Assessment – China And Northeast Asia*, November 7, 2005.

32. Video on CCTV includes footage of the J-10 being refueled by an H-6 tanker. The CCTV video is available at *www.sinodefence.com/airforce/fighter/j10news.asp*, accessed February 23, 2007.

33. Zhang Jinyu and Xu Huidong, "First Chinese Air Force Unit to Be Equipped with the F-10 Fighter Is Now an Operational Unit with Real-War Capabilities," *Jiefangjun Bao*, December 30, 2006, available as Open Source Center, "PRC Air Force's Combat Capabilities Enhanced by New Jian-10 Fighter Planes," CPP20061230704001; and David Lague, "China Builds a Superpower Fighter," *International Herald Tribune*, February 8, 2006.

34. "Chengdu Aircraft Industry: CAC FC-1 Xiaolong," *Jane's All The World's Aircraft*, April 21, 2006.

35. "Advanced Fighter Jet Set for Public Debut in Zhuhai," Open Source Center, July 28, 2006.

36. See "The PLA Air Force Will Equip with Several Hundred 'Xiaolong' Fighters," *Wen Wei Po*, February 8, 2007, available in Open Source Center "WWP: Chinese Air Force to Acquire Several Hundred 'Xiaolong' Fighters," CPP20070208710004.

37. Wang Qing, "ZTS 'Roundup' Sees China Developing Third Generation PLA Fighters," Hong Kong Zhongguo Tongxun She, FBIS CPP20051216001015, December 16, 2005.

38. Evan S. Medeiros, Roger Cliff, Keith Crane, and James C. Mulvenon, *A New Direction for China's Defense Industry*, Arlington, VA: RAND, 2005, pp. 170-174; "PRC Magazine: 'Taihang' Turbofan Will End PLAAF's Reliance on Foreign Engines," *Xian Binggong Keji*, FBIS, CPP20060413318001, April 6, 2006.

39. "Moscow Plant Meets Obligations Over Jet Engine Supplies to China," *Agentstvo Voyennykh Novostey*, in Open Source Center CEP20061106950178, November 3, 2006.

40. Richard Fisher, "PLAAF Equipment Trends," in Stephen J. Flanagan and Michael E. Marti, eds., *The People's Liberation Army and China in Transition*, Washington, DC: NDU Press, 2003, pp. 149-150; "JH-7/A, FBC-1, FIGHTER-BOMBER," *www.sinodefence. com/airforce/groundattack/jh7.asp*, accessed February 23, 2007.

41. Pacific Command Commander Admiral William Fallon visited this unit in May 2006 and sat in the cockpit of an FB-7. Edward Cody, "U.S. Aims to Improve Military Ties with China," *Washington Post*, May 16, 2006, p. A14.

42. "Air Force, China" *Jane's Sentinel Security Assessment-China and Northeast Asia*, November 7, 2005, p. 3; Richard Fisher, Jr., "China's 'New' Bomber," International Assessment and Strategy Center, February 7, 2007, available at *www.strategycenter.net/ research/pubID.146/pub_detail.asp#*.

43. "*Xin Hong 6 Gua Xunhang Daodan [Zu Tu]*" ["Hanging Cruise Missiles on New H-6"], *military.china.com/zh_cn/bbs/11018 441/20051024/12778370_2.html*; "*Zhuang Xunhang Daodan Hong 6*" ["Loading Cruise Missiles Hong-6"], *military.china.com/zh_cn/bbs 2/11018521/20050219/12118043.html*; "*Hongliu Kongtou Zhongxing Xunhang Daodan Jichen Ba Jian Quan Guocheng*" ["H-6 Completely Destroys Ship Target From the Air with Cruise Missile"], *tuku. military.china.com/military/html/2006-07-14/34386.htm*. All sites accessed on March 9, 2007.

44. "Russian Aircraft to Make Appearance in Joint Military Exercise," *The People's Daily*, August 3, 2005, *english.people.com. cn/200508/03/eng20050803_200086.html*.

45. "Russia To Sell Strategic Bomber Planes To China," *Pravda*, January 17, 2005, *english.pravda.ru/print/russia/politics/7605-aviation-0*.

46. Hai Chao, "*Zhongguo ying jinyin jin e zhi zhanlue hongzhaji*" ["Does China Need Russian Strategic Bomber?"] *Shipborne Weapons*, June 6, 2005, p. 14.

47. Dan Jie and Ju Lang, "Feixiang zhongguo de 'nihuo' he 'xiong'" ["Flying towards China's 'Backfire' and 'Bear',"], October 25, 2005, *military.china.com/zh_cn/jzwq/01/11028429/20051025/1278 1919.html*.

48. "Russia To Sell Strategic Bombers To China," FBIS CPP20050305000148, March 5, 2005.3; Jie and Lang, "Fei xiang zhongguo de 'nihuo' he 'xiong'" ["Flying in the Direction of China's 'Backfire' and 'Bear'"] October 25, 2005, at *military.china. com/zh_cn/jzwq/01/11028429/20051025/12781919.html*.

49. "Zhongguo jiang huode nihuo hongzhaji?" ["Should China Acquire Backfire Bombers?"], *World Outlook* June, 2005.

50. "PLA Buys Il-78 Refueler to Solve Compatibility Issues With Su-30s," *Xian Binggong Keji*, FBIS CPP20051123318001, October 1, 2005, p. 19.

51. "Russia Scrambles to Save China Plane Contract," September 15, 2006, *www.defensenews.com/story.php?F=2107706*, accessed February 23, 2007. The authors thank Kenneth Allen for providing this article.

52. "Air Force, China," *Jane's Sentinel Security Assessment-China And Northeast Asia*, November 7, 2005, p. 7.

53. "PLA Buys Il-78 Refueler to Solve Compatibility Issues With Su-30s," FBIS CPP20051123318001, *Xian Binggong Keji*, October 1, 2005; Kenneth Allen, "The PLA Air Force: 2006-2010."

54. See Zhu Baoliu, "*Yanzhi kongzhong yujingji nan zai nali*" ["Problems in the Development of Airborne Early Warning Systems"], *Bingqi Zhishi* [*Ordnance Knowledge*], September 4, 2004, pp. 30-33. The authors thank Richard Fisher, Jr., for providing a copy of this article.

55. Luke G.S. Colton, "A Chinese AEW&C Threat? The Phalcon Case Reconsidered," *Chinese Military Update*, Vol. 1, No. 7, January 2004.

56. "PLAAF Developing KJ-2000, KJ-2 AWACS Simultaneously," *Taipei Ch'uan-Ch'iu Fang-Wei Tsa-Chih*, in FBIS, CPP20050419000171, April 1, 2005.

57. Joseph Kahn, "Crash of Chinese Surveillance Plane Hurts Efforts on Warning System," *New York Times*, June 7, 2006; Michael Sheridan, "China's Hi-Tech Military Disaster: Bid to Copy Israeli Electronics Kills Experts," *London Sunday Times*, June 11, 2006.

58. Kahn, "Crash of Chinese Surveillance Plane Hurts Efforts on Warning System"; Colton, "A Chinese AEW&C Threat?"

59. "PLAAF Developing KJ-2000, KJ-2 AWACS Simultaneously."

60. Office of Secretary of Defense, *Military Power of the People's Republic of China*, Annual Report to Congress, 2004, p. 23.

61. "Russia Scrambles to Save China Plane Contract"; Robert Hewson, "China Boosts Its Air Assets with Ilyushin Aircraft," *Jane's Defence Weekly*, September 21, 2005.

62. International Institute of Strategic Studies, *The Military Balance*, 2006 Edition, p. 268.

63. "Chinese Defence Today Details Y-8 Aircraft, Carries Photos," *Chinese Defence Today*, FBIS CPP2005124075002, December 17, 2004.

64. Xiaoming Peisen, "Article on Y-9 Transport Plane Features Improvements Over Other PRC Aircraft," *Xian Binggong Keji*, FBIS CPP200602227325002, February 1, 2006.

65. It is unclear whether China has established a program akin to the U.S. Civil Reserve Air Fleet program to use civil aviation assets in a military contingency. Some Taiwan sources claim that such plans exist. "PRC to Use Civil Planes to Transport Troops to Attack Taiwan," Open Source Center CPP20000923000035, September 23, 2000.

66. Office of the Secretary of Defense, *Military Power of the People's Republic of China*, Annual Report to Congress, 2006, available at *www.defenselink.mil/pubs/pdfs/China%20Report%202006.pdf*.

67. Office of the Secretary of Defense, *Military Power of the People's Republic of China*, Annual Report to Congress, 2006, p. 18, available at *www.defenselink.mil/pubs/pdfs/China%20Report%202006.pdf*.

68. Medeiros *et al.*, *A New Direction for China's Defense Industry*, pp. 86-91.

69. Office of the Secretary of Defense, *Military Power of the People's Republic of China*, 2006, pp. 30-31.

70. "Missile Systems," *China's Aerospace and Defence Industry, Jane's Defense Weekly*, December 5, 2000; Yihong Zhang, "Upgraded Chinese KS-1 SAM revealed," *Jane's Defense Weekly*, January 9, 2001; "CNPMIEC KS-1/KS-1A low- to high-altitude surface-to-air missile system," *Jane's Land-Based Air Defence*, July 26, 2006.

71. K. Sengupta Prasun, "China's New Air Defence Systems Unveiled," *Chinese Military Update*, Vol. 2, No. 3, September/October 2004, pp. 7-8; Medeiros *et al.*, *A New Direction for China's Defense Industry*, pp. 86-91.

72. James Mulvenon and David Finkelstein, eds., *China's Revolution in Doctrinal Affairs: Emerging Trends in the Operational Art of the Chinese People's Liberation Army*, Alexandria, VA: Center for Naval Analysis, 2004, *www.cna.org/documents/doctrinebook.pdf*.

73. Ken Allen, "Reforms in the PLA Air Force," *China Brief*, Vol. 5, No. 15, July 5, 2005, *www.jamestown.org/publications_details.php?volume_id=408&issue_id=3390&article_id=2369972*.

74. For information on PLAN aviation, see Rick Kamer, "The China Naval Air Force's Re-Emergence," *Chinese Military Update*, Vol. 2, No. 3, September/October 2004, pp. 4-5.

75. See Bernard D. Cole, *The Great Wall at Sea: China's Navy Enters the Twenty-First Century*, Annapolis: Naval Institute Press, 2001, pp. 85-86.

76. Author's interviews with Taiwan Air Force officers, November 2005.

77. State Council Information Office, "China's National Defense in 2004," December 27, 2004, *www.china.org.cn/english/2004/Dec/116032.htm*.

78. Phillip C. Saunders and Jing-dong Yuan, "China's Strategic Force Modernization," in Albert Willner and Paul Bolt, eds., *China's Nuclear Future*, Boulder, CO: Lynn Rienner, 2006, pp. 79-118.

79. Ken Allen, "The PLA Air Force: An Overview," *Chinese Military Update*, Vol. 1, No. 4, September 2003, pp. 1-5.

80. These reforms are summarized in Medeiros et al., *A New Direction for China's Defense Industry*, pp. 28-47.

81. *Ibid.*, p. 157.

82. *Ibid.*, p. 200.

83. "Chronology of the F-22 Program," *www.f22-raptor.com/about/chronology.html*, accessed February 26, 2007.

84. Shambaugh, *Modernizing China's Military*, p. 265; Colton, "A Chinese AEW&C Threat? The Phalcon Case Reconsidered."

85. Medeiros et al., *A New Direction for China's Defense Industry*, p. 202.

86. See Wang Jianjun, "F-10 Displays its Military Might," *Liaowang*, January 8, 2007, available as "PRC Aviation Expert Reviews Jian-10 Fighter Aircraft's Successful Development," Open Source Center CPP20070111710005.

87. Author's interview, Beijing, December 2006.

88. This clear division of labor between the F-15 and F-16 was subsequently muddied by avionics improvements which gave the F-16 ground attack capabilities and the development of the F-15E Strike Eagle variant which had a ground attack mission.

89. See Joe Yoon, "F-22, F-35, & the Hi-Lo Mix," *www.aerospaceweb.org/question/planes/q0216.shtml*, accessed September 26, 2006.

90. Cited in David M. Finkelstein, "Thinking About the PLA's 'Revolution in Doctrinal Affairs'," in James Mulvenon and David Finkelstein, eds., *China's Revolution in Doctrinal Affairs: Emerging Trends in the Operational Art of the Chinese People's Liberation Army*, Alexandria, VA: Center for Naval Analysis, 2004, *www.cna.org/documents/doctrinebook.pdf*.

91. Richard Fisher, "PLAAF Equipment Trends," International Assessment and Strategy Center, October 30, 2001, *www.strategycenter.net/research/pubID.8/pub_detail.asp*.

92. State Council Information Office, "China's National Defense in 2004," December 27, 2004, www.china.org.cn/english/2004/Dec/116032.htm.

93. Kahn, "Crash of Chinese Surveillance Plane Hurts Efforts on Warning System."

94. For a useful discussion, see Keith Crane, Roger Cliff, Evan Medeiros, James Mulvenon, and William Overholt, *Modernizing China's Military: Opportunities and Constraints*, Santa Monica: RAND, 2005.

95. "Air Force, China," *Jane's Sentinel Security Assessment-China and Northeast Asia*, November 7, 2005, p. 4.

96. International Institute for Strategic Studies (IISS), *The Military Balance*, 2006 edition, p. 268; Sid Treventhan, "PLAAF and Civil Aviation Table of Organization and Equipment," Anchorage, Alaska, 2000.

97. "Shenyang Aircraft Corporation J-8I and J-8II," *Jane's*.

98. "Air Force," *Jane's Sentinel Security Assessment — China and Northeast Asia*, p. 5.

99. IISS, *The Military Balance*, 2006 edition, p. 268.

100. Sid Treventhan, "PLAAF and Civil Aviation Table of Organization and Equipment," Anchorage, Alaska, 2000.

101. "Chengdu Aircraft Industry: CAC J-10," *Jane's All the World's Aircraft*, April 21, 2004.

102. IISS, *The Military Balance*, 2006 edition, p. 268.

103. "Aircraft," *China's Aerospace Industry-The Industry and its Products Assessed, Jane's*, April 1997.

104. KnAAPO website.

105. Kamer, "Flankers of the People's Liberation Army," pp. 5-8.

106. "Air Force, China," *Jane's Sentinel Security Assessment-China and Northeast Asia*, p. 5.

107. KnAAPO website

108. Kamer, "Flankers of the People's Liberation Army," pp. 5-8.

109. "Aircraft and Propulsion Systems," *China's Aerospace and Defence Industry-December 2000*, December 5, 2000, p. 27.

110. "Air Force, China," *Jane's Sentinel Security Assessment-China and Northeast Asia*, p. 6.

111. "NAMC, Nanchang Aircraft Manufacturing Company," *Jane's Aircraft Upgrades*, February 10, 2004.

112. IISS, *The Military Balance*, p. 268; Sid Treventhan, "PLAAF and Civil Aviation Table of Organization and Equipment."

113. Shi Jiangtao, "Fighter Jet Puts China on Target to Export Advanced Military Technology," *South China Morning Post*, April 29, 2006.

114. Richard Fisher, Jr., "Report on the 5th Airshow China, Zhuahai, PRC, November 1-7, 2004," *Arms Show Reports*, December 13, 2004, p. 5, *www.strategycenter.net/printVersion/print_pub.asp?pubID=54*.

115. Richard D. Fisher, Jr., "PLA Air Force Equipment Trends."

116. IISS, *The Military Balance*, p. 268.

117. Richard Fisher, Jr., "Report on the 5th Airshow China, Zhuahai, PRC, November 1-7, 2004."

118. "Air Force, China," *Jane's Sentinel Security Assessment-China and Northeast Asia*, p. 6.

119. *Ibid.*, p. 7.

120. "Illyushin, Aviatsionnyi Kompleks Ilmieni S V Ilyushina" ["Aviation Complex Named After S. V. Illyushin"]," *Jane's Aircraft Upgrades*, July 12, 2005.

121. "Chinese Defense Today Details Y-8 Aircraft, Carries Photos," *Edinburgh Chinese Defence Today*, FBIS CPP20051214075002, December 17, 2004.

122. "Article on Y-9 Transport Plane Features Improvements Over Other PRC Aircraft," *Xian Binggong Keji*, in FBIS CPP20060227325002, February 1, 2006.

123. Fisher, "PLA Air Force Equipment Trends," p. 162.

124. IISS, *The Military Balance*, p. 268.

125. "PLAAF Developing KJ-2000, KJ-2 AWACS Simultaneously," p. 116.

126. Kahn, "Crash of Chinese Surveillance Plane Hurts Effort on Warning System."

CHAPTER 9

RIGHT-SIZING THE PLA AIR FORCE: NEW OPERATIONAL CONCEPTS DEFINE A SMALLER, MORE CAPABLE FORCE

Kevin M. Lanzit and Kenneth Allen

> The People's Liberation Army (PLA) is stepping up preparations for military struggle. To meet the requirements of informationalized air operations, the Air Force has gradually shifted from the mission of territorial air defense to that of both offensive and defensive operations. Emphasis is placed on the development of new fighters, air defense, and antimissile weapons, and the means of information operations and automated command systems. Combined arms and multirole air combat training is intensified to improve the capabilities in operations such as air strikes, air defense, information countermeasures, early warning and reconnaissance, strategic mobility, and integrated support. Efforts are being made to build a defensive Air Force, which is appropriate in size, sound in organization and structure, advanced in weaponry and equipment, and possessed of integrated systems and a complete array of information support and operational means.
>
> *China's National Defense*: 2004[1]

Introduction.

The PLA Air Force (PLAAF) is in the midst of a dramatic transformation aimed at transitioning from a benign defensive force to one that incorporates modern defenses and robust offensive strike capabilities. During the next decade, China's Air Force will gain significant additional capabilities in a number of important mission areas.[2] These will come as a result of several programs which are already in, or soon will be, an advanced state of development. In fact, the introduction of new air- and ground-based weapons has already led to major advances in all-weather defensive and, for the first time, offensive operations.

Anticipated future enhancements in command, control, communications, computers, intelligence, surveillance, and reconnaissance (C4ISR) will enable China to significantly improve its ability to perform long-range strikes using stand-off, precision-guided munitions. The emerging capacity of indigenous Chinese defense plants to design and build their own complex weapons can be expected to further accelerate deliveries of high-tech hardware into operational units and wean the PLAAF from its dependence on Russian systems.

In charting the PLAAF's progress, it is tempting to focus on the hardware. However, this would be a mistake since the PLAAF's further advancement will hinge on a broad range of long overdue doctrinal and institutional changes that are now underway. Beginning in the 1990s, the PLAAF embarked on an expansive program of reforms that targeted doctrine, leadership, force structure, organizational structure, and officer and enlisted education, and training.[3] New mission requirements and an emphasis on joint operations are forcing military strategists to rethink old concepts of air doctrine. Force modernization and the drive for joint capabilities have imposed new challenges on Air Force leaders and led to substantial restructuring of command elements. Additionally, the introduction of highly advanced weapons has created the need for the PLAAF to revamp its education and training programs. These and other "software" developments will play a key role in determining the pace and scope of the PLAAF's further modernization and provide important clues about how the force is being "right-sized" for its new missions.

Evolving PLAAF Mission Requirements.

China's National Defense in 2006 states, "The Air Force aims to speed up its transition from territorial air defense to simultaneous offensive and defensive operations. It also aims to increase its capabilities in the areas of air strike, air and missile defense, early warning and reconnaissance, and strategic projection."[4] To "meet the requirements of informationalized air operations," the PLAAF is in the process of a long-term transformation from a territorial air defense force to a modern force capable of conducting short-duration, high-intensity offensive operations against high-tech adversaries.[5] This new orientation for the Air Force is part of a broader Chinese military doctrine that emphasizes mobility, speed, and long-range attack, plus synchronized combined arms and joint operations through the full spectrum of air, land, sea, space, and electromagnetic battlespace, all while relying heavily upon extremely lethal, high-technology weapons.

What is new and strikingly different is the PLAAF's focus on its offensive capacity. The pursuit of a robust set of offensive capabilities became an imperative for the PLA after the 1995-96 Taiwan Strait crisis exposed operational deficiencies and the limited range of military options that could be executed against Taiwan. Since that time, the PLA has striven to develop the capability to carry out a variety of military operations against Taiwan—air and missile attacks, a naval blockade, or even an outright invasion of the island—to block a move towards independence by Taiwan. These capabilities are also intended to deter, delay, and complicate U.S. efforts to intervene on behalf of Taiwan. The successful execution of such military actions will depend upon a broad range of advanced air operations—both offensive and defensive.

Modernization Leads To a Smaller, More Capable Force.

Following a worldwide trend toward smaller, more capable air forces, the PLAAF is downsizing and reshaping the force structure to perform a variety of new support tasks that are required to conduct both defensive and offensive air operations. Since the PLAAF was founded, territorial air defense has remained the highest mission priority for China's Air Force, with successive commanders allocating force structure accordingly. To detect an air attack and direct air defense operations, the PLAAF developed an extensive network of ground-based air warning radars and air defense operations centers. To ward off and defend against attacks, the PLAAF deployed and has maintained fixed antiaircraft artillery (AAA) and surface-to-air missile (SAM) batteries in the vicinity of most major population centers and key installations. Ground-based point defense weapons were augmented by a vast fleet of air interceptors to rove the skies and fill the voids between potential targets and the relatively short-ranged, ground-based weapons.

For PLAAF commanders, the challenge of defending China's airspace became increasingly complex through the 1970s and 1980s as China continued to lose ground to the technologically advanced weapon systems that were entering service in the air forces on its periphery. With their attention concentrated on the air defense mission, PLAAF leaders allocated far less attention and far fewer resources to secondary mission responsibilities, including medium-range nuclear weapons delivery, battle area interdiction, and airborne and airdrop operations.

Throughout its history, the PLAAF's growth and development have experienced challenges from a ground force-dominated PLA opposed to a more independent Air Force. A statement by Liu Yalou, the PLAAF's first commander, aptly makes this point. In 1951, Liu wrote, "The PLAAF must oppose two erroneous tendencies. The first tendency is to believe that the PLAAF is a new service that can disregard the legacy of the Army. The second tendency is to be [cognizant of only] some of the Army's experience. Both of these tendencies are wrong and will impede the PLAAF's development."[6] These points were reiterated in February 1951, when at the first expanded meeting of the PLAAF Party Committee, it was formally affirmed that "the Air Force will be developed [as a part] of the Army."

In the early 1980s, when the PLA began reorganizing the ground forces into group armies, the PLAAF was tasked only to provide defense for group army positions. Specific guidance was given that "each branch and unit of the PLAAF must establish the philosophy that they support the needs of the ground forces and that the victory is a ground force victory."[7] Thus, the PLAAF was still effectively tied to supporting the ground forces rather than acting in its own right as a service with a unique and valuable role. This mindset began to slowly change in the mid-1980s when, almost 40 years after the PLAAF's founding, General Wang Hai became the first aviator to be appointed as the commander.

Changing Capabilities to Meet Mission Requirements.

The Air Force is now attempting to develop the capability to conduct all-weather, day-night, high-intensity simultaneous defensive and offensive operations, while extending its ability to operate beyond the periphery of China's coastline. This major shift in mission orientation has forced the PLAAF to embark on a broad range of new weapons and ancillary programs — air refueling, air defenses, airborne warning and control — required to support offensive operations. Integration of these new weapons and capabilities will require substantial readjustments in the size and composition of the Air Force. In addition to obvious realignments at the tactical level, operational command and control elements will require fundamental restructuring to facilitate the planning and execution of offensive air operations beyond China's borders.

Establishing Acquisition Priorities.

In recent years, the PLAAF has employed a number of approaches to acquiring new platforms. Historically, political, economic, and security considerations have caused the PLAAF to rely on domestic producers for new equipment. Although Chinese defense industries successfully introduced a number of incremental improvements to legacy fighters and air defense systems, it was not until recently that they were able to design and develop new weapons incorporating advanced technologies. As a result, the PLAAF entered the 1990s with a weapons inventory that was more representative of the technology of the 1960s and 1970s.

In May 1990, just prior to Central Military Commission (CMC) Vice Chairman Liu Huaqing's visit to Moscow to negotiate the first SU-27 contract, the PLAAF issued an internal document that laid out a weapons procurement plan addressing doctrinal needs and budgetary constraints.[8] This document argued that it would take many years to attain its needs if China relied solely on indigenous efforts to develop new capabilities. Although self-reliance in designing and developing new equipment remained a strategic goal for the aviation industry, this left the PLAAF with few alternatives for fleet modernization and often led to unmet requirements. The PLAAF adopted four criteria—New, Quality, Modify, and Introduce—to address its modernization challenges:

- *New*—Use the newest weapons and equipment already in the inventory.
- *Quality*—Focus on acquiring and employing weapons and equipment that provide meaningful military capability and possess high operational capability. Maintain aircraft and engines to extend their service lives.
- *Modify*—Use new technology and materials to upgrade existing equipment, thus giving it new life. Designing and developing a new aircraft from the ground up is not considered a feasible option and would consume vast amounts of capital.
- *Introduce*—Acquire and integrate advanced weapons and equipment from abroad.

The PLAAF also announced that it would deploy modern equipment based on the threat facing each war zone. This would ensure rapid mobilization for battle and facilitate training. Finally, the 1990 guidelines argued against spreading new equipment evenly among every unit, on the grounds that this would dissipate strength where it was needed most.

In deciding which weapons and equipment to modernize, the PLAAF determined that it must focus on six combat and combat support capabilities:

- Air superiority,
- Ground attack,
- Transporting troops and supplies,
- Airborne early warning and reconnaissance,
- Electronic countermeasures, and
- Maintenance and logistics.[9]

The 1990 plan also laid out the following general guidelines for proportionally developing its force, although no precise percentages or numbers were specified:

- Fighter aircraft must have the highest priority.
- The proportion allocated for ground attack aircraft must be larger than the portion for bombers, since ground attack aircraft with a refueling capability could be used against rear-echelon targets.
- There must be a certain proportion of bombers, especially strategic bombers.
- Reconnaissance aircraft, jamming aircraft, and airborne early warning aircraft must be supplied in relevant proportions.

- Development of transport aircraft, which have a strategic capability of moving troops and supplies, cannot be slowed.
- Aerial refueling must constitute a certain proportion of combat aircraft as a force multiplier.
- China must pay attention to developing helicopters, especially armed helicopters, for the Army and Navy.
- The Air Force must develop ground-based weapon systems, particularly air defense missiles, radar, and communication systems.[10]

During the past 15 years, the PLAAF has largely achieved the goals established under the 1990 guidelines through the implementation of an equipment modernization strategy that relied heavily on access to Russian military equipment.[11] Major acquisitions of aircraft and air defense systems from Russia include Su-27[12] and Su-30 fighters, Il-76 transports, S-300 SAMs, and Mi-17, Ka-28, and Mi-8 helicopters. The PLAAF has also deployed its first B-6 aerial refueling tankers for J-8II fighters, and is trying to acquire the Il-76 refueling variant for its *Sukhoi* aircraft. Since 2003, the PLAAF has taken delivery on as many as 60 new J-10 fighters, a highly evolved derivative of the Israeli Lavi program, developed in Chengdu, China. Significantly, the PLAAF has not yet achieved its goals in the development of strategic bombers or airborne early warning and control (AEW&C) aircraft.

Force Restructuring.

The PLAAF, along with other branches of China's military, has steadily reduced force structure through the retirement and replacement of outmoded weapons systems with smaller numbers of advanced weapons. In this process, the PLAAF has trimmed personnel and equipment and deactivated units, while adding new capabilities and combat power to its forces. Recapitalization and realignments within the PLAAF follow the global trend toward smaller, more capable air forces. Since the early 1990s, the PLAAF has completed the following force reductions and realignments:

- Decreased total personnel strength from 490,000 to less than 400,000;
- Reduced combat aircraft from more than 5,000 to about 2,000 (and the number is continuing to decrease annually);
- Reduced combat air divisions from 50 to 28;
- Decreased the average number of regiments per air division from three to two; and
- Decreased the number of aircraft per regiment.[13]

As shown in Table 1, the PLAAF, despite a concerted effort to modernize its forces, continues to operate a large fleet of legacy aircraft that are variants of Soviet designs—MiG, *Tupolev, Antonov*—originating from the 1950s. These aircraft slow modernization efforts by consuming operational and maintenance funds while contributing little to China's defense. During the past two decades, the PLAAF has retired nearly 3,000 aircraft, shrinking its combat inventory from roughly 5,000 to approximately 2,000 combat aircraft. This has removed many, but not all, of the vintage airframes from the fleet. Yet, further reductions may prove more difficult to justify and execute because many of the remaining legacy airframes were manufactured as recently as 10 years ago. This creates a number of

PLAAF Combat Aircraft	Numbers
Advanced Fighters	
All Flankers*	240
J-10	50
Other Fighters	
J-6, J-7, J-8	1,200
Attack	
JH-7, Q-5	420
Bombers	
H-5, H-6	100
Reconnaissance	180
Total Combat Aircraft	1,800

Table 1. PLAAF Combat Aircraft.[14]

programmatic challenges for PLAAF leaders as they contemplate modernization schedules in the years ahead. With at least two stealth programs under development, China may elect to restrain near-term acquisitions of nonstealth aircraft in anticipation of more capable airframes becoming available in the near future.

Aircraft replacement rates over the course of the next several years will determine the pace of transition to a truly modern force. At present, the PLAAF's inventory of advanced fighters—300 *Flanker* and J-10 aircraft—remains modest. The newest and most formidable additions to the *Flanker* line-up are the 76 multirole Su-30MKKs stationed in the Guangzhou and Nanjing military regions (MRs) where they are poised to conduct precision strikes against Taiwan and U.S. military bases in Okinawa. Each of these aircraft is currently equipped with an aerial refueling probe.

Once the PLAAF receives the first batch of Russian Il-76 tankers to air refuel the *Flankers*, their range and loiter capability will be further improved.

By many measures China's Air Force remains relatively small. China's size—along with its growing power and influence—dominates the Asian landmass. Yet it borders 15 other nations, thus creating unique challenges for an Air Force tasked to maintain air sovereignty along 22,000 kilometers of national boundary. Although PLAAF basing has been concentrated along the eastern seaboard, China has not neglected other security concerns and vulnerabilities to the northeast and northwest. This can be evidenced in the basing of *Flankers*, which are spread among each of the seven MRs.

DOCTRINAL REFORMS

China's military modernization is underpinned by a new PLA operational doctrine, still evolving, that emphasizes preemption, surprise, and shock based on the concept that the early stages of conflict are crucial to the final outcome. To implement this doctrine, the PLA has assigned priority to modernizing naval, air, and strategic missile forces.[15] This new doctrine reflects China's shift away from its historical predominant reliance on ground forces toward a more balanced defense posture incorporating the full panoply of PLAAF capabilities.

As Table 2 shows, the PLAAF began developing its current doctrine in the mid-1980s, starting with campaigns, then tactics, and, finally, strategy. Note that the doctrine on PLAAF strategy was published in 1995; however, the culmination of the PLAAF's efforts in behalf of its own strategic doctrine did not truly come to fruition until 2004, when the CMC incorporated a PLAAF component into the National Military Strategic Guidelines.

Doctrinal Title	Preliminary Approval	Publication Date
Science of PLA Air Force Campaigns	1984	1988
Science of Air Force Tactics	1989	1994
Science of Air Force Strategy	1992	1995
Introduction to Air Force Military Thought	1998	2006
Science of Integrated Air and Space Operations	2003	2006

Table 2. Chronology of PLAAF Doctrinal Development.

DEVELOPMENT OF THE PLAAF'S STRATEGIC DOCTRINE

The PLA's collective National Military Strategic Guidelines (*guojia junshi zhanlue fangzhen*) have three major components. The first is a strategic assessment of the international environment. The second is the operational component, known as the "Active Defense" (*jiji fangyu*) strategy. The third is "Army Building."

From its founding in 1949 and until 2004, while lacking its own strategy component to establish a broad direction for air operations, the PLAAF relied almost solely on the PLA ground force's "Active Defense" operational component as its strategic-level doctrinal guidance. Chinese military literature indicates that initial discussions on the development of a PLAAF strategic theory may have begun in the mid-1980s. In 1987, PLAAF commander General Wang Hai proposed that the PLAAF be accorded its own operational component in the PLA's "Active Defense" strategy. General Wang introduced "the goal of transforming from defending the country's airspace to building an Air Force capable of simultaneous offensive and defensive operations (*gong fang jian bei*)."[21] Wang emphasized that the combined arms combat environment of the 1980s required a force that could:

- Move quickly over long distances;
- Fight in an electronic environment;
- Have the capability to attack an enemy; and
- Protect the PLAAF from sustaining catastrophic damage from an enemy air attack.

The *China Air Force Encyclopedia* declares that the capability to conduct simultaneous offensive and defensive operations is the "guiding concept for Air Force building."[22] The capacity for simultaneous offensive and defensive operations is closely linked to the PLAAF's organizational structure, weapon systems and equipment, education and training, C4I system, and logistics support system.

Concurrent with General Wang's proposals, National Defense University (NDU) commandant General Zhang Zhen broached the idea of establishing Air Force strategic theory as a specialty within the university.[23] Although this suggestion did not take hold, it apparently led to further discussions on the need to develop strategic guidelines for air operations. In fact, the PLA debated this issue vigorously during the mid-1980s in what could best be described as "turf wars" or "internal politics" within the PLA. According to *Science of Air Force Strategy*, Chinese military theorists of a ground-dominance bent put forward several arguments countering the alleged need for Air Force strategic guidelines:

- China can have only one national military strategy, and that is "Active Defense";
- The PLAAF does not have strategic weapons, so it is not qualified to have an Air Force strategy;
- The PLAAF already has half of an Air Force strategy in that it has the strategic mission of air defense of the nation, but supporting the ground forces does not qualify as a strategic mission;
- Because the PLAAF's command personnel can implement directive guidance only from above, lacking command and decision authority in its own right, there is no Air Force strategy; and
- The PLAAF is a multifaceted service with many missions; therefore, it already has a strategy.[24]

As we saw in Table 2, beginning in 1992 the PLA established a committee of NDU and PLAAF officers to initiate research on Air Force strategy, culminating in the publication of *Science of Air Force Strategy* in 1995. This book laid out an argument, based on international air power doctrine, for the PLAAF to be an "independent" service and to be assigned its own operational component in the PLA's National Military Strategic Guidelines.

The Gulf War and the 1995-96 Taiwan Strait crisis provided additional ammunition for the PLAAF to seek its own strategic doctrine. For example, statements by Chinese Communist Party (CCP) Central Military Commission (CMC) Chairman Jiang Zemin and recently installed PLAAF commander Liu Shunyao emphasized the PLAAF's requirement to fight offensive battles. In 1997, General Liu stressed this new strategic direction in the following words:

The Chinese Air Force plans to build up state-of-the-art weapon systems by early next century, including early warning aircraft, electronic warfare warplanes, and surface-to-air missiles. The PLA Air Force is now able to fight both defensive and offensive battles under high-tech conditions. The Air Force is now capable of waging high-level long-distance combat, rapid maneuverability, and air defense, and is able to provide assistance to Navy and ground forces. The Air Force now sources most of its equipment domestically, fielding a large number of Chinese-designed and produced high-quality fighters, attackers, bombers, reconnaissance aircraft, and special purpose aircraft. Over the next few years, the Chinese Air Force will enhance its deterrent force in the air, its ability to impose air blockades, and its ability to launch air strikes, as well as its ability to conduct joint operations with the ground forces and Navy.[25]

The CMC Approves the PLAAF's Strategy.

In 2004, the CMC approved the PLAAF's "Active Defense" strategy as a component of the National Military Strategic Guidelines for air operations.[26] The PLAAF's strategic component was designated as *"Integrated Air and Space, Simultaneous Offensive and Defensive Operations"* (*kong tian yiti, gong fang jian bei*).[27] According to Hong Kong press reports, the CMC's approval was timed to coincide with the PLAAF's 10th Party Congress in May 2004 and represented a major milestone in China's efforts to build a strategic Air Force.[28] The approval also signaled a fundamental shift in how the PLAAF was to be viewed. The article states that this change is encapsulated in three bold new assertions on the strategic positioning of the PLAAF:

- First, the PLAAF is a national Air Force led by the CCP.

- Second, a modern Air Force must be built to unify aviation and spaceflight, combine defense and offense, and unify information and firepower.
- Third, the PLAAF should be a strategic Air Force standing side by side with the Army and Navy to achieve command of the air, ground, and sea.

Integrated Air and Space.

China's 2004 and 2006 *Defense White Papers* clearly show the growing importance of the PLAAF and its missions. However, although both white papers describe the PLAAF's transition to simultaneous offensive and defensive operations, neither paper references integrated air and space.

Even though the two white papers did not refer to this component, however, the PLAAF has apparently thrown its hat into the air/space ring, having indicated its desire to become actively involved in managing China's military space program with an emphasis on the informationalization aspects. Specifically, in March 2004, the PLAAF published *Air and Space Battlefield and China's Air Force*, following in August 2006 with *The Science of Integrated Air and Space Operations*.[29] Although the first doctrinal book did not provide linkage between space and the PLAAF, the last chapter of the second book, which contains forewords by PLAAF commander General Qiao Qingchen and political commissar General Deng Changyou, lays out six steps for China in establishing a model in which "the PLAAF is the leading organization for 'integrated air and space', the PLAAF is . . . the leading organization to manage China's military space force, and the PLAAF is the primary force for [air and space] combat."[30] However, the book focuses on managing the "informationalization" aspects of the space program, and does not indicate that the PLAAF wants to manage the launch sites, satellite development, and missile program. The six proposed steps are as follows:

- Determine a sound scientific development model for creating a process for the employment of air and space power.
- Establish an Air Force Space organization (*kongjun hangtian jigou*) to use as the base for organizing integrated air and space operations.
- Establish PLAAF space units (*kongjun hangtian budui*).
- Establish information links that provide technology for integrated air and space operations.
- Nurture Air Force space personnel possessing a knowledge of space.
- Expand the PLAAF's overall scope of warfighting power, increasing the PLAAF's air offense capabilities, air defense countermissile capabilities, and airborne troop combat capabilities.

In the introduction to *The Science of Integrated Air and Space Operations*, General Qiao states that under the Party Central Committee's and CMC's leadership, the PLAAF is implementing the transformation from mechanization to informationalization, from a force based on national air defense to one based on simultaneous offensive and defensive operations, from a force based on aviation to one based on integrated air and space, and from a force based on quantity to one based on quality.[31]

In 2006, the PLAAF published *An Introduction to Air Force Military Thought* with opening remarks by PLAAF commander Qiao.[32] The inaugural edition of this new Air Force primer argues that the PLAAF should use informationalization to control the land and sea, and should move toward developing integrated air and space operations.[33] These declarations may be just the beginning of a long turf war within the PLA over managing and employing China's military space assets.

CAMPAIGN THEORY (OPERATIONAL DOCTRINE)[34]

The PLA's *Science of Campaigns* categorizes military operations into 22 distinct types of campaigns. Three among these—air offensive, air defense, and air blockade—are specifically designated as Air Force campaigns.[35] Moreover, PLAAF airborne forces and aircraft are key elements of the joint airborne campaign, and PLAAF AAA and SAM forces can be expected to play a major role in the PLA's joint anti-air strike campaign.

PLAAF Campaign Theory.

Historically, the PLAAF has conducted operations as a series of air campaigns in support of the PLA's overall campaign objectives. In the early days, the PLAAF had little choice but to adopt operational concepts and tactics of foreign as air forces. By the mid-1950s, however, the PLAAF was able to apply operational experiences obtained during both its civil war and the Korean War to create its own adaptations of air campaign theory and tactics.[36] During the mid-1960s, the PLAAF codified its rules and regulations, courses of study, and teaching materials, demonstrating "use the PLAAF as the primary force [during a conflict]" (*yi wo wei zhu*) doctrine.[37]

Beginning in the early 1980s, the PLAAF's research on military theory focused even greater attention on air campaign theory. In 1988, the PLAAF formally published *Science of PLA Air Force Campaigns*, which described the characteristics of operational art, the development of campaign theory, and the mission of the PLAAF's campaign headquarters, and then discussed how these three elements pertain to a unified command organization.[38] Thereafter, the PLAAF published various teaching materials, such as the *Course Material for the Science of Air Force Campaigns* [*Kongjun Zhanyi Xue Jiaocheng*], to guide campaign training.[39] In 1999, the PLAAF revised its Campaign *Gangyao*, which provides the doctrinal basis and general guidance for how the PLAAF will fight future campaigns.[40]

PLAAF Campaign Terminology.

Before we discuss PLAAF campaigns, a brief discussion of key terms is necessary. The term "Air Force campaign" applies to all types of Air Force campaign operations.[41] The PLAAF describes an Air Force campaign as the use of "from one to several campaign *juntuan* (*zhanyi juntuan*) or campaign and tactical *bingtuan* (*zhanyi zhanshu bingtuan*) to carry out the integration of a series of battles according to a unified intention and plan to achieve a specific strategic or campaign objective in a specified time. An Air Force campaign is implemented under the guidance of the national military strategy and the PLAAF's strategy."[42] For the PLAAF, a *juntuan*-level organization refers to the seven MR Air Force (MRAF) headquarters, and a *bingtuan*-level organization refers to division, brigade, or regiment headquarters.

An Air Force campaign is also described as "a campaign conducted independently by an Air Force campaign *juntuan* or with the coordination of other services and branches. An Air Force campaign is guided by the national military strategy and is limited by the PLAAF's strategy. An Air Force campaign involves various air-to-air, air-to-ground, and surface-to-air battles to achieve specific military objectives. The campaign determines the battle's character, goals, missions, and actions, and directly supports the local and overall war."[43]

PLAAF Campaign Categories.

In this regard, the PLAAF has been methodical in the way it has defined its campaign theory and used the theory to provide operational guidance for its forces. PLAAF campaign theory can be categorized into that for aviation (aircraft), for air defense (SAM, AAA, and radar troops), and for airborne troops.[44] Not surprisingly, these three categories reflect the way the PLAAF is organized administratively and operationally.

Characteristics and Objectives. Based on campaign characteristics and objectives, the publication *Science of PLA Air Force Campaigns* identifies three specific types of PLAAF campaigns:[45]

- Offensive air campaigns (*kongzhong jingong zhanyi*);
- Air-defense campaigns (*fangkong zhanyi*); and
- Air-blockade campaigns (*kongzhong fengsuo zhanyi*).

Also based on their characteristics and objectives, *Science of PLA Air Force Campaigns* identifies the following two types of joint-service campaigns where the PLAAF plays a major part:

- Joint anti-air strike campaigns (*lianhe fankongxi zhanyi*); and
- Airborne campaigns (*kongjiang zhanyi*).[46]

Operational Scale. Based on a campaign's operational scale, *Science of PLA Air Force Campaigns* divides PLAAF campaigns into the following three types:

- Multiple war zone (*duo zhanqu*) Air Force campaigns, such as an air defense campaign of the capital (*shoudu fangkong zhanyi*), and Air Force offensive campaigns to destroy the enemy's potential power (*quanli*);
- War zone (*zhanqu*) Air Force campaigns; and
- War zone direction (*zhanqu fangxiang*) Air Force campaigns.

Command Relationships between Services and Branches.

Based on the command relationships and the services and branches participating in the war, PLAAF campaigns can be divided into the following three types:

- Independent (*duli*) Air Force campaigns;
- Combined arms (*hetong*) Air Force campaigns; and
- Jointly executed (*lianhe shishi de*) Air Force campaigns.

THE PLAAF'S TACTICAL DOCTRINE

On its 45th anniversary in November 1994, the PLAAF published *The Science of Air Force Tactics*, an internal military tactics manual which has not been made publicly available. According to a brief explanation provided in the *China Air Force Encyclopedia*, the manual discusses both basic theory (*tactics*) and practical application theory (*techniques and procedures*).[47] The manual identifies weapons and equipment, combat personnel, the battlefield environment, combat command, and combat support as the principal factors influencing the tactical level of conflict. The tactics manual also includes practical application theory (*techniques and procedures*) for aerial combat, air-to-ground combat, and surface-to-air combat.

Training Guidance Concepts.

Instructions titled "military training guidance concepts" (*junshi xunlian zhidao sixiang*) issued by the PLAAF in 2001 downplay safety considerations, focusing rather on realistic and demanding training. Training guidance concepts are issued by the PLAAF Party Committee to unify training ideology, address major challenges, identify training restrictions, and establish overall training objectives.[48] The concepts are reviewed and modified when "situations and mission development change, weapons and equipment are replaced, and new regulations and outlines are implemented.[49] The first set of PLAAF training guidance concepts was published in 1951. Revisions have been issued only seven times—1952, 1954, 1958, 1965, 1974, 1987, and 2001. A comparison of the two most recent sets of concepts, shown in Figure 1, clearly demonstrates the shift in training philosophy that occurred between 1987 and 2001. In 1987, "safety" was the watchword, with little focus on training per se. In 2001, every line points to practical, realistic training.

Concepts issued in 1987
 Adhere to reform (jianchi gaige)
 Enhance effectiveness (tigao xiaoyi)
 Improve steadily (wenbu qianjin)
 Ensure safety (baozheng anquan)

Concepts issued in 2001
 Closely adhere to actual combat situations (jintie shizhan)
 Stress training against opposing forces (tuchu duikang)
 Be strict during training (cong nan cong yan)
 Apply science and technology during training (keji xingxun)

Figure 1. Comparison of 1987 and 2001 Training Concepts.

The Dagang (military training and evaluation outline or program). The PLA published its first training guidance in 1955 under the title *PLA Combat Training Dagang* (*Zhongguo Renmin Jiefangjun Zhandou Xunlian Dagang*), usually denominated simply as the "*dagang*"). The *dagang* established the military training plan for all services and branches of the PLA. Between 1957 and 1980, the PLA revised the basic *dagang* three times. When the *dagang* was revised again in 1989, the General Staff Department (GSD) became responsible for issuing the Army *dagang*, with the PLA Navy, the PLA Air Force, and the Second Artillery each responsible for issuing its own.

The *dagang* provides the general plan for military training. It establishes the "laws governing military training" and the "foundation for organizing and implementing military training."[51] It includes training goals, principles, content, implementation phases and procedures, timing, methods, and quality-control inspection procedures.[52] Each *dagang* is divided into separate volumes according to different objectives and levels, with each volume then further divided into subsections by organization or specialty. For example, in 1997, the Army *dagang* had a total of five volumes comprising 35 subsections.

In 2001, the GSD revised the *dagang* for the ground forces yet again. The English version of the 2002 *Defense White Paper* translated this *dagang's* title as *Outline of Military Training and Evaluation (junshi xunlian yu kaohe dagang)*.[53] It appears this was the first time the word evaluation (*kaohe*) was included.

In April 2002, the PLAAF published its own revised *Air Force New Generation Outline of Military Training and Evaluation (kongjun xinyidai junshi xunlian yu kaohe dagang)*.[54] The PLAAF's *dagang* was divided into several sections, addressing command personnel, headquarters department, branches (aviation, AAA, SAM, airborne, and radar), and all support elements such as the communications troops.

A GROWING LEADERSHIP ROLE FOR THE PLAAF

Historically, Army officers have held all key leadership positions in the four General Departments, the National Defense University, the Academy of Military Science (AMS), and the seven MR headquarters. In recent years, however, this has begun to change slowly. Since 2000, the CMC has steadily assigned PLAAF officers to an increasing number of pivotal leadership positions in Beijing and MR headquarters. These steps indicate a gradual but more concerted effort to implement joint reforms at the highest levels within the PLA.

Over the years, general officer realignments have been observed within the headquarters at the seven MRs. The first changes occurred in the late 1980s, when the seven Military Region Air Force (MRAF) commanders were concurrently appointed as deputy commanders of MR headquarters. Today, some, and possibly all, of the MRAF political commissars are also concurrently appointed as MR deputy political commissars. In 2002, the first PLAAF general officer was appointed to serve as deputy director of the Nanjing MR Operations Department. In late 2003, the PLAAF began to augment each of the seven MR headquarters by appointing a major general to serve as a deputy chief of staff in the Headquarters Department.

In August 2003, Lieutenant General Zheng Shenxia was elevated from the position of Chief of Staff of the PLAAF to the commandantship of the PLA's Academy of Military Science.[55] As the first Air Force officer to hold this post, General Zheng has already brought new emphasis to the integration of air operations into PLA strategic doctrine.

In 2004, the CMC made several significant decisions and appointments affecting the Air Force. During a May meeting, the CMC approved a PLAAF component as part of the National Military Strategic Guidelines and elevated PLAAF Commander General Qiao Qingchen to be a member of the CMC.[56] Although the PLAAF remained subordinate to the four General Departments, the placement of an Air Force commander on the CMC demonstrated a remarkable change in the PLA protocol.[57] Also in 2004, the CMC selected Shenyang MRAF Commander Lieutenant General Xu Qiliang to serve as a Deputy Chief of the General Staff, making him only the second officer in the history of the PLAAF to hold this position.[58] In late 2004, two more PLAAF generals were named as deputies of a General Department.[59] PLAAF Deputy Commander Lieutenant General Li Maifu was appointed as the first Air Force deputy director of the General Logistics Department, and Lieutenant General Liu Zhenqi was named as the first PLAAF officer to hold the position of deputy director of the General Political Department. Another significant appointment occurred in August 2006, when Lieutenant General Ma Xiaotian was appointed as the first PLAAF commandant of the National Defense University,[60] but, surprisingly, no PLAAF officer has as yet been assigned as a deputy in the General Equipment Department.

As a whole, these appointments and the CMC's approval of an Air Force component to the National Military Strategic Guidelines represent a significant break with a past in which the Army retained a stranglehold on senior leadership positions, enabling them to subordinate Air Force interests and potential contributions. These changes in senior officer appointments reflect a significant change in the PLA culture that can be observed in other more subtle ways. For example, up until the late 1990s, irrespective of service or branch, all military personnel assigned to duty within the PLA General Departments, NDU, AMS, or an MR headquarters were required to wear a PLA uniform. Today, personnel assigned to "joint" positions are allowed to wear the uniform of their own service. In 2006, the PLAAF introduced new military uniforms, which, for the first time, were not of the basic Army uniform pattern.

The PLA is at an early stage of transition toward improved joint operational capabilities. The appointment of several Air Force officers to national-level positions is a clear signal of intent and purpose.

However, it remains unclear whether power and authority have shifted in any substantial way to senior PLAAF leaders. Greater PLAAF authority might manifest itself in more Air Force discretion over the direction and management of Air Force weapons development and acquisition programs. It might also be anticipated that the PLAAF will enjoy a larger role in planning and execution of operations. Future advancement along this path toward greater jointness might be evinced by the appointment of Air Force officers to key director or deputy director billets in the MR first-level departments (i.e., Headquarters, Political, Joint Logistics, and Equipment Departments), or the second-level departments such as the Operations Department.

ORGANIZATIONAL REFORMS

The PLAAF has been engaged in a PLA-wide restructuring aimed at achieving "optimal force structures, smoother internal relations, and better quality."[61] Ten rounds of PLA force reductions since 1985 have trimmed nearly 2 million uniformed personnel from the PLA active duty ranks. Under the most recent order, the PLA was directed to eliminate 200,000 active duty positions between September 2003 and the end of 2005, cutting the size of the PLA to 2.3 million. Previous cuts fell on enlisted ranks, resulting in mass demobilizations and unit deactivations. This round targeted the PLA's bloated officer rosters.[62] Approximately 170,000 officers—85 percent of the announced reduction in force—were pared from the top-heavy personnel rosters.[63] Based on a proportional slice, the PLAAF was forced to cut 30,000 officer billets.

Two specific goals of this latest force reduction were to replace junior officers with noncommissioned officers (NCOs) and reduce the number of general officers. For reducing the number of staff officers assigned to headquarters, the solution was to downgrade by two echelons all five air armies (*kongjun jun*) and the five army-level bases to division-level command posts.[64] As a result, the PLAAF currently has a total of 13 command posts, with two in each of five military regions, three in the Lanzhou MR, and none in the Jinan MR.

These structural changes were necessary to reshape the PLAAF's operational command structure, but the changes have also adversely affected morale among officers at all levels whose jobs were eliminated or who have been denied an eventual promotion to the next level to secure their retirement benefits.

The possibility exists for yet another major PLA force reduction by the end of the decade, which could further restructure headquarters staffs. In addition, the PLAAF can be expected to continue the restructuring of air divisions and air regiments. As the PLAAF continues to introduce the Su-30, J-11, J-10, FC-1, and JH-7 into the inventory as replacements for the vintage Q-5, J-6, and J-7, some units will transition to new weapon systems; others will be deactivated.[65] The final restructuring by the end of the decade could leave the PLAAF with just under 30 operational divisions, most of which will have only two regiments each.

PLAAF logistics and maintenance units have experienced significant reorganization and restructuring since the 1990s. Major changes were necessary to accommodate new operational mission requirements as the PLAAF transitioned from a force confined to employing single branches (aviation, surface-to-air missiles, antiaircraft artillery, radar, and airborne troops) and single aircraft types in positional defensive campaigns, to a force capable of combined arms operations in mobile, offensive campaigns.[66] Ultimately, the PLA is striving to conduct joint service operations supported by joint logistics. To achieve these goals, the PLAAF has reconfigured logistics and maintenance systems, which traditionally have not been structured to support mobile, offensive operations. While many of the changes are still underway, some are still only aspirational.

Historically, a single airfield has hosted one regiment fitted to a single type of aircraft. The logistics and maintenance structure was organized to support only that type of aircraft. When aircraft deployed, they flew to a base with the same type of aircraft. Today, however, that situation is changing as a result of the PLAAF's emphasis on achieving new mobility goals. Now, small logistics and maintenance teams deploy, usually by rail or road, along with the deploying aircraft to any type of airfield. Furthermore, efforts are underway at PLAAF airfields to instruct specialized maintenance teams in the cross-servicing of multiple aircraft types.

NEW PERSONNEL PROGRAMS

Significant changes in the PLAAF's recruiting and training of conscripts, NCOs, and officers (cadre) have taken place since the late 1990s and will continue through the end of the decade.

Enlisted Force.

Prior to 1999, the PLAAF's enlisted conscripts served for 4 years. At the end of that period, they could remain on active duty as a "volunteer" for an additional 12 years. In 1999, China revised its Military

Service Regulations, which reduced the conscription period for all of the PLA's services and branches to 2 years.[67] All conscripts report for duty on November 1 and are demobilized 2 years later on October 31. During the early 2000s, the PLAAF began recruiting civilian college students who had not yet completed their studies to join as enlisted troops. The goal is to have them remain on active duty as NCOs at the end of their initial 2-year service.

The revised service regulations also established a formal NCO corps, whose members can now serve until they have 30 years of service or until they reach age 50. The PLAAF must now provide housing for them and their families as well. In terms of education, some PLAAF NCOs can attend an officer academy to receive a technical degree before returning to their unit as an NCO.

Although the PLA does not announce specific figures for the number of troops by rank and specialty, it appears that the number of conscripts has gradually been reduced, while the number of NCOs has increased accordingly to provide greater experience and stability to the overall enlisted force.

Officer Corps.

The PLAAF has also begun to reform the way it recruits officers. Historically, most officers were recruited from high school graduates or the enlisted force. Once they join the PLAAF, they attend a PLAAF academy, where they receive a 3- or 4-year degree and are commissioned as an officer. Today, the PLAAF is recruiting civilian college graduates and providing them with 3 months of basic training before they are commissioned as officers.

The most significant changes have taken place in recruiting pilots, which historically relied on high school graduates and enlisted personnel selected for officer pilot training. In 2000, the PLAAF began to recruit its pilots from graduates who have a 4-year bachelor's degree in specific areas from one of the PLA's academies, including the Army, Navy, and Second Artillery.[68] In 2003, the PLAAF extended the program to civilian college graduates with specific bachelor's degrees.[69] These graduates receive 2 years of flight training at a PLAAF flight academy and 1 year of transition training before being assigned to an operational unit. As a result, the first group of pilots selected from PLA college graduate began entering the operational force in 2003. The first group of pilots selected from civilian college graduates entered the operational force in mid-2006.

ENHANCEMENTS TO EDUCATION PROGRAMS

The total number of PLAAF schools and academies has expanded and contracted over the years in response to policy changes regarding training objectives or war preparations. At one point, the PLAAF had over 30 academies, including as many as 17 flying schools during the Cultural Revolution. Today, the PLAAF has less than 20 academies, including eight flying academies and one NCO school.

In 1986 the PLAAF upgraded its officer schools to academies and began offering master's degrees in certain subjects. In 1999, three schools in the Xian area were combined administratively to become the PLAAF Engineering University, so that the first 2 years of basic training could be conducted in a single location, and doctorate degrees could be offered. The PLAAF's Antiaircraft Artillery Academy also expanded its curriculum to include training airborne officers for the first time. In 2004, three additional academies in the Changchun area were combined into the Air Force Aviation University. The trend of consolidating academies into universities and expanding the curriculum will most likely continue through the end of this decade.

The PLAAF has always placed great emphasis on training officers to be proficient in tactics and technical skills, but did not begin focusing on officer education until the mid-1980s. Whereas the schools before the 1980s taught officers to fly, maintain, and support aircraft, these technically oriented schools did not spend much time on the theory of warfighting at the campaign and strategic levels.

In 1996, the PLAAF's official magazine, *China Air Force*, carried an article written by the PLAAF's Command College that discussed the lack of adequate combined arms and joint training characterizing the PLAAF officer corps in the early 1990s.[70] The article described the PLAAF's commanders at the regimental to MRAF headquarters levels as "lacking knowledge, having poor concepts, and being incompetent in joint operations." As part of the reforms to produce trans-century commanders, the Command College began focusing on theories such as joint combat operations, mobile warfare, information warfare, and electronic warfare, and updated its combat theory.

Not only was the PLAAF concerned about its commanders being unable to command combined arms and joint forces, it was also concerned about their inability to use high-tech systems effectively. According to a *People's Daily* article in May 2000, the PLAAF began requiring all of its officers at and above the regimental commander level to receive high-tech training within

1 year.[71] The new course includes modern air combat theory, development trends in modern fighter aircraft, weaponry for modern combat, equipment for high-tech warfare, and command automation devices.

On the occasion of the PLAAF's 56th anniversary in 2005, commander Qiao Qingchen and political commissar Deng Changyou stated, "We should continue to step up the training of Air Force pilots, new equipment operators, combined arms force commanders, and high-level scientists, technicians, and experts, and gradually create a sufficient number of outstanding young and middle-aged qualified personnel."[72]

By the end of this decade, the PLAAF will most likely still be concerned about its officers having the ability to command at the combined arms and joint level due to the dearth of PLAAF officers assigned to joint positions in the seven MR Headquarters and four General Department headquarters. However, the computer skills of its officer corps should be much better as younger officers who have grown up with computers move into command and staff positions at the regiment and division level.

SUMMARY

While the PLAAF has made impressive progress towards comprehensive force modernization, most Western observers have concluded that it will require an additional 10–15 years before the process is complete. Several obstacles stand in the way. The most visible impediments are the lingering hardware deficiencies. China's Air Force continues to face significant shortfalls in key weapon systems and other hardware—advanced fighters, airborne warning and control (AWACS), aerial tankers, and C4ISR infrastructure—that are essential for the conduct of high-intensity, offensive air operations. Chief among the PLAAF's challenges is a large inventory of obsolete aircraft that contribute little to capabilities and will require substantial additional time and resources to maintain and replace. Modernization has also been hampered by lengthy delays in fielding command and control and air surveillance aircraft, two systems that are essential for the Air Force to extend its reach beyond the shoreline.[73]

PLAAF modernization also rests on its ability to introduce the full measure of reforms that are currently underway within the ranks. The force structure is being radically reshaped to accommodate the introduction of advanced new weapons and the logistics support required to sustain these systems. In addition, Air Force strategists are actively engaged in the development of new operational concepts and doctrine, tasks made doubly difficult by the PLAAF's lack of recent combat experience. Significant changes are also underway in the training and educational programs to ensure that Chinese airmen have the skills and knowledge required to operate advanced weapons in a complex operational environment. Many of these changes are just now taking shape, and another dozen years will be needed before today's lieutenants mature into seasoned mid-grade professionals.

ENDNOTES - CHAPTER 9

1. *China's National Defense in 2004*, Beijing, State Council Information Office, December 27, 2004, *www.china.org.cn/english/ MATERIAL/116010.htm*.

2. China's air forces include the PLA Air Force, PLA Navy Air Force and PLA Aviation (helicopter) units. The focus of this chapter is on the PLA Air Force.

3. For the PLA, reforms (*gaige*) encompass three components: modernization (*xiandaihua*), regularization (*zhengguihua*), and revolutionization (*geminghua*). Within the PLA's context, modernization refers primarily to equipment and weapon systems, or what can be considered the PLA's hardware; regularization is the software component, which includes leadership, personnel, education, training, organizational structure, housing, and funding, etc.; and revolutionization is the political component.

4. See *China's National Defense in 2006,* December 29, 2006, www.china.org.cn/e-white/.

5. Hua Renjie, Cao Yifeng, and Chen Huixiu, eds., *History of Air Force Theory (Kongjun Xueshu Sixiang Shi),* Jiefangjun Publishers, Beijing, 1991, pp. 294-331.

6. *Ibid.*

7. Chengdu Military Region Campaign Training Office, *Air Force Utilization During the Campaign to Defend Group Army Field Positions (Jituanjun Yezhan Zhendi Fangyu Zhanyi Kongjun de Yunyong),* February 1982, p. 1. This ground force domination is not surprising, since every PLAAF commander and deputy commander until the late 1980s had their roots in the ground forces. It was not until 1973 that the PLAAF even had its first aviator as a deputy commander and 1985 as the commander. Even now, the Army still selects the PLAAF senior officers.

8. Teng Lianfu and Jiang Fusheng, ed., *Air Force Operations Research (Kongjun Zuozhan Yanjiu),* Beijing: National Defense University Publishers, May 1990, pp. 276-282.

9. Teng and Jiang, pp. 296-298.

10. *Ibid.*

11. Richard Fisher has written extensively on the PLAAF's acquisition of Russian weapon systems.

12. The PLAAF operates three separate versions of the SU-27 *Flanker* aircraft, including the Russian-built single-seat SU-27SK and tandem-seat SU-27UBK, and a domestically-produced single-seat model that has been designated the J-11.

13. The information for this comparison comes from the International Institute for Strategic Studies (IISS), *Military Balance* for 1990 and 2005, plus data from the U.S. Department of Defense, Annual Report to Congress on the Chinese military.

14. See *www.sinodefence.com/airforce/fighter/default.asp*. Estimates vary for exact numbers of operational combat aircraft in the PLAAF. This list appears reasonably accurate, although lower than the numbers found in the DoD Report which credits China with a combined total of 2,300 PLAAF and PLAN combat aircraft, including 1,525 fighters and 775 attack/bombers.

15. *China's National Defense*, 2004.

16. *Science of PLAAF Campaigns* (*Zhongguo Renmin Jiefangjun Kongjun Zhanyi Xue*), Beijing: Liberation Army Publishers, November 1988. Approval for research to begin on this publication occurred in June 1984.

17. Science of Air Force Tactics (Kongjun Zhanshu Xue), Beijing: PLA Press, November 1994.

18. Dai Jinyu, ed., *Science of Air Force Strategy* (*Kongjun Zhanlue Xue*), Beijing, Guofang Daxue Publishers, July 1995. According to the book's introduction, General Zhang Zhen broached the idea of establishing Air Force strategic theory as a specialty within the university in 1986, but no action was taken until the early 1990s.

19. Min Zengfu, ed., *An Introduction to Air Force Military Thought* (*Kongjun Junshi Sixiang Gailun*), Beijing: PLA Press, January 2006, p. 42. This book was published as part of the PLAAF's Military Theory Research 10th Five-Year Plan. In 1998, PLAAF Headquarters approved the basic concept of the book so that research could begin.

20. Cai Fengzhen and Tian Anping, eds., *Kong Tian Yiti Zuozhan Xue* (*The Science of Integrated Air and Space Operations*), PLA Press, August 2006.

21. Hua Renjie, Cao Yifeng, and Chen Huixiu, eds., *History of Air Force Theory*, pp. 294-331; Gao Rui, ed., *Zhanlue Xue* (*Science of Strategy*), Academy of Military Science, October 15, 1987, p. 114. Since its inception in 1987, the concept of simultaneous offensive and defensive operations has gradually evolved from a "goal" (*mubiao*) or "model" (*moshi*), indicating the aspirational nature of this desired capability.

22. Yao Wei, ed., *China Air Force Encyclopedia* (*Zhongguo Kongjun Baike Quanshu*), Beijing: Aviation Industry Press, November 2005, p. 39.

23. Dai Jinyu, ed., *Kongjun Zhanlue Xue* (*Science of Air Force Strategy*), Beijing, Guofang Daxue Publishers, July 1995.

24. *Ibid.*

25. Hua Renjie, Cao Yifeng, and Chen Huixiu, eds., *History of Air Force Theory*, p. 312; Oliver Chou, "President Calls for Hi-Tech Push by Air Force," *South China Morning Post*, March 3, 1999; Sun Maoqing, "PLA Commander on Modernizing Air Force," *Liaowang*, April 14, 1997, No. 15, *Foreign Broadcast Information Service (FBIS)*, pp. 20-21; Hua Chun, Chang Tun-Hua, and Kuo Kai, "Air Force Trains Crack Units Troops for Offensive, Defensive Operations, Interview, Lieutenant General Liu Shunyao, PLA Air Force Commander," Hong Kong *Ming Pao*, August 2, 1997, *FBIS-CHI-97-226*, August 2, 1997. See also speech by Liu Shunyao, "Comprehensively Push Forward PLA Modernization Building," *Jiefangjun Bao*, December 24, 1998 *(FBIS)*. The timing of the first comments on an offensive capability came from Liu Shunyao as he took over the commander's position in December 1996 and as Taiwan began final preparations to receive the first squadron of 150 F-16s in April 1997.

26. David M. Finkelstein, "China's National Military Strategy," Chapter 2 in this volume. In 1985, the CMC approved "Offshore Defense" as the PLA Navy's component of the "Active Defense" strategy within the National Military Strategic Guidelines.

27. Yao Wei, ed., *China Air Force Encyclopedia*, p. 57. At the same meeting, Jiang Zemin stepped down as the Chairman and the CMC approved elevating the commanders of the PLA Navy, Air Force, and Second Artillery as CMC members, but it was the second time a PLAAF commander was a member of the CMC. Zhang Tingfa, who was the PLAAF commander from 1977 to1985, served on the CMC from August 1977 to September 1982. Zhang was also a member of the Chinese Communist Party Central Committee Politburo from August 1978 until September 1985.

28. "China Plans To Build Strategic Air Force, Acquire Long-Range Bombers," *Hong Kong Feng Huang Wang*, June 28, 2004.

29. Cai Fengzhen and Tian Anping, eds., *Air and Space Battlefield and China's Air Force (Kong Tian Zhanchang Yu Zhongguo Kongjun)*, PLA Press, March 2004; Cai Fengzhen and Tian Anping, eds., *The Science of Integrated Air and Space Operations (Kong Tian Yiti Zuozhan Xue)*, PLA Press, August 2006. At the time of the first book (2004), Cai was the commandant of the PLAAF Engineering University in Xian. At the time of the second book (2006), he had moved

up to be one of the deputy chiefs of staff in the Headquarters Department at PLAAF Headquarters. Tian is an instructor at the PLAAF Engineering University.

30. Cai Fengzhen and Tian Anping, eds., *The Science of Integrated Air and Space Operations (Kong Tian Yiti Zuozhan Xue)*, PLA Press, August 2006, pp. 299-301.

31. *Ibid.*, p. 2.

32. Min Zengfu, ed., *An Introduction to Air Force Military Thought*, pp. 1-3.

33. *Ibid.*, p. 42.

34. An imperfect fit exists between the lexicons of Western and Chinese military planners and theorists. Western planners refer to "strategies," "doctrine," and "tactics, techniques, and procedures (TTP)" to address the concepts at the strategic, operational, and tactical levels of military operations, respectively. Chinese military planners are in agreement with Western counterparts at the strategic level, but prefer to use the term "theory" to describe operational and tactical level concepts.

35. The remaining 19 campaigns are as follows: ground force campaigns—mobile warfare, positional offensive, urban offensive, positional defensive, and urban defensive; naval forces campaigns—sea blockade, sea lines of communications (SLOC) destruction, coastal raid, antiship, SLOC defense, naval base defense; Second Artillery campaigns—nuclear counterattack, conventional ballistic missile campaigns; and joint service campaigns—blockade, landing, antiair raid, airborne, and antilanding. Wang Houqing and Zhang Xingye, eds., *Zhanyi Xue (Science of Campaigns)*, Beijing: National Defense University Publishers, May 2000, Chapter 13, pp. 350-351.

36. Hua Renjie, Cao Yifeng, and Chen Huixiu, eds., *History of Air Force Theory*, pp. 294-331.

37. *Ibid.*

38. *Zhongguo Renmin Jiefangjun Kongjun Zhanyi Xue (Science of PLAAF Campaigns)*, Beijing: Liberation Army Publishers, November 1988. Approval for research to begin on this publication occurred in June 1984.

39. Hua Renjie, Cao Yifeng, and Chen Huixiu, eds., pp. 294-331.

40. See *big5.xinhuanet.com/gate/big5/news.xinhuanet.com/zilian/2004-07/26/content_1649800.htm*.

41. Teng Lianfu and Jiang Fusheng, eds., *Air Force Operations Research*, May 1990, p. 157.

42. *Ibid.*, p. 152; Wang Houqing and Zhang Xingye, eds., *Science of Campaigns*, p. 346.

43. Wang Houqing and Zhang Xingye, eds., *Science of Campaigns*, p. 346.

44. Teng Lianfu and Jiang Fusheng, eds., *Air Force Operations Research*, p. 155.

45. This is in contrast to the 1988 version of *Science of PLAAF Campaigns*, which discusses air defense, air offensive, and combined arms campaigns, but does not mention air blockade, airborne, or joint service campaigns. The 1988 book also devotes separate chapters to air campaigns relative to command of the air, electronic countermeasures, and operations under chemical, biological, and nuclear conditions.

46. Of note, the 2000 *Science of Campaigns* book lists airborne campaigns as a joint campaign. However, a new PLA book published in 2006 has shifted airborne campaigns from joint status to being an Air Force campaign. See Zhang Yuliang, ed., *Zhanyi Xue (Science of Campaigns)*, Beijing, National Defense University Press, September 2006.

47. Yao Wei, ed., *China Air Force Encyclopedia*, pp. 108-110.

48. Zhu Rongchang, ed., *Kongjun Da Cidian (Air Force Dictionary)*, Shanghai: Shanghai Dictionary Publishing House, September 1996, p. 779, under the heading, *Principles of Flying Training of Air Force*, and p. 180, under the heading, *Guide Thought for Air Force Military Training*.

49. Ken Allen in discussions with senior PLAAF officers.

50. See *web.wenxuecity.com/BBSView.php?SubID=military_best&MsgID=1661*.

51. Yang Changlin, ed., *Dangdai Junguan Baike Cidian (Contemporary Military Officer Encyclopedia-Dictionary)*, Beijing: PLA Publishers, July 1997, p. 92. The term *"dagang"* can be translated in different ways. Various PLA dictionaries and encyclopedias published during the 1990s have an entry for the *dagang*, translating it as the Training Outline (*xunlian dagang*) or Programme of Military Training (*junshi xunlian dagang*). For

example, the PLA's *Air Force Dictionary*, which was published in 1996, translates *dagang* as the Programme for Military Training. See Zhu Rongchang, ed., *Kongjun Da Cidian* (*Air Force Dictionary*), Shanghai: Shanghai Dictionary Publishing House, September 1996, p. 180; Zhang Xusan, ed., *Haijun Da Cidian* (*Navy Dictionary*), Shanghai: Shanghai Dictionary Publishing House, October 1993, p. 236.

52. Zhu Rongchang, ed., *Air Force Dictionary*, p. 180.

53. *China's National Defense in 2002*, English Version, Beijing: Information Office of the State Council of the People's Republic of China, December 2002.

54. Based on multiple sources. The PLAAF's *dagang* consists of at least 15 chapters, each of which is divided into multiple sections.

55. Zheng Shenxia was one of 15 officers elevated to the rank of general or admiral in June 2004. See *english.peopledaily.com.cn/200406/20/eng20040620_146943.html*.

56. The only other PLAAF commander to serve on the CMC was Zhang Tingfa (from 1977 to 1982). Zhang was also a member of the Party's Politburo from 1977 until 1985, when he retired. For information on the PLA's grade structure and how it fits into the promotion system, see Kenneth Allen and John Corbett, "Predicting PLA Leader Promotions," in *Civil-Military Change in China: Elites, Institutes, and Ideas After the 16th Party Congress*, U.S. Army War College, Carlisle, PA, October 2004, pp. 257-278. Qiao is only the ninth commander since the PLAAF was established in 1949. It is highly likely that Qiao will be replaced at the 17th Party Congress in 2007, at which time he will be 68 years old.

57. Interviews.

58. The first PLAAF officer to be appointed to the position of Deputy Chief of General Staff was former PLAAF commander Wu Faxian in the 1960s.

59. "Four General Departments of the PLA Announce Personnel Changes," *Hong Kong Wen Wei Po*, January 12, 2005.

60. "Ma Xiaotian Promoted to Office of the President of Defense University," *Hong Kong Ta Kung Pao*, August 18, 2006. Prior to his appointment to NDU, Ma was one of the PLAAF's four deputy commanders.

61. *China's National Defense in 2004*.

62. Officers reportedly comprise over 30 percent of the PLA; officer ratios are even higher in the PLAAF, the PLA Navy, and 2nd Artillery. See Cheng Ying, Li Xuanliang, and Liu Fengan, "Transformation of PLA's Military Training Strategy: An Exclusive Interview with Zhang Baoshu, Director of the Military Training and Arms Department of the PLA General Staff Headquarters," *Shanghai Liaowang Dongfang Zhoukan*, No. 32, August 10, 2006, pp. 28-29.

63. Chiang Hsin-hsien, "China to Complete Force Reduction of 200,000 Troops in 2005," *Hong Kong Wen Wei Po*, October 6, 2005.

64. *Ibid*. In addition, see *www.chinamil.com.cn/site1/ztpd/2004-09/09/content_11222.htm*. Some of the command posts, such as the Wulumuqi Command Post, may have been downgraded by only one grade to a deputy army/corps-leader grade organization.

65. The last F-6 was produced in 1979. See *www.china-military.org/units/divisions.htm* for information on each division.

66. This description is a composite of information taken from Hong Heping and Tian Xia, "Head to the New Century," *Zhongguo Kongjun* (*China's Air Force*) 1996-95; and Wen Guangchun, ed., *Jidong Zuozhan Houqin Baozhang* (*Logistics Support for Mobile Operations*), PLA General Logistics Department Headquarters Department, PLA Press, January 1997. This is one of six books under the title *Gaojishu Tiaojian Xia Jubu Zhanzheng Houqin Baozhang* (*Logistics Support for Local Wars under High-Tech Conditions*) that the General Logistics Department commissioned the National Defense University and all logistics organizations to compile in 1995.

67. PRC 2004 *Defense White Paper*.

68. See *www.pladaily.com.cn/item/flying/content/1620.htm*.

69. See *www.pladaily.com.cn/item/flying/content/1634.htm*.

70. Hong Heping and Tian Xia, "Head to the New Century," *Zhongguo Kongjun* (*China Air Force*), October 1, 1996.

71. *People's Daily Online*, english.people.com.cn/english/20005/16/print20000516_40895.html.

72. Sun Maoqing and Ren Bo, "PLA Air Force Commander and Political Commissar Discuss Air Force Development," *Xinhua*, November 11, 2005.

73. These and other hardware issues have been documented extensively in *Jane's Defence Information* and other aviation journals and publications.

PART V:

THE PLA NAVY

CHAPTER 10

THE STRATEGIC AND OPERATIONAL CONTEXT DRIVING PLA NAVY BUILDING

Michael McDevitt

INTRODUCTION

The objective of this chapter is to place the ongoing development of the People's Liberation Army Navy (PLAN) in the context of China's overall strategy. Because building and sustaining an up-to-date navy capable of conducting a "modern war under high-tech informationalized conditions" is a very expensive proposition, to understand why the leadership of the People's Republic of China (PRC) is willing to commit the resources to "navy building" is central to answering the question of what "drives" the force posture, size, and capabilities of the PLAN.[1]

That the PLAN has been introducing capable new ships, submarines, and weapons over the past 15 years means that a compelling strategic case has convinced a national leadership unschooled in things maritime to dedicate the resources necessary for naval development. PRC leaders have thus come to believe that the strategic interests of the state can be secured only with a robust naval force—which is a historic departure from the dominant strategic traditions of China.[2]

This chapter postulates five separate but interrelated factors that animate, or drive, the leadership to actively support the development of the PLAN: first, what the PLA calls the "major strategic direction," which essentially means the compass direction from where potential threats to Chinese interests originate; second, a maritime strategy that comports with the continental strategic tradition of China; third, the need to deter Taiwan's independence and, if necessary, to deter or defeat a U.S. Navy relief force if the PRC elects to attack Taiwan; fourth, the historically novel situation in which international seaborne trade is what drives the economic growth of China; and fifth, the increasing dependence of the PRC's economic development on oil and natural gas delivered to the PRC by ships.

This chapter also will speculate about future PLAN developments regarding out-of-region presence deployments and will briefly comment on the possibility of the PLAN developing a robust sea-based leg for its strategic nuclear deterrent.

THE FIRST DRIVER: MAJOR STRATEGIC DIRECTION

Recently the PLA's Academy of Military Science published in English an important work on PLA strategic thought. Entitled *The Science of Military Strategy*, it has nearly 500 pages worth of important insights into how the PLA thinks about strategy.[3] One of the chief among them is the significance that strategic planners attach to determining the "major strategic direction." This determination is central to translating strategy into real operational plans and concepts because "the major strategic direction" forms the basis *on which operational plans are then developed and appropriate forces are procured, postured, and trained*.[4]

The Academy of Military Science defines the major strategic direction as "the focal point of the struggle of contradictions between ourselves and the enemy . . . in the overall strategic situation, it is the vital point of *greatest importance*" (emphasis added). In other words, the major strategic direction is where China's most important interests are either threatened or unresolved. *The Science of Military Strategy* goes on to say, "The major strategic direction is basically determined according to the national strategic interests and the fundamental international and domestic strategic situation."[5]

Once established, the major strategic direction tends to remain fixed. Since it involves fundamental national interests, and forms the basis for procuring specific military capabilities, it is not subject to arbitrary changes. In *The Science of Military Strategy*, this point is explicit:

> For example, the main strategic direction of China has seen three major adjustments since the establishment of the PRC. . . . In the mid-1950s the Party Central Committee and the Central Military Commission [CMC], in light of the strategic encirclement of China by foreign forces led by the United States and the serious situation of a possible strategic offensive launched against China, specified the southeastern coastal area of China as the main strategic direction. Between the 1960s and 1970s, as Sino-Soviet relations broke up and the Soviet Union deployed a million troops along the Sino-Soviet border and posed an increasingly serious military threat to China, the leadership changed the main strategic direction decidedly to the three northern regions. In the 1980s, they once again adjusted the main strategic direction according to the new international situation.[6]

This means that there have been only three iterations of the main or major strategic direction since 1949. The authors of *The Science of Military Strategy* are coy about its current direction, but it is not hard to discern since the authors point out that it has to do with the current international situation.

In analyzing the "current international situation" from the perspective of Beijing, it is clear that over the past 15 years PRC party leaders and diplomats have done a good job of advancing the national interest of stability in the area around China. They have secured the PRC's land frontiers by resolving or mitigating territorial disputes with Russia, Vietnam, Kazakhstan, Kyrgyzstan, and India. They have also negotiated "strategic partnerships" with most of these countries, and, in the case of the "stans" and Russia, have knitted them into the fabric of a regional security relationship called the Shanghai Cooperation Organization (SCO).[7] As a result, the PRC does not face a credible military threat from its continental neighbors in the near to middle time frame, nor does it have a territorial dispute with them that could be the pretext for military action. In terms of simple military capability, Russia is an exception to this judgment because of its still substantial strategic nuclear force; however, that threat has been moderated by good political relations enshrined in the "Sino-Russian Good Neighborly Treaty of Friendship," which went into effect on March 1, 2002.[8]

However, while its land frontiers are stable, looking east from Beijing beyond its eastern seaboard the situation is more strategically problematic. The PRC's maritime approaches are replete with unresolved sovereignty issues and genuine vulnerabilities. Strategic vulnerability from the sea is not a new issue for China. Weakness along its long maritime frontier has been a problem for Beijing since at least 1842, when the Treaty of Nanking ended the first Opium War. This 3-year conflict with Great Britain exposed Imperial China's military weakness, and ushered in the so-called "Century of Humiliation." The repeated military and diplomatic humiliations and defeats that

China suffered were inflicted by Western powers, including Japan that came mainly *from the sea*.[9]

The difference today, is that the PRC has the resources and political coherence necessary to address the reality that the vast majority of China's outstanding sovereignty issues and unresolved strategic problems are *maritime* in nature. For military strategists and planners, this makes establishing the major strategic direction a reasonably straightforward proposition. Consider the following issues, which are aspects of "the focal point of the struggle of contradictions between ourselves and the enemy":

- With Taiwan being an island, it is the combination of Taiwan's air defense and the threat of intervention by the U.S. military (primarily the U.S. Navy) that effectively keeps the Taiwan Strait a moat rather than a highway open to the PLA.

- Perhaps as strategically significant to a PLA planner as Taiwan is the geostrategic reality that the PRC's economic center of gravity is its east coast. Because it is a "seaboard," it is extremely vulnerable to attack from the sea—a military task the United States is uniquely suited to execute.

- Territorial disputes with Japan over islands and seabed resources in the East China Sea have become more serious, representing a potential flashpoint where Sino-Japanese interests are contested. Each state is emphasizing its claims by the periodic deployment of naval and coast guard vessels. The entire issue is maritime in nature.[10]

- Unsettled territorial disputes, and their concomitant resource issues, remain with respect to the Spratly Islands and the South China Sea. Again, this problem is maritime in nature.

- China's entire national strategy of reform and opening depends largely upon maritime commerce—i.e., trade. The PRC's economy is driven by the combination of exports and imports which together account for almost 75 percent of PRC gross domestic product (GDP). This trade travels mainly by sea.[11]

- Finally, there is the issue of energy security—or, as one commentator put it, "energy insecurity." It has become commonplace to observe that the PRC will increasingly depend upon foreign sources of oil and natural gas, most of which come by sea.[12]

Beijing's primary military competitor is the United States, the world's foremost naval power, which has maintained—for the past 50 years—a significant naval presence on "China's doorstep." Should the PRC elect to use force to resolve either the reunification dispute with Taiwan or outstanding maritime claims, the United States is the one country that could militarily deny success. Also by its air and naval presence in the region, it could stymie any Chinese attempt to use the growing capability of the PLA to settle these issues by *force majeure*. The United States is also becoming even more closely allied with China's historical antagonist Japan, which itself has an excellent navy and a formidable maritime tradition.[13]

Because of these factors, and especially because China's economic health depends upon unimpeded access to and use of the high seas, Beijing has been forced to think more seriously about how to deal with its maritime frontier. In the past, China could simply surrender ground to an invader without being defeated. But now, although the threat of invasion is long past, being complacent about its maritime frontier is no longer a viable strategic choice.

Given the maritime nature of all the PRC's outstanding strategic issues and its dependence upon trade for continued economic development, there is little question that the PRC's "main strategic direction" is eastward toward the central Pacific Ocean, and southeast toward the South China Sea and the shipping lanes from the Middle East. This judgment is reinforced by the December 2004 Chinese *Defense White Paper*, which breaks with the tradition of land force dominance, clearly stating that the PLAN, the PLA Air Force (PLAAF), and the ballistic missile force—the Second Artillery—are to receive priority in funding. Further, it explicitly lays out its ambitions for the PLAN, PLAAF, and Second Artillery in these words:

> While continuing to attach importance to the building of the Army, the PLA gives priority to the building of the Navy, Air Force, and Second Artillery force to seek balanced development of the combat force structure, *in order to strengthen the capabilities for winning both command of the sea and command of the air, and conducting strategic counter strikes.* (emphasis added)[14]

It is noteworthy that the authors of *The Science of Military Strategy* date the current (i.e., toward the sea) strategic direction from the 1980s. This seems to coincide with Deng Xiaoping's "strategic decision" in May–June 1985. At an enlarged meeting of the CMC during that period, he stated flatly that while there were still dangers of wars and conflicts in the world, the possibility of war with the Soviets was remote, as was that of war with the United States. My colleagues, Dr. Paul Godwin and Dr. David Finkelstein, have argued that as a result of this announcement, almost every aspect of China's national security orientation (political, economic, and military) shifted "from continent to periphery." The shift to the periphery included looking toward the sea. This time frame also coincided with a shift in the strategic focus of the PLAN, from coastal defense to offshore defense.[15]

GEOGRAPHY IS STRATEGIC DESTINY

Throughout China's long history, its strategic orientation could be categorized as continental, and hence its strategic tradition — its way of thinking about and framing strategic issues — has been largely focused on land war.

Today, however, the risk of cross-border aggression has moderated. The combination of globalization, democratic governance and the resulting "democratic peace," international norms of state behavior, and the deterrent value of nuclear weapons have substantially lowered the likelihood of cross-border aggression. The threat of invasion — the primary worry of Chinese, or indeed most Eurasian strategists for many centuries — has all but disappeared. As globalization proceeds, economic growth increasingly depends on trade, most of which is carried in containers loaded on ships. As a result, security on the high seas is becoming a growing preoccupation of countries that historically were not strategically focused on the maritime domain. The PRC is in the midst of this evolving strategic *zeitgeist*. Of course, this does not mean that PLA strategists have totally abandoned their land warfare strategic traditions.[16]

The PRC's maritime strategic outlook is securely nested in the continental tradition of using maritime power in a defensive strategic context—which, in the PRC's case, translates to protecting offshore sovereign interests and denying other nations the use of the high seas as an avenue for attacking China. This perspective dates back to August 1985, when then-CMC Vice Chairman Yang Shangkun addressed a meeting of the PLAN Party Committee and directed that the concept of "offshore defense" become the strategic concept that guides naval modernization. In effect, the PLAN was told to become more than merely a coastal defense force.[17] As former Navy Commander Vice Admiral Shi Yunsheng put it, "Following the . . . Central Military Commission meeting in 1985, we established the Navy's strategy of offshore defense . . . and defined the strategic mission of the Navy in the new period."[18]

THE SECOND DRIVER: DEFENSE-ORIENTED MARITIME STRATEGY

Arguably, a major reason the PLAN has fared so well in the internal competition for resources is that it presented a compelling strategic rationale for Navy building that managed to fit comfortably within a decisionmaking milieu dominated by a continental and ground force-oriented strategic culture. Thus Navy building circumvented that culture's fixation on the Navy as a defensive force, as opposed to a power-projection force.

The PLAN's notion of offshore defense is based on how another continental power thought about maritime strategy—the Soviet Union.[19] The Soviets developed a defensive maritime strategy with spaced, roughly parallel sea lines of defense (so-called "thresholds") at varying distances from the Soviet Union's coasts, with each succeeding line defended by weapon systems and tactical schemes appropriate to its location. This linear ground combat approach to thinking about maritime defense, or what might be thought of as "layered" defense, was used to rationalize the operational capabilities Soviet naval and air forces required to deny the use of the sea to its canonical threat, the United States. The high point of the Soviet approach to maritime defense was realized by the mid 1980s, when the Kremlin had in place a force of about 270 attack submarines, 280 major surface combatants, and over 1,300 naval aircraft allocated between the North Atlantic, eastern Mediterranean, and Pacific maritime approaches to the Soviet Union. The difference between the Soviets and the PRC approaches is that PLA has elected to define distance-related thresholds in terms of "island chains."

The parallels between the Soviet and PLA approaches to coastal defense almost certainly have much to do with continental strategic culture and Soviet mentorship, but it is also a very sensible approach to addressing the operational problem of defending the homeland against a force approaching from the sea. In today's U.S. Defense Department jargon, what the Soviets planned to do and what the PRC is planning to do are called "anti-access" — keep U.S. forces from getting close enough to the Chinese mainland to attack the PRC itself, or to interfere in a PLA attack on Taiwan.

The Soviet template considered the waters closest to the mainland out to approximately 200 nautical miles to be an area that Soviet naval forces and land-based air forces must be able to "control." Beyond this threshold, moving further to sea (to a range of about 1,000 nautical miles) the Soviets' strategy was to "deny" those waters to the U.S. Navy.[20] In other words, the military requirement is sea control close in, and sea denial as the distances from the mainland increase.

Overlaying this template on a map of East Asia results in a requirement for the PLAN to "control" the Yellow Sea, much of the East China Sea, the Taiwan Strait, the very northern portion of the South China Sea, and the Tonkin Gulf. Not surprisingly, this sea control area also closely approximates the PRC's exclusive economic zone (EEZ) and also generally follows the contour of the so-called "first island chain" that stretches southeast from Japan, through the Ryukyus, Taiwan, and the Pratas and Paracel islands in the northern portion of the South China Sea.

If the entire South China Sea is included within the first island chain threshold, the "sea control" zone runs beyond 200 nautical miles, greatly increasing the degree of difficulty in executing the mission. But by doing so, it encompasses the entirety of Beijing's territorial claims in the South China Sea, thus creating a "requirement" to improve the military potential of disputed islands as bases or outposts in the South China Sea. Whether or not all of the South China Sea is in or out of the sea control area, the PLAN faces an enormous challenge in controlling it. First, it is a vast space, and the waters are heavily traveled. Sea "control" implies a requirement to keep track of all the ships and craft at sea in the area to be controlled. To actually control the sea in time of conflict would require very thorough around-the-clock surveillance and control of the air space above the surface. These are a capabilities the PLA does not yet possess.

The expanse beyond the sea control threshold to the PLA's second island chain threshold is considered to be the sea "denial" area. This second threshold, approximating the Soviet 1,000 nautical mile (NM) line encompasses the enormous strip between 200 to 1,000 NM from shore. This is the area in which use of the seas would be "contested." The PLA ambition would be to deny it when necessary to U.S. forces.

Such considerations are not as arcane as might seem on first blush. These thresholds establish requirements for specific PLAN capabilities and as such are a driver of PLAN requirements. By establishing specific distances and areas where certain military effects are deemed to be necessary, it becomes simpler for "scientific strategists" to then stipulate precise operational characteristics for specific weapon systems, thus to determine how many ships, submarines, and aircraft are required to accomplish the intended missions.

LAYERED DEFENSE – ANTI-ACCESS

The first and most important requirement of a layered defense of the seaward approaches to China is an effective surveillance system that covers ocean approaches. Finding and locating ships on the high seas are very difficult because of the vastness of the oceans. Moreover, since ships move, simply determining the location of a ship at a particular point in time is quickly perishable information. One must keep track of moving ships by constantly updating the surveillance plot. In addition, a surveillance system must be able to distinguish between merchant ships and oil tankers, on one hand, and warships on the other.

Without effective surveillance, it is impossible to position offensive weapon systems or intercept moving naval task forces. The Soviets built an integrated surveillance system composed of radio direction-finding, electronic "spy ships" that could locate electronic signals, and space-based satellites designed to detect either electronic or infrared emissions from ships. It is worth noting that surveillance satellites are in relatively low orbits around the earth and therefore pass overhead relatively quickly. Constant, around-the-clock coverage of any geographic area requires a large constellation of satellites. That is why high-altitude drone aircraft have become such important new surveillance tools: they can loiter over a specific area for a long time.

The second element in the Soviet's layered defense system was land-based long-range aircraft that could be employed *en mass* to fire long-range antiship cruise missiles. The Soviet *Backfire* bomber remains the prototypical example of this capability. The Soviet tactic was to send aerial raids composed of two regiments (approximately 46 aircraft) against each enemy carrier battle group to ensure that enough bombers would survive the defensive screens to get within range to launch ship-killing cruise missiles.

It was this tactical threat that drove the U.S. Navy to develop the well-known *Aegis* radar-based air-defense system. The system was built specifically to permit missile defense ships to shoot down barrages of cruise missiles. China does not have anything equivalent to the *Backfire*, and this aspect of its layered defense is therefore not especially capable. The closest aircraft it has to the *Backfire* are the FB-7 fighter-bomber and the Chinese variant (B6H) of the venerable Soviet *Badger* bomber. Neither of these aircraft has the range of the *Backfire* or carries long-range cruise missiles.[21]

The third facet of the Soviet layered strategy was the use of submarines directed to their targets in much the same way that German U-boats were dispatched toward transiting convoys: they were vectored by commands from shore, based on surveillance information. The Soviet variant of this old operational concept was to intercept carrier battlegroups by the use of nuclear-powered submarines especially equipped with large magazines of cruise missiles. The PLAN is adapting this approach. It has focused on more modern, high-performance, conventionally-propelled submarines, which, while lacking the time on station and submerged speed of nuclear-powered submarines, are much more difficult to detect. But because conventionally-powered submarines do not have sustained endurance, they depend relatively more on accurate surveillance to help them locate targetable ships.

The Soviets recognized the vulnerability of their surface ships to both U.S. submarines and U.S. carrier aircraft, both of which could attack before the Soviet ships reached their cruise missile firing range. As a result, the Soviets intended to use their surface ships in roles closer to shore, either to defend against air raids headed toward the Soviet mainland or as last-ditch defenses. PLAN surface combatants suffer from the same vulnerability. It is likely that the PLAN would opt for the same solution as the Soviets: use surface warships closer to shore. In the PLAN's case, this would mean keeping them within the first island chain as last-ditch defenders, or to search for enemy submarines, or to fight the Taiwan Navy if the scenario included an attack on Taiwan.[22]

The operational template that the Soviets developed and the PLAN has adopted is a classic response of a continental strategic culture more interested in defending itself from attack from the sea than in using the ocean as a highway to attack another country.

This being the case, it is reasonable to inquire, given that the PLAN is optimized to deploy a layered defense against an expeditionary force or a force responding to a PLA attack on Taiwan, How does a layered defense contribute to an attack on Taiwan that is an offensive, not a defensive, undertaking?

THE THIRD DRIVER: TAIWAN

Anyone who has serious discussions with PRC uniformed or civilian officials will almost inevitably, at some point, be informed about how important Taiwan is to China as a matter of national sovereignty and territorial integrity.[23] Taiwan is the remaining unresolved territorial issue from China's Century of Humiliation. What is less frequently discussed is the geo*strategic* importance of Taiwan to the PRC. PLA strategists claim that Taiwan can have either a negative or a positive impact on the "survival and development of the Chinese nation and the rejuvenation of the great nation of China in this century."[24] The following argument is typical:

> Taiwan is located in the southeast of our sea area and is in the middle of the islands surrounding our coastline. It is in the key area of sea routes of the Pacific Ocean, and is thus crowned as "the key to the southeast coastal area of China," and "the fence to the seven provinces in the center of China." The sea routes from the East China Sea to the South China Sea, from Northeast Asia to Southeast Asia, as well as the route from the West Pacific to the Middle East, Europe, and Asia pass here.
>
> It is where we can breach the chain of islands surrounding us in the West Pacific to the vast area of the Pacific, as well as a strategic key area and sea barrier for defense and offense. If Taiwan should be alienated from the mainland, not only our natural maritime defense system

> would lose its depth, opening a sea gateway to outside forces, but also a large area of water territory would fall into the hands of others. What's more, our line of foreign trade and transportation which is vital to China's opening up and economic development will be exposed to the surveillance and threats of separatists and enemy forces, and China will forever be locked to the west side of the first chain of islands in the West Pacific.[25]

This remarkable assessment makes clear that in PLA strategic thought, Taiwan in the hands of the PRC provides an important element in the seaward defenses of mainland China, while Taiwan in unfriendly hands constrains China's access to the open ocean and could provide a base for attacks against the PRC.

During much of the Cold War, when China's military potential was focused on a threat from the Soviet Union, or was consumed by the "Cultural Revolution" and remained wedded to a doctrine of "people's war," the PLA did not possess the wherewithal to surmount the barrier posed by the Taiwan Strait to the application of PLA power against Taiwan. During this time, when the PRC threatened Taiwan with military punishment, its threats were largely empty. The PRC itself was "a paper tiger."

In retrospect, after the 1950s this did not matter much. Mao could trigger a Cultural Revolution, and Deng could focus on the Soviets because there was little threat that Taiwan would be permanently lost to China. The political leaders on both sides of the strait sought the same end: eventual reunification of the island and mainland. The argument was over what party would be in charge of the "uniting," not over whether to have one Taiwan and one China. Beijing displayed little urgency in improving its capability either to credibly deter Taiwan's independence or to field the means to capture it.[26]

This situation changed during the early 1990s, when democracy and notions of a *de jure* independent Taiwanese state began to resonate politically in Taiwan among the electorate. In turn, Beijing made policy pronouncements on the use of force to prevent the permanent separation of Taiwan from the mainland. Taiwan became an operational *idée fixe* for the PLA, which sought to field capabilities that would lend credibility to these anti-reunification pronouncements. In this whole process of fielding capabilities that could deter a declaration of independence by Taiwan, the PLAN has *not* played a central role.[27]

The PLA's single-minded focus on the operational problem of Taiwan has resulted in weapons and military capabilities that allow the PLA to "reach out and touch" Taiwan in a way that was not possible in earlier decades. This capability has translated into two PLA focus areas: putting hundreds of ballistic missiles in the hands of the Second Artillery; and purchasing excellent Russian tactical aircraft systems, which have allowed the PLA to credibly begin to match Taiwan's heretofore qualitatively better aircraft. The two strands of development go hand in hand: the missiles will punish Taiwan, destroy its command and control, and ground its air force, while the tactical aircraft will exploit this effort by gaining and sustaining air superiority (or "air control") over the strait and perhaps over Taiwan itself. Control of the air over the Taiwan Strait is the main prerequisite for an invasion of Taiwan.[28]

In a campaign to invade Taiwan, the PLAN has the seemingly prosaic but vital mission of getting the Army across the strait once air superiority has been achieved. Presumably, it is also responsible for dealing with Taiwan's small Navy, either at sea or by sealing it within its naval bases by mining the entrances. The requirement to get the Army to Taiwan is a driver for one aspect of Navy building, resulting in a steady growth of small, specially-designed amphibious warships. The PLAN also has at its disposal the substantial and modern PRC merchant fleet, plus a mobilizable fishing fleet.[29] Everything hinges on the PLA Air Force's ability to execute its mission. If that force can achieve and sustain air superiority over the strait, getting the Army to Taiwan would be within the capability of the PLAN.

But getting the PLA Army to Taiwan is not the most difficult problem for the PLAN. Its most important and most difficult mission is to stop the U.S. Navy from intervening. The PLAN must deter or defeat the hypothesized actions by U.S. Navy carrier strike groups to keep them out of the fight long enough for the combined forces of the Second Artillery, the PLA Air Force, and the Army to succeed. "Success" means creating the circumstances necessary to cross the strait (establishing air superiority), get ashore, and establish a defensible foothold on Taiwan, and subsequently cause the government in Taipei to surrender or flee. Any one of these enablers can be upset if the United States is able to intervene effectively.

In other words, the PLAN has an important role in a joint strategic mission to keep the most disruptive element of U.S. power at bay long enough for the actual assault to be effective. *This is a primary driver for the PLAN* and is in harmony with the PLA's doctrinal emphasis on what is called "key point strikes."[30]

In our discussion thus far on drivers, the conceptual approach to denying access through a layered defense was discussed within the context of a Soviet operational template. The translation of the conceptual approach into a specific operational task for the PLAN highlights both strengths and shortcomings in the PLAN as well as areas where it is reasonable to expect the PLA to focus its future efforts. The most obvious shortfalls in the PLA (that is, the entire PLA, since the Air Force and Second Artillery have important roles) are its weakness in the area of surveillance of the open oceans and its shortage of land-based aircraft to attack enemy warships before those ships launch aircraft that could interfere with the Taiwan attack or conduct attacks against mainland China.

For reasons of economy, internal development, prestige, and defense requirements, the PRC has focused on space-based systems. Because space-based systems are so important in open-ocean surveillance, it is reasonable to expect a continued emphasis in this area. Space-based surveillance is not a direct driver for the PLAN per se, but without surveillance the Navy's ability to execute its anti-access mission would be severely impaired. According to open sources, the PRC currently has seven satellites in orbit that can contribute to ocean surveillance. Significantly, in April 2006, Beijing launched its first radar satellite. It carries synthetic aperture radar, which can probably inspect objects as small as 20 meters in length and is thus excellent for identifying ships.[31] This may be the first in the constellation of radarsats necessary to maintain around-the-clock coverage.

The land-based air component of the layered defense consists of both PLAAF and PLAN Air Force aircraft. Based on open-source information, the only PLAAF bomber complement with antiship missiles is a single regiment (about 20 aircraft) of the *Badger*-variant B6H bomber. According to PLA airpower expert Ken Allen, these aircraft have been practicing over-water missions and antiship attacks since around 2002. The

PLAAF also has one regiment of FB-7 fighter-bombers and two of the new Russian-built Su-30MKK multirole regiments that could be used in antiship roles. PLAN organic aviation has one bomber regiment, one FB-7 regiment, and one Su-30MKK regiment that are capable of launching antiship cruise missiles. Again, each regiment has about 18–20 aircraft.[32]

In sum, the PLAAF and naval aviation force can field about seven regiments of aircraft with cruise missiles to attack approaching warships—perhaps 130–140 aircraft. Based on a metric of two aircraft regiments to oppose each enemy carrier battlegroup, the PLA could muster enough aircraft to attack a three-carrier force. But it has not fielded a long-range, air-launched cruise missile that would permit these aircraft to launch while remaining outside the surface-to-air missile envelope of U.S. warships. As a result, the aircraft would be vulnerable to U.S. Navy air defenses.

The PLA has apparently decided upon its submarine force as the most important element in its layered defense. This makes sense, given the inherent difficulty for the U.S. Navy or, for that matter, any navy, to locate very quiet modern submarines. The PLAN gets the most "bang for the buck" from submarines because they are very difficult to find, and hunting them will take a large number of USN ships, airplanes, and submarines. In the 10 years between 1995 and 2005, the PLAN commissioned 31 new submarines, but only two are nuclear-powered. As previously mentioned, because the vast majority of the PLAN submarine force is conventionally powered, it has significant operational drawbacks—limited endurance and speed.[33] Nonetheless, it is today an imposing force, and there is every expectation that it will continue to improve as it adds more nuclear-powered subs.

Operationally, submarines may have to be stationed as far away as 750 nautical miles from the PRC coast for sea denial so they can concentrate and attack enemy carrier forces before carrier aircraft can be involved in numbers in the air battle over the Taiwan Strait. If the intent is to delay the U.S. Navy, and perhaps even deter it from proceeding toward Taiwan, the PLAN will have to mass submarines in large numbers once carrier forces have been located in order to raise the risk to U.S. surface ships to the point where commanders might elect to stay outside the denial area until it clear of PLAN submarines. This deterrent task may take as many as six or more submarines per approaching carrier strike group.

Assuming that three to four U.S. carriers were assembled to respond to an attack against Taiwan, the PLAN would need at least 18 to 24 submarines on station. Its ability to sustain that posture would be a function of how often submarines rotated home and how long it would take to transit between homeport and patrol station. A rough estimate is that 60 modern submarines would be required for the anticarrier mission, if we assume the need to relieve on-station boats on a sustainable basis. In other words, it is reasonable to expect the PLAN to continue to grow a modern submarine force if it is to execute an anti-access strategy with confidence.

The PRC has added a new element to the layered defense—one that is uniquely Chinese and exploits one of the PLA's most effective capabilities. This new wrinkle is to use ballistic missiles to attack moving surface warships.[34] Traditionally, ballistic missiles were considered a poor weapon to use against ships at sea: ships move, and once the missile is fired, the aimpoint of a ballistic trajectory, by definition, would not be altered to account for target movement.

However, the PLA is apparently trying to place seekers in high-explosive missile warheads that will activate as the warhead descends into the target area, and then steer the warhead to the moving ship. I am informed that this is a difficult but not impossible technical task that depends on accurate surveillance (once again) plus missile warhead maneuvering technology that can slow down the warhead when it reenters the atmosphere so the seekers are not incinerated by the heat of reentry.[35] If the PLA can master and field this weapon system, it will be able to present a challenge to the U.S. Navy as serious as the one presented by Soviet *Backfire*-launched cruise missiles before the introduction of the *Aegis* radar system.

The foregoing discussion of how the PLAN might operate in support of a Taiwan invasion scenario is not based on any special insight into PLAN's plans. Rather, it is based on a good understanding of how the Soviet Union thought through the very same operational problem—defense against attacking carrier forces.[36] Since the principles of war on the high seas and the employment of warships and submarines do not greatly differ from one body of water to another as long as the operational scenario is similar (the high seas have no unique terrain features), this discussion is intended to respond to the question of what would drive PLAN force structure and capabilities.

THE FOURTH AND FIFTH DRIVERS: MARITIME TRADE AND IMPORTING ENERGY BY SEA

For discussion's sake, these two drivers can be combined since the issues are similar as far as the PLAN is concerned. In the case of both international maritime trade and the importation of oil by tankers, PRC reform and opening up have created new Chinese dependencies and therefore new problems that China has not confronted before. These dependencies create a probable requirement for the PLAN to contribute to the safe passage of ships bound for China with oil and natural gas or trade goods in times of crisis or conflict. This is a future requirement, in my judgment, because in a practical sense today the PLAN could not have much effect one way or another on the safety of merchant ships sailing to or from China in time of war with the United States.

China never had this problem in a strategically significant way because of its traditional economic "independence." Until Deng Xiaoping's decision to reform economically and "open up" China, the economy was largely autarchic. In 1793, George, Earl Macartney, headed the first British mission ever to be received by the Chinese court. His objective was to initiate relations so that trade could be established. The Qianlong emperor's oft-cited response that "China possesses all things and does not need European trinkets" is worth remembering if only to remind ourselves of what a unique turn-about China is facing today, when such a huge amount of its economic life is bound up in international trade.[37] Today, about 75 percent of the PRC's annual GDP is based on international trade (imports and exports, including both goods and services). Hong Kong port facilities alone process something close to 25 million containers annually.[38]

That China's economic life was not centered on trading abroad does not mean that Chinese did not venture to sea to fish and trade: these activities were always pursued. The difference is that they were not vital economic activities. One reason why China has so seldom focused on maritime issues is obvious: its security threats historically came from "barbarians" to the north and west. But it is also because the economic life of the country was not dependent on maritime commerce. This is not unusual among great states: over the course of history very few nations—"ten or fewer," in Colin Gray's calculation—have had a commerce-driven maritime strategic outlook.[39]

According to the International Monetary Fund (IMF), China's integration into the world economy is a landmark event with huge implications for both global and regional economies. This historical evidence, together with the still substantial development potential of the country, suggests that China could maintain relatively strong export growth for a number of years, provided that its growth momentum is not upset by the prevailing economic and political vulnerabilities. In other words, if all goes well, trade will continue to be an important aspect of China's growth.[40]

Dependence on foreign sources of oil is an irreversible fact of life for China—a fundamental feature of its energy profile and of the global energy market for years to come. West Africa and the Persian Gulf are both natural magnets for China in this regard. Indeed, Gulf suppliers are already vitally important energy partners for China. Since 1996, about 60 percent of China's crude oil imports have come from the Middle East.[41] Because most of this oil comes to China by sea via the Strait of Malacca, the PRC leadership reportedly became concerned about the possibility that in a conflict over Taiwan, the United States might try to block the strait and cut off PRC oil imports that come by that conduit.[42]

Clearly, PRC leaders who worry that the United States will block the Malacca Strait need to look at a map so they can understand that if this vital strait were closed, there still would be other deep-water passages through the Indonesian archipelago. Even if all these passages were somehow closed, maritime traffic from Africa and the Middile East could sail around Australia and proceed to East Asia via the central Pacific. The point is that the oceans of the world are seamless, and stopping traffic once it is operating on the high seas is very difficult. Oil travels to China in ships flying the flags of many nations. Today, only some 10-12 percent of China's oil imports are carried by Chinese flag tankers. The wide variety of carriers would complicate any attempt to identify and isolate tankers bound for China from those bound for Japan or Korea. Ironically, Beijing wants to reduce its dependency on foreign flag carriers, and over the next 15 years has set the objective of shipping 75 percent of its oil imports in Chinese flag carriers, making it easier for the United States to determine which ships are carrying oil to Chinese ports. [43]

Protecting commerce at sea from a determined opponent has been a mission that only a handful of Western navies have ever done successfully. Preying on commerce has been going on for centuries. Historically, it has taken two forms: piracy by independent actors, and state-sponsored attacks. In fact, the U.S. Navy was established because President John Adams needed to protect American trade plying the Mediterranean from the depredations of North African state-sponsored commercial warfare.

Historically, trying to cut off a country's maritime trade by intercepting ships on the high seas has been very difficult to do. The most successful attempts at cutting off maritime trade have been either at the point of origin or at the destination. (The Royal Navy's tight blockade of Europe during the Napoleanic Wars or of the United States during the War of 1812 were relevant examples of blockades at the destination, as was the U.S. success in isolating Japan during World War II.) Although PRC oil dependency has become an issue long after the "first island chain" was described as including the entire South China Sea, sea lane security for oil will now provide an additional driver for PLAN capabilities to exercise control over the sea lane from Singapore to China.

At the other end of the oil sea lane, tankers carrying Persian Gulf oil to East Asia, including China, must pass through the Strait of Hormuz, the entrance to the Persian Gulf, a choke point the U.S. Navy already dominates. Unlike Malacca, where alternative transit options are available, there are no other alternatives to getting large amounts of oil out of the Gulf except via Hormuz. If the PRC were serious about protecting Gulf oil bound for China, it either would have to be on the scene with a significant naval capability or would need to depend upon an ally in the area capable of acting efficaciously on Beijing's behalf.

So far, there have been no indications that the PLAN is actively planning to maintain a naval presence in this region, but it would be foolish to rule out the possibility. We should recall that throughout the 1980s the Soviet Union maintained a small naval force of submarines and surface ships in the Northern Arabian Sea, using facilities provided by Yemen. This squadron had the mission of demonstrating a Soviet presence, showing the flag in a region that the Soviets considered important. That the Soviets normally included a submarine in this mix ensured that U.S. Naval units in the area were always in a high state of readiness. One cannot rule out the possibility of similar facilities in either Iran or Pakistan being made available for a PLAN squadron. Such a squadron would be valuable both for a PRC peacetime presence and in crisis situations, but would probably be lost if an all-out conflict with the United States over Taiwan broke out.[44]

Deployed peacetime presence squadrons are one of the missions that the growing PLAN destroyer and frigate force could perform. It is not entirely clear that maintaining ships on distant stations is something the PLAN is planning to do, but it would not be a surprise if in time it elected to do so. For example, today most North Atlantic Treaty Organization (NATO) countries—as well as Asia powers Australia and Japan—maintain a more or less permanent naval presence in the Persian Gulf/Northern Arabian Sea region.

The model could become a driver of force structure. PLAN surface combatants can act as valued symbols of China as a great power with global interests, capable of operating and exercising around the world. Showing the flag is an important peacetime mission. These ships also have the capability to perform escort roles and would have some functions in a Taiwan scenario, operating against the Taiwanese navy. However, in a conflict against the U.S. Navy, the inadequate antiaircraft and antisubmarine defenses on these ships would make them very vulnerable to U.S. tactical air or U.S. and allied submarines. The PLA has spent considerable time studying the United Kingdom's 1982 Falklands campaign, and is aware of the sinking of the Argentine cruiser *Belgrano*, a dramatic illustration of surface ship vulnerability to submarine attack.[45]

The best way for the PRC to protect its sea lanes and commercial traffic is to maintain good diplomatic relations with trading partners and littoral states adjacent to them. The only country that could seriously disrupt merchant traffic destined to or from China is the United States, and it is not clear to this observer what the PLAN could do about it. It will be many years before the PLAN is able to operate surface ships independently at sea in the face of a hostile United States. That is not to say that maintaining distant squadrons lacks high utility in peacetime and in periods of crisis. Whether such utility results in a demand for surface ships remains to be seen.

A LOOMING DRIVER: THE MILITARY IMPLICATIONS OF OVERSEAS INTERESTS AND OF BEING A RESPONSIBLE "STAKEHOLDER"

Theorists of naval strategy have always drawn a close connection between a nation's far-flung economic interests and a strong navy. It seems clear that China has bet its future on globalization and its ability to succeed in the global system. The latest PRC *White Paper on National Defense* explicitly makes the point that "China has never been so closely bound up with the world as it is today."[46] What this means in practical terms is that the PRC is developing global interests that are derived from its global trade and quest for energy security. As discussed above, China has reaped tremendous economic benefits from opening up and becoming an integral part of the global economy.

According to Assistant Foreign Minister Lu Guozeng,

> After more than 20 years of development, China is now enjoying closer and closer relations with other countries. There were very few Chinese-funded businesses overseas at the beginning of reform and opening up. In contrast, by the end of 2003, China has had an accumulative foreign direct investment (FDI) of USD 33.4 billion with

3439 companies covering 139 countries and regions. The statistics released by United Nations Conference on Trade and Development (UNCTAD) recently show that China is expected to surpass Japan to become the 5th largest source country of FDI. Therefore, China's overseas interests are on the rise.[47]

In a more recent speech, PRC Vice President Zeng Qinghong stated, "Overseas investment by Chinese companies has increased by over 20 percent annually, with 80 percent of it made in Asia. In 2005, Chinese made 31 million overseas visits. Asia is the top choice of a large number of Chinese tourists. All this has played and will continue to play an important role in promoting economic growth in Asia and the world."[48]

According to the PRC Consulate in Houston, Texas, Chinese workers overseas face greater danger than ever before. An article posted on the consulate's website raises the question of how China should protect its citizens overseas. There were some 30,000 Chinese workers in Iraq before the war and some 230,000 spread among Israel, the United Arab Emirates, Jordan, and Egypt. Chinese workers were attacked in Afghanistan, and about 5,000 are working on various projects in Pakistan. The point of the article was that, as more and more Chinese go abroad, the danger of Chinese citizens being harmed or killed increases.[49]

Recently, a car bomb detonated harmlessly near an oil refinery in Nigeria, and the group claiming responsibility warned, "We wish to warn the Chinese government and its oil companies to steer clear of the Niger Delta. Chinese citizens found in oil installations will be treated as thieves. The Chinese government by investing in stolen crude, places its citizens in our line of fire." The correspondent reporting this episode raises an interesting rhetorical point: How did China find itself in the crosshairs of the Movement for the Emancipation of the Niger Delta? The answer, of course, is that China is active around the world, especially in search of energy.[50]

Just as China is economically engaged globally, it is also increasing its global military presence by increasingly participating in United Nations (UN) peacekeeping missions. The latest is the Lebanon mission: Prime Minister Wen Jiabao pledged 1,000 PLA soldiers in response to UN Resolution 1701. The BBC report editorializes that this pledge is evidence that China is starting to intensify "its diplomacy in areas it had not seen as vital." This is the sort of behavior one could expect from a responsible stakeholder.[51]

That the PRC is going global is well recognized and the subject of frequent commentary, which need not be repeated here. By going global, Beijing is translating economic engagement into political interests. Historically, economic and business interests, when paired with concerns about the safety of citizens, have translated into the employment of naval forces on distant stations to safeguard those interests and to respond to local crises and disturbances. In previous eras, this was often dubbed "gunboat diplomacy." While this form of coercive diplomacy is no longer routinely practiced, the modern counterpart of dispatching naval forces when instability threatens a nation's economic interests or the lives of its overseas nationals is still very much alive. Over the past few years, Great Britain, France, and the United States have deployed forces to various crises in West Africa. The recent evacuation of foreign nationals from Lebanon by French, Italian, and U.S. naval forces is just the latest example.

Clearly, officials in the PRC are considering the implications of its global interests in terms of responding to threatening events. When combined with Beijing's worries about energy security and sea lane protection, it is likely only a matter of time before we witness periodic deployment of PLAN ships or small task groups designed to show the flag and maintain presence in areas where the PRC has economic interests or large numbers of citizens abroad.

This driver is reinforced by pressure from Washington for the PRC to become a responsible stakeholder. Responsible stakeholders participate in UN-sanctioned peacekeeping missions, as the PLA is increasingly doing. If this trend continues, it will certainly create a demand for a PLAN that can support these UN missions. Such erstwhile responsible stakeholders as the United States, the United Kingdom, and France routinely deploy naval forces abroad to perform these sorts of missions. The other member of the UN permanent five, Russia, used to do so during Soviet days—and now that its energy export-driven economic resurgence is taking hold, there are signs that its still large Navy is stirring back to life. If the former tendency resumes, it will leave the PRC as the only member of the permanent five not to have a globally deployable naval force.

If the PLAN begins to conduct distant peacetime presence operations, and I suspect it will, it will be accomplished by ships with modest expeditionary capabilities. The current trend in naval construction in Europe, as well as East Asia, is for 12,000 to 17,000 ton multipurpose amphibious (or expeditionary) ships that can carry a few hundred soldiers or marines, several helicopters, good medical facilities, and the wherewithal to establish good command, control, and communication centers. These are the sorts of ships that are useful in missions such as humanitarian relief, disaster relief, evacuation of people in danger, or simple presence to signify a willingness to protect interests in jeopardy.

A NON-DRIVER: TAKING STRATEGIC NUCLEAR WEAPONS TO SEA

Another potential driver would be the desire on the part of the PRC to take advantage of the vastness of the open ocean to enhance the survivability of its nuclear deterrent against the United States and potentially circumvent U.S. missile defense by being able to launch intercontinental ballistic missiles (ICBMs) from submarines along azimuths outside the engagement zones of antiballistic missile (ABM) systems. Should the PLA elect to pursue this course of action with ballistic missile submarines (SSBNs), it would need to overcome a potentially serious vulnerability by making certain that its SSBN force was so acoustically quiet that it could not be tracked by U.S. attack submarines.

Russian advisors to the PLAN may have discussed Cold War vulnerability issues related to the Soviet Navy's own SSBN force. These issues were so serious that the Soviet Navy had to cluster its ballistic missile submarines in heavily protected maritime enclaves (called bastions) to ensure that its boats survived in case of war with the United States.

The combination of close contacts with the Russian navy and the growing body of unclassified studies on Cold War naval operations[52] must have made it abundantly clear to PLA planners that unless PLAN SSBNs can operate undetected by U.S. forces, it would be risky to make substantial investments in a sea-based leg of their nuclear retaliatory capability. Noisy SSBNs would be vulnerable on the high seas, and would become a resource black hole if the PLA had to create a Soviet-like "bastion" defense to protect them.[53]

The PLA has an alternative basing option that can take advantage of the vastness of the Chinese mainland, in which the Second Artillery's new solid-fuel road-mobile systems are far more survivable. A far more likely PLAN option, and one that its modernization program suggests it is pursuing, is to arm nuclear attack submarines with nuclear-tipped cruise missiles. These multimission submarines can be employed in a wide range of operational tasks, and at the same time provide a hedge in support of China's avowed "nuclear counterattack" doctrine.[54]

CONCLUSION

The PRC is investing in Navy building for the straightforward reason that, without a capable Navy, it has serious strategic vulnerabilities otherwise unaddressable—especially in the case of Taiwan. Without a credible naval establishment, it can threaten Taiwan with punishment but not seizure.

Given that the strategic case for Navy building emerged some 20 years ago, the CMC and PLA had to choose what sort of Navy to build. The choices were relatively clear. One was the historical model of the Imperial Japanese Navy (IJN). The IJN is tangible proof that a Western-style blue water Navy was possible in an Asian context. But developing such a Navy would have meant a departure from China's continentalist strategic tradition. Besides being countercultural to an Army-dominated PLA, such a blue water Navy would have been expensive and very difficult to make credible in terms of training and technology. China's only attempt to field such a Navy met with disaster in 1895.

The PRC's early relationship with the Soviets provided the second, more obvious, template for the PLA. The geostrategic circumstances facing the Soviet Union and China were similar when it came to threats from the sea, and the defensive Soviet-style anti-access model was also less expensive and easier to build because the PLA could capitalize on Soviet-developed technology and operational concepts. Finally, this approach to Navy building fitted within the continentalist worldview at the highest levels of military and party decisionmaking.

As it turns out, this approach to Navy building also fits well with the political message that Beijing has been sending to the world: China's rise will be peaceful and nonthreatening. Fielding an obviously defense-oriented Navy would be tangible evidence that the PRC was not going to become an expeditionary or power-projection threat. Exceptions to this assessment of the PRC as nonthreatening are the cases of Japan and both Koreas. They are within or adjacent to the PLAN sea-denial area—the first island chain.

The PLAN submarine force in particular is a capability-based threat to Japan's economic lifelines of maritime trade that Japan cannot, and probably will not, ignore. For the rest of Asia, an avowedly power projection PLAN would be counterproductive to China's broader strategic objectives of not creating powerful enemies in the region, especially since such a naval force would not be essential to satisfying the PRC's strategic objectives. In this context, the PLAN's focus on commissioning many more diesel submarines than nuclear submarines also helps reinforce the positive diplomatic message of a peaceful rise. They are quieter, are very hard to find, and create the image of *being defensive in nature*. They fit within the template of East Asian naval developments that feature South Korea, Singapore, and Malaysia joining Japan, Taiwan, and Australia as nations with conventionally-powered submarines.

It is unlikely that when Liu Huaqing developed his "island chain" approach to maritime strategy, he foresaw the tremendous growth in China's global trade and quest for natural resources (especially energy). Nor is it likely that he foresaw the PRC's growing international role in UN peacekeeping. The idea that thousands of PRC citizens would be working or traveling abroad did not seem likely to any student of China 20 years ago. That those citizens might need protection from terrorists or criminals was equally unanticipated, if similar American failures in long-range thinking are any guide.

A combination of such factors, plus the pressure from the United States to become a responsible stakeholder, is creating demand signals for a PLAN that can support UN-sanctioned missions, protect PRC interests abroad with a show of force, protect or evacuate PRC citizens in jeopardy, protect sea lines of communication, respond to natural disasters, and demonstrate PRC resolve in support of embattled friends in Africa and along the South Asia littoral. But today these are issues that Beijing and the PLAN are, I believe, just beginning to think about seriously. It is not enough simply to think about wartime employment concepts, the PLAN, unique among all of the PRC's military services, must also now consider distant, prolonged peacetime operations as part of its core mission set.

These combinations of potential missions will, I believe, require the PLAN to learn how to deploy and sustain surface combatants, amphibious ships, and support ships on distant stations for long periods of time. Also, it will almost certainly create a sound rationale for having some sort of an aircraft carrier, since helicopters are particularly valued in most of these missions.

This means that the PLAN probably faces another addition to its core mission in its future. It will continue to maintain a defensive strategy for the defense of China and its possessions, but it will also deploy a force whose primary utility will be to provide peacetime presence, sea lane monitoring, and crisis response. This force will probably not be particularly valuable in case of a real war with the United States, but such a war is not likely. This next-generation Navy will be useful to the PRC in furthering its own interests while also demonstrating that it too can be a responsible stakeholder among military forces of the community of nations.

ENDNOTES - CHAPTER 10

1. PRC *Defense White Paper*, December 2004, Information Office of the State Council of the PRC, December 2004, Beijing, at *english. people.com.cn/whitepaper/defense2004*.

2. U.S. Department of Defense, *Annual Report to Congress on the Military Power of the People's Republic of China 2005*, Washington, DC: Office of Secretary of Defense, released May 2005. This and the preceding annual reports are the most authoritative open-source references to PLA modernization. In terms of major warships, since 1995 the PLAN has commissioned about 31 new submarines and 26 new surface combatants (destroyers and frigates). It has also fielded sundry modern missiles and torpedoes with which to arm these vessels.

3. This entire section is drawn largely from Peng Guangqian and Yang Youzhi, eds., *The Science of Military Strategy,* Beijing: Military Science Publishing House, Academy of Military Science of the People's Liberation Army, 2005, pp. 230–234. The title is suggestive, reminding us that the PLA approach to strategy purports to be scientific. As in the Soviet forces, military science in the PLA takes as one of its major purposes the identification of laws of war and principles of armed conflict. The process seeks to describe objectively the relative capabilities of adversaries and to assess them by means of historical (experience) and scientific

analysis based on Marxist-Leninist principles. See the schematic on p. 36.

4. *Ibid.*, p. 231.

5. *Ibid.*, p. 232.

6. *Ibid.*, pp. 233–234.

7. This is my interpretation. For the latest official assessment, see "China's National Defense in 2006," PRC *White Paper on National Defense, www.opensource.ove/portal/server.ptgateway/PTARGS*. The opening section of the latest PRC *Defense White Paper* discusses Beijing's assessment of the security environment.

8. *People's Daily on Line,_english.peoples.com.cn/200202/28/eng 20020229_91188.html.*

9. Dean Cheng, *Shedding the Century of Humiliation: China's Place in Asia,* a paper prepared for the Sino-Japan Strategic Rivalry Workshop, which took place at CNA Corporation headquarters, Alexandria, VA, April 2006.

10. The Center for Naval Analyses has partnered with the Institute for Defense Analyses, National Defense University, and Pacific Forum/Center for Strategic and International Studies (CSIS) on an 8-month project that examines all aspects of the current state of Sino-Japanese relations. The study is paying particular attention to the disputes in the East China Sea. A final report was completed in December 2006, and is available from this author. See "Chinese Warships Make Show of Force at Protested Gas Rig," *Japan Times,* September 10, 2005," *search.japantimes.co.jp/print/nn20050910a1.html.*

11. Discussion with Daniel Rosen, President, China Strategic Advisory, LLC, on January 24, 2007.

12. Dan Blumenthal and Joseph Lin, "Oil Obsession: Energy Appetite Fuels Beijing's Plans to Protect Vital Sea Lines," *Armed Forces Journal,* June 2006. Posted on the AEI website,_*www.aei.org/publications/pubID.24499,filter.all/pub_detail.asp,* p. 10.

13. A recent expression of PRC angst over the strengthening of the U.S.-Japan Alliance and Japan's evolution toward becoming a more normal major power is found in the 2006 PRC *Defense White Paper* in the first section dedicated to a discussion of "The Security Environment."

14. PRC *Defense White Paper*, December 2004, Information Office of the State Council of the PRC, December 2004, Beijing, english.people.com.cn/whitepaper/defense2004.

15. Paul H. B. Godwin, "From Continent to Periphery: PLA Doctrine, Strategy, and Capabilities Toward 2000," *China Quarterly*, Vol. 146, June 1996, pp. 464–487. This section of the present chapter also draws from David M. Finkelstein, "Reform and Modernization of the People's Liberation Army Navy: From Marginalization to Major Importance," unpublished paper in author's possession, presented at the Center for Naval Analyses Conference on the PLAN, April 6–7, 2000.

16. *The Science of Military Strategy*, pp. 3–13. These pages contain a very interesting but fulsome discussion of China's history of strategic thought.

17. Chiang Shang-chou, "China's Naval Development Strategy — Building an Offshore Defensive Naval Armed Force," *Kuang Chiao Ching*, December 16, 1998, *Foreign Broadcast Information Service* (FBIS). See also Paul H. B. Godwin, "Force Projection and China's Military Strategy," paper presented at the Sixth Annual Conference on the PLA at Coolfont, VA, June 1995, p. 4; Dr Nan Li, "Reconceptualizing the PLA Navy in Post-Mao China: Functions, Warfare, Arms, and Organization," Institute of Defense and Strategic Studies (IDSS), Singapore, Working paper No. 30, August 2002, p. 6.

18. Chen Wanjun and Zhang Chunting, "Shouldering the Task of a Century-Straddling Voyage — Interviewing Newly Appointed Navy Commander Vice Admiral Shi Yunsheng," *Liaowang*, cited in David M. Finkelstein, "Reform and Modernization of the PLA Navy: From Marginalization to Major Importance," CNA Corporation: April 2000.

19. Bernard D. Cole, *The Great Wall at Sea: China's Navy Enters the Twenty-First Century*, Annapolis, MD: U.S. Naval Institute Press, 2001, pp. 163–164. See also an excellent discussion of Soviet influence in You Ji, "The Evolution of China's Maritime Combat Doctrines and Models: 1949-2001," IDSS, Singapore, Working paper No. 22, May 2002, pp. 2–8.

20. Notions of sea control and sea denial date back to the writings of Alfred Thayer Mahan and his near contemporary, Julian Corbett. They stem from concepts of "command of the sea." A search of their writings will yield often contradictory definitions of these terms. My understanding is that "sea control"

means having the capability to prevent an enemy from using some segment of maritime geography for as long as one wishes. In other words, one party can use the sea at its pleasure, while an opponent cannot. *Today, this is hard to accomplish in practice unless one also controls the air above the water in question.* "Sea denial," on the other hand, means temporarily controlling an area of water, with the recognition that control will be contested and that neither side has complete freedom to use the sea as it wishes. Many books talk about and around this topic. For the best extended discussion, see Colin S. Gray, *The Leverage of Sea Power: The Strategic Advantages of Navies in War*, New York: The Free Press, a division of Macmillan, Inc., 1992, pp. 19, 274.

21. Cruise missile range is important if the attacker hopes to be successful. The ideal occurs when the attacking aircraft can launch its cruise missiles before it enters the defensive umbrella of the surface ships, so that the defender is confronted with having to shoot at cruise missiles themselves—a difficult target. It is much easier for modern surface-to-air missiles to intercept aircraft. The colloquial characterization of this tactical problem is to "shoot the archer before he launches his arrows."

22. John B. Hattendorf, *The Evolution of the U.S. Navy's Maritime Strategy, 1977–1986*, Naval War College Newport Paper 19, Newport, RI: Naval War College Press, pp. 101–181. The referenced pages reprint the National Intelligence Estimate, 11-15-82, which discusses the Soviet approach to what is now known as an anti-access maritime strategy. On p. 170, a map displays Soviet sea denial and sea control areas in the Pacific. This entire section on the Soviet approach to layered defense is also informed by Hattendorf's personal experience in the 1980s, when he spent 4 years conducting strategic studies and program assessments oriented toward understanding and defeating Soviet maritime strategy.

23. The PRC *White Paper on National Defense 2006*, p. 3, uses these very words. Electronic version in English found at www.opensource.gov/portal/server.pt/gateway/PTARGS.

24. The *Science of Military Strategy*, p. 442.

25. Ibid., pp. 442–443.

26. Harry Harding, "Speech before the Asian Affairs Committee of the Association of the Bar of New York on November 1, 2000," Reprinted in the National Committee for U.S.–China Relations newsletter, March 2001.

27. While Taiwan is a priority for campaign planning and the PLA's most likely contingency, it would be a mistake to consider Taiwan as the only reason that the PLA is modernizing. See James Mulvenon and David Finkelstein, eds., *China's Revolution in Doctrinal Affairs: Emerging Trends in the Operational Art of the PLA*, Conference Report, Washington, DC: CNA Corporation, December 2005, p. 12.

28. See DOD Report to Congress, pp. 3-6, for a discussion of the forces postured opposite Taiwan.

29. *Ibid.*, pp. 4, 30, for a discussion of PLAN amphibious and expeditionary forces.

30. "Key point strikes" call for the concentration of the PLA's most powerful capabilities to destroy or degrade *the enemy's best capabilities in* order to (1) level the technological playing field at the inception of hostilities, and (2) disrupt the enemy's campaign before it can achieve operational momentum. The PLA's approach rests upon the correct selection of "enemy vital targets" and "key point application of force" against those targets. David Finkelstein, *Evolving Operational Concepts of the Chinese People's Liberation Army: A Preliminary Exploration*, Alexandria, VA: The CNA Corporation, 2001, p. vi.

31. Conversation with Mr. Dean Cheng, CNA China analyst and a leading expert on Chinese space activities. See also his conference report, *China's Space Program: Civilian, Commercial, and Military Aspects*, A CNA Conference Report, Alexandria, VA: The CNA Corporation, May 2006.

32. Discussion with Ken Allen, September 6, 2006.

33. Ronald O'Rourke, *China Naval Modernization: Implications for U.S. Navy Capabilities – Background and Issues for Congress*, CRS Report for Congress, updated August 29, 2006. CRS website, Order Code RL 33153, p. 8. This is the single best open-source compilation of information on the PLAN available to scholars and research specialists.

34. DOD Report to Congress, p. 4. The report says, "China is exploring the use of ballistic missiles for anti-access/sea-denial missions."

35. Interview of technical experts in the U.S. Department of the Navy, June 2006. See also O'Rourke, CRS China Naval Modernization, p. 5.

36. In the 1980s, the U.S. Navy developed a maritime strategy involving a 600-ship force conceived to defeat the Soviet strategy. See Hattendorf, *Evolution of U.S. Navy's Maritime Strategy*.

37. F. W. Mote, *Imperial China: 900-1800*, Cambridge, MA: Harvard University Press, 1999, pp. 913.

38. Data extracted from a presentation given by Dr. David Denoon, Professor of Politics and Economics at New York University, at a CNA workshop held at CNA headquarters, Alexandria, VA, in July 2006. Denoon calculates the combined total of China's exports and imports in 2004 as being approximately $1,125 billion. He posits China's gross domestic product (GDP) in 2004 as being $2,000 billion.

39. Colin Gray, *Leverage of Seapower*, p. xii.

40. Eswar Prasad, ed., "China's Growth and Integration into the World Economy: Prospects and Challenges," IMF Occasional Paper 232, Washington, DC: International Monetary Fund, 2004, *www.imf.org/pubs/cat/longres.cfm?sk=17305*, p. 9.

41. John Calabrese, "China's Global Economic Reach and US primacy," testimony before the U.S.–China Economic and Security Review Commission, August 3, 2006. Paper in author's possession.

42. Blumenthal and Lin, "Oil Obsession," p. 1. According to the state-run *China Youth Daily*, June 15, 2004, Hu Jintao raised the issue of energy security at the closing session of the Central Economic Work Conference held on November 29, 2003. Hu is reported to have stated that more than half of China's domestic oil imports came from the Middle East, Africa, and Southeast Asia and that roughly four-fifths of its crude oil imports were transported through the Straits of Malacca. Hu expressed concern that "certain major powers have always been encroaching and attempting to control the transport routes along the Straits of Malacca."

43. Gabe Collins, "China Seeks Oil Security with New Tanker Fleet," *Oil and Gas Journal*, October 9, 2006. This is an excellent discussion of this entire issue by a member of the U.S. Naval War College faculty.

44. Personal experience during deployed operations in this region.

45. LTC Wu Jianchu, "Joint Operations—the Basic Form of Combat on High-Tech Terms," *Junshi Kexue*, Vol. 4, 1995; Zhou

Xiaoyu, Peng Xiwen, and An Weiping, *New Discussions on Joint Campaigns*, Beijing: National Defense University Publishing House, January 2000, pp. 21-22.

46. PRC *Defense White Paper 2006*, Preface, p. 1.

47. Lu Guozeng, *Seminar on Diplomacy and Economics*, December 18, 2004. See *www.fmprc.gov.cn/eng/topics/desenglish/t176097.htm*.

48. PRC Vice President Zeng Qinghong, Speech to Open the 2006 Annual Meeting of Boao Forum for Asia, April 22, 2006. *www.fmprc.gov.cn/eng/zxxx/t248674.htm*.

49. Consulate General of the People's Republic of China in Houston, TX, News Release, *houston.china-consulate.org/eng/zt/ljzg/t140336.htm*.

50. Ian Bremmer, "As China Rises, So Does Its Risk," *Fortune Magazine*, June 8, 2006, *money.cnn.com/magazines/fortune/fortune_archive/2006/06/12/8379268/index.htm*.

51. BBC News, September 19, 2006, *news.bbc.co.uk/go/pr/fr/-/2/hi/asia-pacific/5355128.stm*.

52. Sherry Sontag and Christopher Drew, *Blind Man's Bluff: The Untold Story of American Submarine Espionage*, New York: Harper Paperbacks, 1999.

53. Hattendorf, *Evolution of U.S. Navy's Maritime Strategy*, pp. 30–36.

54. PRC *Defense White Paper 2006*. p. 5.

CHAPTER 11

RIGHT-SIZING THE NAVY: HOW MUCH NAVAL FORCE WILL BEIJING DEPLOY?

Bernard D. Cole

INTRODUCTION

This chapter evaluates China's plans for building naval forces based on estimates of need by Beijing. Addressing this question and associated issues involves exploring the current state of China's naval forces and strategy. A brief assessment of how this force compares to the missions laid down in the national defense white paper will be offered as well. The current state of the People's Liberation Army Navy (PLAN) will be described in terms of both numbers and capabilities. Future developments will be estimated out to 10 years. The PLAN in 2016-17 will almost certainly be operating in an international environment different from that prevailing today. How different the Chinese government believes that environment might be will in turn go far toward determining the character of the naval forces Beijing estimates it needs.

That decision, in turn, will depend on Chinese national security aspirations and concerns. Implementing that decision—or rather a whole range of associated decisions—will be empowered by the economic and personnel resources available within a framework of national industrial and research capabilities. The geography of maritime Asia will remain the same, of course, but the interests, resources, and intentions of the nations therein will not. How these are interpreted by China, by her allies, and by her opponents will strongly influence Beijing's decisionmaking process with respect to future Navy building. Hence, this chapter will use maritime scenarios as a vehicle for evaluating situational parameters that are likely guiding, and will guide in the future, China's decisions on how strong and capable a PLAN will be deployed.

China's naval ambitions and possible employment will be measured in accordance with three scenarios. While somewhat arbitrary in selection, these scenarios are based on several plausible possibilities. Each scenario will be developed within the following five parameters. First, each describes an existent (2006-07) conflict of interests between China and at least one other nation. Second, none of these conflicts appears to be susceptible of near-term resolution (that is, before 2016-17), although this is admittedly the most arbitrary of the selected parameters. Third, each scenario is inherently maritime in character in terms of geographic location and access to naval forces, and would dictate a PLAN deployment against significant naval power should Beijing elect to implement the naval instrument of statecraft.

Fourth, the scenarios are not predetermined to lead to the employment of violent naval force, although that would be an attractive option in each case. Resolution through diplomacy, through third-party arbitration, or via multilateral negotiation might very well serve as a decision path, but will not receive the focus that employment of the PLAN will receive. Fifth, third-party naval intervention will be considered, but will not form an integral part of the discussion of each scenario.

The navies of Japan and South Korea are formidable, modernizing forces, but the only likely navy the PLAN is considering in its active planning process is that of the United States. Despite the U.S. Navy's current and likely future ability to exercise overwhelming naval power anywhere in maritime Asia, Chinese naval planners are doubtless working to develop a way to avoid, counter, or perhaps even co-opt that power.

The process of evaluating the PLAN will focus on its force structure, to include surface combatants, aircraft, submarines, and shore establishment. An attempt will also be made to assess the state of PLAN training, from the individual to fleet and joint levels, in terms both of theory and practice. This assessment will lead in turn to a discussion of PLAN doctrine, the goal being to evaluate its origins and its linkages to force structure, perceived threats, and possible future developments. Finally, China's *2006 White Paper on National Defense* will be used as a baseline from which to measure PLAN capability today and in 2016: Can the Navy meet its mission requirements, as spelled out in the *2006 White Paper*?[1]

CURRENT PLAN COMPOSITION

Evaluating China's naval capabilities will begin with the size of the force. This is the most obvious first indicator in evaluating naval capabilities, and it is the strongest possible indicator of national intentions: What proportion of national treasure is being devoted to naval strength? This may not necessarily serve as a gauge of national belligerency or of the likelihood of Beijing choosing to employ naval force. Available naval resources do provide a strong indicator, however, of Beijing's propensity to utilize the PLAN as an instrument of statecraft in a crisis involving perceived threats to vital security interests.

The PLAN will be analyzed in Western naval terms, that is, as "communities" or classes. These are the surface, subsurface, and aviation warfare communities.

Surface Forces.

Surface ships form the oldest, most visible, and most vulnerable naval community. Almost all are "warships," but these are usually classified as either combatant or noncombatant. The former are built to do battle, and in the PLAN include destroyers, frigates, and patrol craft. The first two types of ships are intended to serve as multimission platforms; they are armed, equipped, and crewed with the goal of enabling them to conduct operations in all primary naval warfare areas.

The first area is antisurface warfare (ASUW).[2] This area refers to operations conducted to detect, localize, target, and attack surface ships, typically carried out with radars, guns, missiles, or torpedoes. ASUW missions may also be carried out by submarines, aircraft, or shore batteries.

The second is antisubmarine warfare (ASW) operations, conducted to detect, localize, target, and attack submarines. These are typically carried out by sonar and torpedoes, both tube-launched and rocket-assisted. ASW is best carried out by submarines but is also assigned typically to aircraft as well as to surface ships.

The third primary surface ship warfare area is antiair warfare (AAW), conducted to detect, track, target, and attack manned aircraft and unmanned air vehicles. The usual means of conducting AAW are radar, missiles, and guns. AAW is a primary mission assigned to aircraft, as well as to shore batteries.

Less prominent but not unimportant naval warfare areas include mine warfare (MIW), which—with sonar, unmanned vehicles, divers, and mammals—includes operations both to install mine fields and to detect, localize, and destroy or remove mines planted by opposing forces. MIW is also carried out by aircraft, both fixed and rotary wing, as well as by submarines. There is also information warfare (IW), consisting of information operations, operational security (OPSEC), psychological operations (PSYOPs), military deception, and electronic warfare (EW). EW in turn includes operations to utilize the electronic spectrum for detection and warning, while denying an opponent the ability to do so. Finally we have amphibious warfare (AMW), referring to operations designed to move ground forces ashore, usually with small seaborne landing craft, air cushion vehicles (LCACs), helicopters, or vertical take-off and landing (VTOL) fixed-wing aircraft. Navies typically operate with marine forces for this mission, but most armies also have units dedicated to AMW. Indeed, the PLA has divisions stationed in Fujian and Guangdong Provinces that are dedicated to the amphibious mission.[3] AMW includes several categories, ranging from the full multidivision assault against an opposed beach to small raids conducted by special operation forces (SOF).

China's surface ship force is riding on the leading edge of current PLAN modernization. New ships have been launched every year since 2000, following a more deliberate but well-funded ship design and commissioning program during the 1990s. Two ship types are conspicuous—destroyers and frigates—

both armed with very capable antisurface ship cruise missiles (SSM). All of these ship classes are designed to be multimission-capable, which means they are assigned missions across the spectrum of naval warfare areas, especially ASUW, ASW, and AAW.

Destroyers of two classes and frigates of a single class were commissioned during the 1990s. The two *Luhu*-class and one *Luhai*-class destroyers are essentially identical in capability; the latter is larger — displacing 6,000 rather than the former's 4,600 tons — primarily because China was forced to buy Ukrainian-built gas turbine engines as the ship's main propulsion plant. This resulted from the U.S. embargo on sales of military equipment to China following the 1989 Tiananmen Square massacre: prior to that event, the United States had sold China five General Electric-built LM-2500 gas turbine engines, four of which power the two *Luhu*-class destroyers.[4] The Ukrainian engines are significantly larger than the U.S. machinery, hence the *Luhai*'s hull had to be increased in size. The 2,250 ton displacement *Jiangwei*-class frigates, at least 12 of which have now been built by China, are smaller versions of the destroyers, powered by German-designed diesel engines.

These 1990s shipbuilding programs were tentative in nature, as the PLAN built small types of ships, trying different combinations of mostly foreign-built or at least foreign-designed weapons, sensors, command and control, and propulsion systems. The decade was also one in which PLAN budgets had yet to begin benefiting in a major way from the significant defense budget increases that began in the early 1990s and have come to fruition in the 21st century. Such increases will no doubt continue, as China's remarkably expanding economy continues to yield greater revenues to the national government. However, none of the ships built during the last 10 years of the 20th century were capable of operating successfully in a modern fleet environment with opposition from U.S., Japanese, South Korean, or perhaps some other navies.

The PLAN surface force that deployed in 1999—*Luda*-class destroyers and *Jianghu*-class frigates in addition to the ships discussed above—was characterized by a formidable anti-surface ship capability, but only weakly equipped to conduct ASW and AAW operations. In fact, none of these ships is capable of operating at less than very great risk to themselves in an environment subject to attack by land- or carrier-based aircraft. And they would have to rely on luck to work against well-operated submarines successfully.

Two ship classes have bridged the gap between 20th and 21st centuries for the PLAN, however. China has continued to modify the *Jiangwei*-class frigates, of which three subclasses now exist. The second class (*Jiangwei II*), featuring a Chinese-built copy of the *Jiangwei I*'s French-built AAW missile system, was not successful. The *Jiangwei III*, at least one of which is in commission, appears to differ from its earlier sisters primarily in its improved command and control capabilities.[5]

The second cross-century combatant is the *Sovermenny*-class destroyer, four which China has purchased from Russia. This 8,000 ton displacement ship was designed by the Soviet Union specifically to target U.S. aircraft carriers, with its long-range, heavy-warhead *Sunburn* antiship cruise missile. That the ships were designed to operate as part of a multimission-capable task force is indicated by its marginal ASW and AAW capabilities. The *Sovermennys*' steam plants also have a problematic history. The PLAN will have to employ the boats conservatively, so as to avoid exposing them to the air attacks against which they are so vulnerable.

With the dawning of the 21st century, the PLAN launched shipbuilding programs reflecting a new confidence and technological expertise by the Chinese warship construction industry. By the middle of the new century's first decade, China has already launched three new classes of destroyers and a new class of frigate. All have continued the PLAN's emphasis on very capable antiship cruise missile batteries and, while still equipped with problematical ASW systems, are armed with the most advanced AAW system yet put to sea by China. The *Luyang I*, *Luyang II*, and *Luzhou*-class destroyers are all gas-turbine powered ships designed with some stealth characteristics and intended to provide the PLAN for the first time with ships capable of area AAW defense.[6]

Which of these ships will be a class leader, and how many hulls will be built in each class are unanswered questions. The *Luyang II* is the most intriguing of this new class of destroyers, since it is equipped with an antenna array characteristic of the U.S.-designed *Aegis* AAW system. Probably two of these classes of ships — *Luyang I*, *II*, *Luzhou* — are armed with a highly capable Soviet-designed antiaircraft missile system known as the *Grizzly* in NATO parlance.

China's frigate force is now led by the diesel-powered *Jiangkai*-class, three of which have reportedly been commissioned. This ship appears to be a larger version (3,500 ton displacement) of the *Jiangwei*-class, the chief difference being a hull and superstructure design exhibiting "stealthy" characteristics. In fact, with its sleek rounded surfaces and reported radar absorbent coatings, the *Jiangkai* bears a strong resemblance to the French-designed *Lafayette*-class frigates operated by Taiwan's Navy. A less obvious but significant advantage is the *Jiangkai*'s size; with half again the displacement of the *Jiangwei*, the *Jiangkai* will offer a significantly more stable platform for weapon and sensor systems, and will be a better sea-keeping ship.

These modern ship classes—*Luhu, Luhai, Luyang I* and *II, Luzhou, Jiangwei, Jiangkai*—augment China's older *Luda* and *Jianghu* combatants. These older ships are armed with capable antisurface ship cruise missiles, although being even less ASW and AAW capable then their newer fleet mates.

Since 2000 the PLAN has commissioned several new ships in other mission areas as well, including at least 17 amphibious warfare ships. Most significant of these are the 4,800 ton displacement *Yuting I* and *II-*class landing ship-tanks (LSTs), which are designed to beach themselves for direct off-loading of troops and vehicles. Although lightly armed—as are all amphibious assault ships—the *Yutings* are equipped with helicopter landing platforms, which increase their flexibility by allowing the vertical transport of embarked troops and equipment. An additional 10 *Yunshu*-class landing ship-mechanized (LSMs) have also been commissioned since 2003. These are smaller versions of the *Yuting*-class, displacing 1,460 tons. Even more significant is a much larger amphibious ship, displacing between 18,000 and 25,000 tons, that was launched in late 2006. This ship looks almost identical to the U.S. *San Antonio*-class Landing Platform Dock (LPD); it will offer the PLAN a platform capable of deploying at least four helicopters and four air-cushion landing craft, and embarking at least 400 troops.[7] It will be the first Chinese naval vessel capable of force projection as defined by Western navies.

The PLAN mine warfare force remains limited to one dedicated mine laying ship (few, if any, of the world's navies dedicate vessels to this single mission) and a force of aging Soviet-designed minesweepers. The PLAN certainly is not ignoring this warfare area, however, since new MIW technology is being acquired and exercised.[8] MIW would occupy a prominent place in any scenario involving PLAN action against Taiwan.

Discussion of the Chinese Navy's future plans often poses questions about an emergent blue water navy or talk about power projection. Both of these expansionist courses depend on logistical support for viability. Hence, the PLAN's inventory of ships capable of replenishing combatants and amphibious ships at sea while underway is a critical indicator of China's naval ambitions. Until 2005, the PLAN included just three such ships, and only one of these, the ex-Soviet *Vladimir Peregudov* (renamed *Nancang* by China), is large enough for fleet operations, at 37,000 tons displacement. It is also the only PLAN replenishment ship with a helicopter deck and hangar.

Two replenishment vessels of the *Fuqing*-class displace just 21,000 tons. In 2005, however, China built and commissioned two new *Fuchi*-class replenishment-at-sea (RAS) ships, each displacing 28,000 tons and capable of supplying the fleet with fuel, ordnance, food, and other supplies. If Beijing uses these new RAS ships as replacements for the two smaller units, it will indicate a continued lack of blue water ambition. If, however, the *Fuqings* are retained until replaced by larger ships and each of China's three naval fleets—North Sea, East Sea, South Sea—grows to include two or more large RAS ships, then the PLAN will be capable of more long-range deployments. This capability will also indicate Beijing's more ambitious intentions for its Navy.

Submarine Force.

True long-endurance submersible warships became feasible only with the advent of nuclear propulsion in the U.S. Navy in the mid-1950s. Nuclear-powered attack submarines (SSN) are expensive, however, and modern conventionally-powered submarines have developed into extremely capable antisurface ship and mine-laying platforms. Germany and Russia have led the way in designing such boats, and China has benefited directly from both.

The PLAN first built SSNs in 1980, with the earliest of the five-ship, *Han*-class. These boats were built along the lines of the 1950s vintage Soviet-designed *November*-class SSN. They are "noisy" and have experienced significant maintenance problems during their lifetime; in fact, no more than four and perhaps just three of the *Han*-class remain operational.[9] China is currently building and deploying a new class of SSN, however, the Type-093 or *Shang*-class. Two of these boats are operational, with at least one more under construction. They strongly resemble the 1980s Soviet-designed *Victor III*-class SSN, although no doubt much modernized. This similarity almost certainly reflects Russian assistance in China's construction of this new generation of SSN. The *Shangs* have been compared to the U.S. *Los Angeles*-class SSNs which, although now 30 years old, continue to form the bulk of the American submarine force.[10]

The PLAN has never succeeded in deploying a nuclear-powered submarine armed with intercontinental ballistic missiles (ICBMs) carrying nuclear warheads. The *Xia*-class fleet ballistic missile submarine

(FBM or SSBN) was constructed in 1987, but apparently never regularly patrolled, probably due to engineering problems.[11]

China is building a new FBM, the Type-094 or *Jin*-class. This seems surprising, given the nation's already successful DF-31 and 31A land-based, road-mobile, solid fuel ICBMs, which are much less expensive, much more secure, and much more controllable than are sea-based missiles. It may be that Beijing is determined to have more than one leg to its nuclear deterrent force. A PLAN desire to have a role in this mission may also be a factor, although FBMs are subject to the command and control of the Chinese national command authority, operating through the Second Artillery and not directly with the PLAN.

China already owns the world's most formidable force of conventionally-powered submarines (SS). The oldest part of this force consists of almost 60 *Romeo*-class boats, copies of an early 1950s Soviet design. The PLAN probably operates no more than a dozen of these submarines, however, due to high maintenance requirements and the lack of crew personnel. More useful are the 17 boats of the *Ming*-class, an updated version of the *Romeo*, which began entering active service in 1975.[12] This submarine offers only slightly improved capabilities over the *Romeo*, and China is well into a large-scale construction program for its next-generation conventionally-powered attack boat, the *Song*-class.

At least 12 *Songs* have been commissioned or are in production. The first two of these boats demonstrated serious shortcomings in China's ability to design and construct advanced submarines, but these problems have apparently been overcome, and the *Song* appears to be the PLAN's indigenously produced, conventionally-powered submarine of choice for the first three decades of the 21st century.

China has also purchased 12 Russian-built *Kilo*-class boats, hitherto one of the very best SSs in the world. A thus far new class of submarine, named the *Yuan*, was unveiled in the summer of 2004; it may be a follow-on to the *Song* class, but given the *Yuan*'s relatively small size and lack of significant follow-on production, it is more likely an experimental boat of sorts, perhaps an attempt to reverse engineer the *Kilo*-class.

China has yet to incorporate air-independent propulsion (AIP) in any of its conventionally-powered submarines. An AIP plant allows a boat to remain submerged for up to 14 days instead of the 4 associated with conventionally-powered boats (when only slow speeds are called for). The technology is not thoroughly proven, however, and Beijing may be waiting for further Russian (or other foreign) developments in AIP engineering before purchasing the plants.

Naval Aviation.

PLAN aviation is the Navy's weakest branch, although progress is being made. All fixed-wing aircraft are based ashore, including approximately 48 of the Su-30 fighter-attack aircraft that China has purchased from Russia. This is the PLAN's only truly modern tactical aircraft, although the 18 JH-7s and 120 J8-IIs are the result of indigenous attempts to produce a contemporary fighter.

China's naval aviation force also deploys Soviet-designed B-6 bombers capable of firing antiship cruise missiles (ASCM), its primary tactical role. The PLAN's patrol and ASW aviation force is relatively weak, with approximately 24 H-5 and H-6 aircraft operating.

The PLAAF apparently continues to provide China's primary air-to-air refueling and electronic warfare aircraft for maritime missions.

The PLAN's main aviation strength lies in its shipborne helicopter fleet. Of either French or Russian design, about 60 aircraft are deployed, mostly onboard ships. Each of China's new destroyers and frigates is capable of hangaring and operating one helo, although only the four or so newest ships appear capable of digital linkage with its aircraft in flight. This computer connection permits automated flight control and engagement information to be passed both ways between ship and aircraft, and is crucial for prosecuting engagements.

Personnel and Training.

The PLAN is currently coping with personnel shortages, a common issue among almost all modernizing military forces. These shortages have a basis in the booming Chinese economy and the growing technological sophistication of the weapons, sensor, and engineering systems with which new PLAN ships, submarines, and aircraft are being equipped. The traditional Chinese serviceman of Mao Zedong's construct—an uneducated "man of the people" who could put down his hoe and pick up a rifle—is no longer adequate material for molding a proficient modern sailor. Whether draftee or volunteer, the new PLAN enlisted man must possess both the native intelligence and formal education necessary to maintain and operate complex electronic and mechanical systems.

To this end, during the past decade and a half, the PLAN has significantly revised its system of educating and training enlisted technicians. While the draft remains in place, its usefulness has been restricted by the reduction of obligated service to just 24 months. Hence, promising enlisted conscripts must agree to an extended period of service, perhaps 4 years at a minimum, before justifying a PLAN investment in extensive education and training. The needs of naval modernization and operational requirements have also persuaded the PLAN that a well-developed corps of noncommissioned officers (NCOs) who are both proficient technicians and effective leaders is needed. Such a corps of technicians and NCOs is indeed emerging in the PLAN.[13]

A similar phenomenon is occurring in the PLAN's officer corps. The days of poorly educated, even illiterate, officers is past; the Navy simply cannot afford officers who do not possess the education necessary to understand how to maintain and employ the complex technological systems with which their ships, submarines, and aircraft are equipped. To expand the available base of qualified officer candidates, the PLAN has in recent years established several officer accession programs similar to the U.S. Naval Reserve Officer Training Corps (NROTC).[14] These programs are oriented toward attracting civilian students majoring in engineering or the sciences, reflecting the PLAN's awareness of the need for technologically competent officers.

However, the PLAN's training paradigm remains based on the calendar year, a relatively rigid construct that inhibits effective operational training and focuses maximum operational readiness on a narrow period of time. One moderating development, however, allows for ships to enter the training program during the calendar year when the timing of unexpected or extended maintenance periods so dictates, rather than being restricted to the beginning. The PLAN follows a Navy-wide training program that at least on paper proceeds from individual personnel training to team, crew, multiship, and finally to joint training on a significant scale, sometimes involving units from all of China's three fleets, the Army, and the Air Force.

Maintenance is a topic seldom addressed by observers of naval strength. This is a crucial failing, since naval forces are only as effective as their state of readiness, and their readiness is more dependent on effective materiel maintenance than any other factor, even personnel proficiency. Of course, the two are directly connected: personnel must be effectively trained to conduct effective maintenance on their assigned equipment and platforms.

For example, an aircraft with an improperly maintained engine or inaccurately aligned fire control radar obviously will not be able to carry out its mission. At sea, if a ship's propulsion plant is not properly maintained, it will operate neither efficiently nor reliably. A destroyer designed to cruise for 2,000 kilometers (km) at 16 knots (kts) may as a result be able to cruise for only 1,600 km at that speed; or a ship designed to operate in combat at 28 kts may be able to attain a speed of only 25 kts, which could be fatal.

Another example is improperly performed or falsified maintenance on a shipboard guided missile fire control system, which would lead to unsatisfactory daily system operability. In time of need, that system would either not operate at all or would perform at a level well below design specifications, again a possibly fatal shortcoming.

The most pointed example of the importance of proper maintenance applies to that performed on almost any of a submarine's many pumps. As noted above, submarines depend for stealth above all other factors for their operational effectiveness, indeed, for their very survival at sea. The primary factor in a submarine's ability to operate stealthily is the lowest possible noise signature. This requires the most detailed attention to maintenance procedures, to ensure that all equipment onboard, from the galley food mixer to the main engine, is operating as quietly as possible.

If, for instance, the prescribed procedures for repacking a bearing on a particular pump calls for Grade "A" lubricant, but because of an onboard shortage of that product the maintenance personnel use Grade "B" lubricant, than no immediate deleterious effect will be observed. Over time, however, the substitute will break down more quickly than would the proper lubricant, and the pump bearing will begin to operate at a higher than designed noise level. This means that the submarine will be generating more noise and will be more detectable by opposing submarines, ships, or listening devices.

How effective are PLAN maintenance programs, especially when considered in relationship to PLAN personnel training? Anecdotally, the PLAN does not have a good reputation for the detailed attention to maintenance demanded of an effective navy.[15] First, PLAN ships deploying on long cruises, to the Western Hemisphere for example, have been assigned additional, specially trained maintenance personnel and special spare parts allocations, a clear admission that normal maintenance protocols are not up to the task of trans-oceanic deployments. Second, the loss of the submarine *Ming 361* in 2003 certainly reflected an unsatisfactory training paradigm for maintenance personnel.[16] Third, two PLAN senior captains embarked as observers on several American warships during the 1998 *Rimpac* exercise conducted in Hawaiian waters. Part of their extensive report of that experience appeared in the military press in China: one of the factors most attracting the PLAN officers' attention was that U.S. sailors continued performing equipment maintenance during underway operations, indicating that this was not the practice in the PLAN.[17] Fourth, another apparent lesson learned by the PLAN from the U.S. Navy is the absolute requirement to have an NCO corps able to assure effective maintenance with a minimum of officer supervision.

THE FUTURE: 2016-17

The PLAN understands the importance of personnel education and training, as it does that of systematic training ranging all the way from individual to unit to fleet to joint service levels. Improvements in these areas began a decade ago and will almost certainly continue during the next 10 years, resulting in substantial increases in operational competence. The loss of *Ming 361* resulted in the relief of the responsible chain of command, extending from PLAN commander Admiral Shi Yunsheng down to the senior captain responsible for the maintenance failures that contributed to the loss of the submarine's crew.[18] In other words, PLAN personnel in 2016-17 will be better educated and more thoroughly trained than their predecessors, and at least as patriotically dedicated to their mission.

Meanwhile, platform and materiel modernization is occurring across all PLAN communities: aviation, surface, and subsurface. The latter has clearly been selected by Beijing to serve as China's primary instrument of naval force, however. The force of 6 SSNs and 25 modern SSs cannot be ignored by any potential maritime opponent, be it Taiwan, with its almost negligible undersea force of two boats, or the United States, with the world's largest and most capable Navy. And it is these two forces that Beijing has in mind as it deploys and increases its submarine fleet.

Detecting, localizing, targeting, and sinking submarines—the essence of ASW—remains the most difficult of naval warfare areas. China's PLAN development program in 2006 is proceeding not to challenge any particular foreign navy directly, but rather to serve as an effective instrument of national will in specific strategic scenarios. Three of these scenarios are most likely and thus most predictive of Beijing's naval plans for the next decade: Taiwan, the East China Sea, and the Straits of Malacca.

Taiwan.

Taiwan is China's number one geostrategic concern. More than that, ensuring Taiwan's reunification with the mainland is a matter of revolutionary ardor and has been elevated by Beijing to symbolize Chinese nationalism. Since Beijing consistently refuses to discount the possibility of employing military force against Taiwan should the island's government cross any one of several thresholds, PLAN would be one of the military instruments of choice in that case.

Options for employing maritime forces against the island range from various levels of restricting seaborne trade to full-scale amphibious invasion. The PLAN would presumably play a prominent role in other applications of military pressure as well, including special operations, decapitation, and blockade. The Navy's most important role in a Taiwan scenario, however, would be to isolate the battlefield by deploying submarines to prevent or at least delay intervention by other countries' naval forces. This means the U.S. Navy, of course, although the Australian Navy and Japanese Maritime Self-Defense Force (JMSDF) could conceivably assist American intervention in the face of a large-scale Chinese assault on Taiwan.

Such U.S. intervention would almost certainly be built around aircraft carriers and other surface ships; effective intervention would not require these ships to enter Taiwanese harbors, or even be close to 200 nautical miles of the island. However, some 20-30 Chinese submarines deployed north, east, and southeast of Taiwan would cause American and other naval commanders to proceed very cautiously.

If China were able to maintain even a dozen submarines on station in the East China Sea for 1 month in the face of the U.S. Navy's approach, it would likely provide an uncomfortable Taipei government with enough time to decide that negotiating was preferable to fighting essentially by itself. There is no reason to expect Beijing to waver in its stated resolve to employ military force to prevent Taiwan from achieving *de jure* independence. Hence, the PLAN will continue to be a primary vehicle for pressuring Taiwan, a role that will likely end only with Taiwan's accession—to a degree acceptable to both—to China's governance.

The East China Sea.

The East China Sea is China's front porch and thus vital for national defense. It contains the nation's most important fishing grounds and possibly rich energy deposits, and is the scene of a sovereignty dispute with Japan. This dispute concerns the Daoyutai (in Chinese) or Senkaku (Japanese) Islands, a cluster of barren, uninhabited rocks claimed by both nations. Although they lie equidistant (170 km) from the Japanese Ryukyu Islands and Taiwan, the Daoyutai are located on China's continental shelf as it is defined in the UN Convention on the Law of the Sea (UNCLOS), some 350 nautical miles from the coast.[19]

China and Japan both appear to have a respectable legal argument, however, and the conflict is most unlikely to be voluntarily settled by the two disputants themselves. More important than any possible economic gains resulting from undisputed possession of these barren land features, however, is the national hubris of China and Japan. Moreover, Taiwan further unsettles the pot with its own claim to the Daoyutais, echoing China's. During the past decade, groups of private citizens from Taiwan, as well as from Hong Kong, mainland China, and Japan, have all conducted forays among the islands, attempting to establish their nation's sovereignty. The groups from China-Taiwan-Hong Kong have all failed, sometimes at the cost of life, while at least one Japanese group temporarily succeeded in establishing a lighthouse on one of the islets.[20]

Despite their doubtful material value, the Daoyutais-Senkakus might serve as a *casus belli* as manifested in naval conflict between the JMSDF and the PLAN. A recent four-session study by several greater Washington, DC, analysts concluded that such a conflict was not unlikely, although it would almost certainly be of short duration. Nonetheless, any shooting incident between Japan and China risks unintended escalation into a serious conflict, one that might well involve the United States by virtue of its Mutual Defense Treaty with Japan.[21]

China has not made overt threats of military action to enforce its claims to either the Daoyutais or the natural gas deposits in the East China Sea ocean floor. This latter issue does bear significant economic gravitas, given the not insignificant oil and natural gas reserves that may lie in two to four sea bottom fields, perhaps as much as 200 billion barrels of the former and 7 trillion cubic feet of the latter.[22] Of current concern is the Chunxiao (Chinese) or Shirakaba (Japanese) natural gas field, currently being extracted from by both China and Japan, while still in dispute. Resolution of the field's ownership by the disputants themselves is no more likely than that concerning the Daoyutais-Senkakus.

Beijing and Tokyo have both used military forces to establish a presence in the area, with China employing ships and aircraft, and Japan using its Coast Guard and aircraft.[23] This patrolling could, as is the case with forays around the Daoyutias-Senkakus, lead to unintended escalation. China has been conducting extensive sea bottom surveys during the past 5 years; such exploration serves both to ascertain the presence of mineral deposits and to map the ocean bottom so as to enhance submarine operations.

Six sessions had been held by Beijing and Tokyo by the summer of 2006 in an attempt to reach a diplomatic resolution of their dispute over East China Sea resources. They failed to reach resolution, but the two sides have established "two expert groups to help settle [their] dispute."[24]

Presumably, any strong move by Beijing in the East China Sea, whether over the Daoyutais-Senkakus or the disputed oil and gas fields, would be conducted by surface combatants, supported by long-range aircraft and submarines. Similar forces would likely be deployed by Tokyo. While China and Japan would likely curtail a naval conflict immediately, the JMSDF's significantly more advanced naval capabilities would, if employed, almost certainly cause the destruction of PLAN units, with significant loss of life. Given China's sensitivity to sovereignty issues, its dramatically rising nationalism, and its historical enmity toward Japan, any such losses at sea would be difficult to accept or relegate to negotiation.

PLAN improvements to better cope with the JMSDF will be a spur to its modernization programs during the next decade. These will likely focus not so much on equipment as on doctrine: joint and integrated operations are one way to describe part of the intent of "netcentric warfare," an American concept which the PLAN undoubtedly aspires to master.

The Chinese Navy of 2016-17, given even moderate progress, will be able to operate in an East China Sea scenario with commonly accepted tactical doctrine, with surface and air forces that have trained and exercised together, and with effective communications among units and shore stations using integrated systems. The continuing submarine modernization program in which Beijing is so heavily investing will enable the East China Sea to be divided into submarine operating areas with each patrolled by at least 24 modern submarines armed with highly effective cruise missiles capable of submerged launch.

Malacca.

Speaking in 2004, President Hu Jintao took note of China's "Malacca dilemma." He was referring to both "indigenous" problems such as piracy, but also to the possibility of the United States having a "choke hold" on China's seaborne energy imports, 80 percent of which flows through Malacca.[25] Since the Malacca and Singapore Straits connect the South China Sea with the Indian Ocean, Hu's concern is justified by the geography and the problems in those waters (this point may be argued—see Chapter 10 by Michael McDevitt).

Six nations claim all or some of the land features that dot the South China Sea, if one counts Taiwan (which echoes Beijing's claims). China is the only claimant, however, that apparently delineates the entire South China Sea—water areas as well as land features—as sovereign territory.[26] The South China Sea's value is both realized and potential, including fisheries, oil and natural gas deposits, national hubris, and most important the fact that more of the world's seaborne traffic uses the area than any other comparable body of water in the world.

All of these elements require a second look, however. First, while oil and natural gas are already being extracted from the seabed underlying the northern and southern South China Sea, the central area is untapped and may or may not contain significant reserves. With the signing of the February 2005 agreement by China, the Philippines, and Vietnam to jointly explore the area, the level of tension associated with the various sovereignty declarations has been very much reduced. Significantly, however, Beijing has not displayed any willingness to compromise on its own sovereignty claims.[27]

Second, not only do the fisheries belong to no single nation, but the stocks in the South China Sea are being over-fished by all claimants; at the present rate, and with the bordering nations' inability to control even their own fishermen, the dispute may soon be moot.[28] Third, while national pride is not usually assuageable by the balm of diplomacy, sovereignty claims in the

South China Sea are susceptible to resolution based on other criteria affecting more substantive issues.

Fourth and finally, what are the threats to the SLOCs in and surrounding the South China Sea that might elicit the use of naval power by one of the claimants? In fact, presently existing threats are of the nonstate variety—piracy and transnational crime, terrorism, and environmental degradation. These may worsen over the next decade, but will best be solved or at least ameloriated through international cooperation, as is indeed occurring today. There is little evidence that the next decade will witness a breakdown of the cooperative international approach to safeguarding these sea lanes. All concerned nations benefit from free and safe sea lanes. Obviously, however, if a truly ominous dispute ever breaks out between the United States and a risen, revanchist China, all bets are off.

Should Beijing decide that the PLAN must be capable of defending South China Sea LOCs and the Malacca Strait, it would have to make extremely large investments in materiel and personnel resources, since it is incapable today of carrying out such a mission. The Navy would need to have an increase in its number of state-of-the-art warships from the less than 20 it currently deploys to at least triple that number. A similar increase in RAS ships would also have to occur, from five to perhaps ten, to support those surface ships during the relatively long periods at sea required to guard against international threats.

If Beijing decides it will deploy the PLAN against any future U.S. and allied interference with the Southeast Asian SLOCs, China's submarine force will have to continue to increase the number of its 22 most modern boats (*Song, Kilo, Shang*), again to approximately three times that number. Most importantly, the PLAN will have to increase its aviation capability to be able to support surface ship task groups operating more than 1,000 nautical miles from home base. This will require not only the construction of bases on disputed, difficult-to-defend South China Sea and Andaman Sea islands, but also creation of a defensive system effective enough both to protect the bases and to afford the degree of protection necessary to allow conduct of offensive missions.

West of the Malacca and Singapore Straits, the Andaman Sea is not subject to the sovereignty disputes of the South China Sea, but it is subject to competing Indian and Burmese interests. Burma, in fact, is currently the subject of a veritable economic invasion by China that has strong political and military facets. Barring the overthrow of the well-established Burmese military dictatorship, this trend will almost certainly continue—despite Indian attempts to establish a contravening influence in the country. By 2016-17, the world may witness PLAN support facilities, if not outright bases, on Burma's coast and islands.[29]

Such facilities, if matched by the Chinese-modernized port at Gwadar, Pakistan, would in theory for the first time provide the PLAN with the logistic infrastructure to conduct extended operations in the Indian Ocean and North Arabian Sea. Even with the three-fold PLAN expansion noted above, however, such distant operations may not be feasible. First, the Indian Navy is a formidable force, and one that will continue to modernize and expand during the next decade. Second, Pakistan and Burma are two of the world's most unstable nation-states, and are as likely as not to have fallen into anarchy by 2016-17.

Finally, two factors argue against Beijing making such a decision for distant operations and basing. First, despite increasing energy needs, including growing dependence on foreign sources, China currently relies on seaborne imports for only 10 percent of its total energy needs. Furthermore, this percentage is likely to decline rather than increase during the next 10 years, as Beijing invests increasingly in pipelines and alternative (nonfossil) energy sources. Second, the United States is more than capable of countering any PLAN moves to resolve the Malacca dilemma or to dominate the Indian Ocean.

2006 DEFENSE WHITE PAPER

The *2006 White Paper on National Defense* continues the *2004 White Paper's* theme of increased PLAN stature and perceived value in the eyes of Beijing's decisionmakers.[30] The first section of the 2006 version, "Security Environment," notes "growing complexities in the Asia-Pacific," highlighting potential dangers to China posed by the strengthened U.S.-Japan alliance. Other threats listed—"territorial disputes, conflicting claims over maritime rights and interests"—are also maritime in character. One significant difference from the 2004 paper is that while the 2006 version criticizes continued U.S. weapons sales to Taiwan, it does not dwell on the island's threat to China, but simply notes the importance of "promoting cross-Straits relations toward peace and stability."

The national military strategy of "active defense" includes a role for the Navy, which is described as "aim[ing] at gradual extension of the strategic depth for offshore defensive operations and enhancing its capabilities in integrated maritime operations and nuclear counterattacks." This last phrase is of particular interest, perhaps attesting to Beijing's determination to deploy ICBM-armed submarines, despite the apparently more attractive option of concentrating on land-based, road-mobile missiles such as the DF-31/31A.

The PLAN is described as benefiting directly from Beijing's determination to instill in its military the benefits of a revolution in military affairs (RMA), including "informationalization" as a goal. To this end, the PLAN is afforded priority in the "development of firepower, mobility, and information capability" to "strengthen its comprehensive deterrence and warfighting capabilities." This last priority might explain in part Beijing's apparent determination to build a sea-based nuclear deterrent force with the *Jin*-class FBM.

According to the *2006 White Paper*, the PLAN will not experience the previously planned personnel reductions, but in fact will continue to claim a larger percentage of overall PLA personnel strength. The Navy is cutting headquarters personnel and reorganizing the shore establishment and some fleet units, but these moves are aimed at smoothing the path for the Revolution in Military Affairs (RMA).

The Navy is charged with "build[ing] itself into a modern maritime force . . . of combined arms with both nuclear and conventional means of operation." It will take "informationalization as the goal and strategic focus," giving "high priority to the development of maritime information systems." The PLAN is directed to emphasize building "mobile maritime troops" capable of "operations in coastal waters, joint operations, and integrated maritime support." Finally, it will continue to "improve and reform training programs . . . in joint integrated maritime operations . . . exploring the strategy and tactics of maritime people's war under modern conditions."

The PLAN is described as being integral to China's efforts to improve logistics by establishing across the armed services an integrated system of "materials procurement and management." Similarly, the Navy's reserve and militia forces will be strengthened "to pursue the principle of self-defense by the whole nation," with a Navy focus on making national "border and coastal defense unified, effective, solid, and informationized."

Of particular interest is the *2006 White Paper's* extensive discussion of national law promulgated in support of China's interpretations of various articles of the UNCLOS. These maritime laws reflect a continuing legal codification of Beijing maritime interests and claims.

At the strategic level, Beijing is awarding the PLAN a first-rank role, especially in dealing with sovereignty and international issues. Operationally, China is determined to continue naval modernization across the spectrum: ships, submarines, aircraft, and personnel. The emphasis on improving amphibious and surface combatant forces underlines China's concern with the Taiwan situation, while the importance accorded to improving joint operational and long-range precision strike capabilities implies direct concern with possible U.S. intervention in that situation.

For the Navy, then, China's *White Paper* is not mere public posturing, but accurately describes the developments already underway in naval modernization. Its intentions are not transiet, but rather will continue to guide PLAN developments for the foreseeable future.

CONCLUSION

China has built formidable navies on several occasions during its history. The Yuan Dynasty employed powerful naval forces in coming to power and striving for national security objectives in the 13th century, while the Ming Dynasty in the early 15th century deployed the world's most advanced and capable Navy. Both dynasties allowed their navies to atrophy, however, once specific strategic objectives had been accomplished. The Qing Dynasty's mid-19th century efforts to build and deploy a modern navy were less successful, first quickly coming to naught before the guns of the French fleet, and then that of the Japanese.

The Chinese People's Navy has since its founding in 1949 labored as an adjunct of the Army. It has been only since the end of the Cold War and the removal of the Soviet threat that Beijing has felt moved to direct significantly increased defense resources to modernizing what has throughout its existence been a marginally effective coastal defense force. The *2004* and *2006 White Papers on Defense* illustrate the new emphasis Beijing and the PLA are placing on modernizing the Navy. The PLAN is accorded heightened importance strategically, operationally, and doctrinally in China's national security paradigm.

By 2006, China had already deployed a Navy with the ships, submarines, aircraft, and systems ready to serve in pursuit of specific national security objectives, with Taiwan at the head of that list. This process of modernization includes improved personnel education and training, further doctrinal development, and a coherent maritime strategic view from Beijing, a process that will almost certainly continue throughout the next decade. By 2016-17, China will have available as an instrument of national power a Navy capable of carrying out ambitious assigned missions. The Taiwan imbroglio may still head that list, but the PLAN a decade hence will also be capable of denying command of the East and South China Seas to another power, and of commanding those seas for discrete periods. In other words, the PLAN of 2016-17, at three times its present size, will dominate East Asian navies, with the possible exception of the JMSDF—we cannot rule out a major Japanese rearmament in this time frame—and will offer a very serious challenge to the U.S. Navy when it operates in those waters.

This relatively higher PLAN status will not result from a failure of either Japan or the United States to pursue its own naval modernization, but does take account of Japan's constrained defense budget and personnel pool, and reflects the continuing reduction in American naval numbers and the U.S. operational overstretch from increasingly widespread and marginal missions in Southwest Asia and in the Global War on Terrorism. By 2016-17, present trends indicate that the Chinese Navy will allow Beijing to exert hegemonic leverage in maritime East Asia.

ENDNOTES - CHAPTER 11

1. China's *2006 White Paper on National Defense* may be found at "Full Text: China's National Defense in 2006," *Xinhua*, Beijing, December 29, 2006, in *Foreign Broadcast Information Service* (FBIS)-CPP2006122968070.

2. The warfare abbreviations in this chapter are those of the United States and its NATO allies; the PLAN at least informally uses them as well, as evidenced in the author's conversations with various senior Chinese officers.

3. Dennis J. Blasko, *The Chinese Army Today: Tradition and Transformation for the 21st Century*, London: Routledge, 2005, p. 188.

4. The LM-2500 is widely used in the U.S. Navy, propelling FFG-7, DD 963, *Aegis* cruisers, and *Burke*-class combatants, among others. The fifth engine purchased by China is most likely used for shore-based training, although it may have been destroyed in an accident or used to provide spare parts.

5. Author's tour of *Jiangwei II* and discussion with the ship's commanding officer, May 2006.

6. These ships' engineering plants also include back-up diesel engines for cruising at low speeds; hence, CODOG, or "combined diesel or gas turbine ," is the name for this system.

7. See report and pictures at *www.chinadefense.com/forum/ index.php?showtopic*=436. Additional pictures are at *25472778d 26665f4891c92ji3.jpg*; *2547277a3079954ce149ddfx6.jpg*; *3_48705_ 288bd2b113483cc.jpg*; and *321 KB)3_4729_78cf70af27cfae5.jpg*.

8. See Wang Shi K'o, "Cross-Strait Underwater Warfare: A Comparison of Mine Deployment and Minesweeping Strength," *Ch'uan-ch'iu Fang-wei Tsa-chih* [*Defense International*], March 2006, in FBIS-CPP20060504103001, May 4, 2006.

9. All vessels generate self-noise from operating machinery and simple passage of the hull through the water. This noise is detectable by an opponent's sonar; hence—and especially for a submarine, which depends on stealth for its very existence—the lowest possible sound "signature" is desirable.

10. This comparison is drawn in "*Shang* Class" at *www. sinodefence.com/navy/sub/type093shang.asp*.

11. A second *Xia*-class may have been constructed, but lost to an accident before commissioning. See Bernard D. Cole, *The Great Wall at Sea: China's Navy Enters the 21st Century*, Annapolis: Naval Institute Press, 2001, p. 27, n. 46.

12. An 18th boat, *Ming* hull number 361, suffered the loss of its entire crew in a 2003 accident; two additional *Ming*s may be under construction, although this report must be considered very doubtful, given ongoing construction of the *Song*-class submarines, which are generally believed to be the successor to the *Ming*. See *www.fas.org/man/dod-101/sys/ship/row/plan/index.html*.

13. The best information on this subject is provided by Dennis J. Blasko, *The Chinese Army Today: Tradition and Transformation for the 21st Century*, London: Routledge, 2005.

14. Author's conversations with senior PLAN officers. Also see Blasko, pp. 58-59; "Nation to Recruit More College Students in Military Conscription," *Xinhuanet*, October 30, 2003, at www.chinaview.cn.

15. *Ibid*.

16. Author's conversations with senior U.S. Navy officers, 2003-04.

17. Jiang Yuanliu, "China's Master-Degree Captain Watches US Naval Exercise, *Jiefangjun Bao*, October 22, 2000, p. 5, in FBIS-CHI-98-316, citing Senior Captains Mao Zhenggong and Jia Xiaoguang.

18. "CMC Chairman Jiang Zemin Denounces PLA Navy for Errors Behind Submarine Accident," *Kuang Chiao Ching*, No. 371, Hong Kong, August 15, 2003, p. 15, in FBIS-CPP20030815000047.

19. This point is disputed by Tokyo, which argues that the Asian continental shelf should be demarcated by the Nansei-shoto Trench well to the East of the Daoyutais, placing them on the same continental shelf as the Ryukyus. The grouping is categorized as five islands and three rocks, although international law is not clear on the question between the two. According to the UNCLOS, "an island is a naturally formed area of land, surrounded by water, which is above water at high tide," while a rock "cannot sustain human habitation," which implies at least the lack of a natural supply of potable water. Yet the definition of an island says nothing about "human habitation."

20. See, for instance, "Japan Chases Off Taiwanese Boat," *AFP*, July 30, 2006, at www.channelnewsasia.com/stories/afp_asiapacific/view/225307/1/.html.

21. See "Sino-Japanese Rivalry CNA/IDA/INSS/Pacific Forum CSIS Workshop Series" reports, especially Brad Glosserman, "Workshop Four: Implications for the U.S.," September 29, 2006, in e-mail: pacforum@hawaii.rr.com, October 18, 2006.

22. J. Sean Curtin, "Stakes Rise in Japan, China Gas Dispute," *Asia Times Online*, October 19, 2005, at www.atimes.com, identifies the Chungxiao/Shirakaba, Duanqiao/Kusunoki, and Tianwaitian/Kashi fields (giving both Chinese and Japanese names), but notes that they may actually be part of the same complex.

23. For instance, see "Chinese Warships Make Show of Force at Protested Gas Rig," *The Japan Times*, September 10, 2005, and

"Japan and China Face Off Over Energy," *Yomiuri Shimbun*, condensed in the *Asia Times*, July 1, 2005, at *www.atimes.com/atimes/printN.html*.

24. "China, Japan to Set Up Expert Groups to Solve Gas-Field Row," *The Financial Express*, July 9, 2006; and "China, Japan End 6th Round of East China Sea Talks: Wide Gaps Remain," *People's Daily Online*, July 9, 2006, at *english.people.com.cn*.

25. Quoted in David Zweig and Bi Jianbai, "China's Global Hunt for Energy," *Foreign Affairs*, Vol. 84, No. 5, September/October 2005, p. 34; and "Ji Xiaohua, "It Is Not Impossible to Send Troops Overseas to Fight Terrorism," *Sing Tao Jih Pao*, Hong Kong, June 17, 2004, p. A27, in FBIS-CPP20040617000054.

26. Author's conversation with senior legal adviser at PRC Embassy, Washington, DC, 2002. Also see Cole, pp. 39-40.

27. The agreement was reported in "Philippines, China, and Vietnam Agree to Explore South China Sea Areas," *Xinhua*, Beijing, March 14, 2005, in *Alexander's Gas & Oil Connections* [referred to hereafter as *Alexander's*], Vol. 10, No. 7, April 6, 2005, at *www.gasandoil.com/goc/news/nts51490.htm*. For varying estimates of energy reserves in the central South China Sea, see Cole, pp. 58-60.

28. See Cornelia Dean, "Study Sees 'Global Collapse' of Fish Species," *New York Times*, November 3, 2006, p. A21: "If fishing around the world continues at its present pace, more and more species will vanish, marine ecosystems will unravel, and there will be 'global collapse' of all species currently fished, possibly as soon as mid-century, fisheries experts and ecologists are predicting."

29. See Nyi Nyi Lwen, "Economic and Military Cooperation Between China and Burma," September 2006, at *www.narinjara.com/Reports/BReport.ASP*.

30. The *2004 Defense White Paper* may be found at *www.fas.org/nuke/guide/china/doctrine/natdef2004.html#5*.

PART VI:

THE WRAP UP

CHAPTER 12

THE "RIGHT SIZE" FOR CHINA'S MILITARY: TO WHAT ENDS?

Ellis Joffe

Any attempt to discuss the "right size" for China's military immediately raises an integrally connected question: The "right size" for what purpose? "Size," it should be emphasized, refers not necessarily to the quantity of men and weapons, but to the overall capabilities of the People's Liberation Army (PLA) — its force structure, technological levels, and organization. And "right" refers to the appropriate size of the PLA as determined by the Chinese alone, in accordance with their interests.

Most analysts of Chinese military affairs agree that the purpose of China's military buildup during the past decade or so has been to acquire a capability that would enable China to coerce Taiwan into accepting a "one-China" solution to the Taiwan problem, or at least to prevent it from moving toward formal independent status. In fact, the title of the wrap-up chapter in the volume preceding this one was "China's Military Buildup: Beyond Taiwan?" Its main point was that although China's post-Mao military modernization was driven by several factors, the chief reason for the accelerated buildup that began in the mid-1990s and increased after 1999 was the emergence of the Taiwan issue in a form that threatened the "one China" solution and was thus unacceptable to the Chinese.

This issue provided the impetus for the acceleration of military modernization. Its purpose was to give the Chinese leadership the military clout needed to prevent the separation of Taiwan—preferably by intimidating the Taiwanese or, failing that, by military moves that could culminate in an invasion of the island. Since the Chinese believed that in this latter event the United States would intervene militarily, an integral purpose of the buildup was to deter or delay American intervention by raising its costs and, in the worst-case possibility, to increase China's chances of overcoming it. This buildup provided the Chinese with collateral capabilities that raised questions about their intentions "beyond Taiwan," but these questions have been peripheral to the central issue of Taiwan itself.

By 2007 the overwhelming significance of the Taiwan issue has diminished dramatically, primarily because the specter of a major war no longer hovers over the Taiwan Strait, even if it has not disappeared entirely. This has been due to a combination of factors— the readiness of China's leaders to acquiesce in the status quo rather than to push for unification as long as Taiwan refrains from declaring formal independence; their belief that economic and other ties will advance the chances for peaceful reunification; the political difficulties of independence-seeking Taiwan President Chen Shuibian and China's expectation that he will not be reelected; and the new determination of the United States to restrain Taiwan from provocative actions.[1]

The more relaxed attitude of the Chinese presumably derives from an additional reason: confidence in their increased capability to coerce or conquer Taiwan, while making it more difficult for the United States to intervene. At the same time, the Chinese have no illusions about narrowing the gap between their overall military strength and that of the United States, nor about their chances of defeating the United States in an all-out war over Taiwan.² These assessments must have raised questions among China's leaders about the future objectives of military modernization and the right size for their armed forces. Now, then, is an appropriate time to take a closer look at the objectives—beside Taiwan—which will shape the PLA in the next stage of its development.

TO MAKE CHINA A GREAT POWER

The purpose of building up the Chinese armed forces is, of course, to provide military backing for the foreign policy objectives of the leadership. However, these objectives differ in the priority attached to them by the leadership and in the time frame for their attainment. They also differ in the capability of the military to support them. Therefore, their influence on the development of the armed forces varies in accordance with the importance of the objective and the connection between its attainment and military force.

The most basic long-range and unalterable objective of the Chinese leadership has been to obtain recognition for China as a great power and to gain the appropriate respect and standing in the eyes of the international community, especially of other great powers, that come with this status. However, although the objective has remained constant, the strategies for attaining it have changed radically over the years—from revolutionary strategies aimed against the great powers and designed to destabilize the international system, to diplomatic strategies aimed at cooperating with the great powers and working within the system.

The Chinese have never publicly articulated in print their vision of this objective, and it is not at all certain that they themselves have thought out the specifics. Nonetheless, inferences from pronouncements and actions make it clear that they have been motivated by two fundamental principles: the preservation of China's independent position in global politics and the upholding of its national honor; and membership in the exclusive group of major powers that make the key decisions defining the workings of the international system. However, if the specifics are vague, the powerful forces driving the objective are not.

The first of those forces consists of China's physical attributes—territory, population, and geographic location—that together endow it with an overwhelming presence and provide an underpinning for the claim of its leaders that China is entitled to a prominent global status. More important is the political and emotional significance of modern Chinese nationalism that has its roots in the grandeur of the ancient Chinese empire, and its power in the determination of Chinese leaders to avenge past imperialist-inflicted humiliations by restoring China to a position of international prominence. Most important is China's economic surge that has catapulted it to the front rank of the global economy and to a position of major political influence. It has also provided China with the economic strength that constitutes one of the two essential pillars of great power status.

The other pillar—military force—is nowhere near a level commensurate with great power status. Although there are no mandatory international criteria that qualify a country for such status, it seems axiomatic that there are at least three conditions: a large population and territory; a credible nuclear capability; and sizable advanced conventional forces that enable it to project military power for long periods far from its borders.

At a minimum, such forces would presumably need to include aircraft carriers; long-range combat and combat support aircraft; transport aircraft and ships for moving large numbers of troops and supplies; air and sea refueling capabilities; global communications systems; and bases in friendly countries. Although China is working on developing some of these capabilities, it still lacks most of them.

Because such capabilities will be out of China's reach for generations, its leaders have never set their attainment as a realistic objective. And for good reasons. First, their global aspirations are a political and emotional goal, not a strategic one, and the absence of these capabilities does not put China's security at risk. Second, this is a long-range goal and the Chinese can move toward it without the urgency that would require an immense and draining military effort. And third, the military effort required to bring China closer to great power levels is so immense that movement toward it has to be incremental in any case, and can begin by focusing on China's short-term objectives.

For these reasons, China's global aspirations have not until now determined the pace and scope of China's military modernization. From the start of post-Mao modernization, the military component of China's great power aspirations remained dormant while Beijing focused on realistic near-term objectives—first, an upgrading of its backward armed forces, and then a rapid buildup after the Taiwan issue burst on the scene. This focus has greatly increased China's military power and has vastly enhanced its regional and international standing—to say nothing of generating at times exaggerated fears and concerns—but it has not moved China much closer to the level of great power capabilities.

How to move China toward such capabilities has by 2007 likely become a subject of discussion, if not debate, among China's leaders, perhaps in connection with a review of the PLA's future direction. A new phrase, reflecting either a consensus among the leaders or a line of argument in a debate, has appeared in an Army newspaper article: "It is a matter of great importance to strive to construct a military force that is commensurate with China's status . . . so as to entrench China's international status." A similar view was expressed by PLA Navy Rear Admiral Yang Yi, director of the Institute for Strategic Studies at the National Defense University: "As a responsible power, China needs to establish a military force that is commensurate with its international position and this is needed . . . to safeguard world peace." Admiral Yang emphasized, however, that "because of insufficient investments over a long period of time . . . the gap between China and the developed counties in the military realm has not shrunk, but rather is continuing to grow."[3]

From the logic of the situation and from fragments of data, it is reasonable to conclude that China's global aspirations will continue to drive PLA modernization into the future. However, because of the PLA's relative backwardness and the gap between it and advanced armies, these aspirations alone will not determine the speed of modernization and the resources that will be allocated to it. It does not make sense for the Chinese leadership to devote resources needed elsewhere for an objective that is both remote and only partially attainable at best in the far-off future. It makes much more sense to focus on more relevant near-term objectives which, over the long haul, will also advance its global aspirations.

SEEKING A PARAMOUNT REGIONAL POSITION

Although the Chinese leadership has been explicit regarding its desires or demands on specific regional issues, its broad objectives have been just as vague as its global aims. However, from inferential evidence it is possible to identify these objectives. The primary one is to gain a paramount position in the East Asian region—a position from which China will have the final say about what does or does not go on in its extended neighborhood. The most important example of what, from China's standpoint, should not be permitted to go on is the conclusion of strategic alliances between countries in the region and the United States.

These objectives are driven by the same powerful forces—physical presence, nationalism, and economic power—that motivate China on the global scene. However, additional considerations are at work in the region which make the attainment of China's objectives more imperative. The first and most important is security. Whereas China's global aspirations are relevant to its prestige and political standing, its regional objectives are directly connected to the defense of the homeland. The Chinese undoubtedly want Asian countries to acknowledge China's paramount position by virtue of its economic strength and political influence, and to act accordingly. This is probably the main reason why they have in recent years pursued policies designed to make friends of influential leaders in Asia. However, there are exceptions—Japan's tougher stance toward China and its closer strategic relations with the United States are one outstanding instance. Whereas China can presumably use its new economic leverage to put pressure on Asian countries, in the end it is only military strength that can protect its interests and ensure its national security. Moreover, unlike the global situation, building a military force for limited regional objectives is within China's reach.

Nonetheless, for more than a decade after the start of modernization, the Chinese felt no urgency about building such a force. During that period, military modernization was limited primarily to the nontechnological aspects of the Army's capabilities and, with a few exceptions, was marked by upgrading old weapons rather than acquiring new ones. The Chinese had good reasons to adopt this policy.

First, the United States did not loom as a military threat, and whatever danger they still perceived from the Soviet Union was remote and required no rapid improvements beyond the progress made by upgrading weapons and other reforms. Moreover, the cost of buying new weapons in large quantities was prohibitive and compounded by the difficulties of assimilation and the reluctance of the Chinese to become dependent on foreign suppliers. From its neighbors, the Chinese faced no military threat and were presumably confident they could carry out limited military actions beyond China's borders after the adoption of the "local, limited wars" doctrine in the latter 1980s. Although modernization was stepped up in the early 1990s, primarily due to the availability of advanced weapons from the former Soviet Union, it was still relatively slow because it lacked the impetus of a strategic focus and sense of urgency.

The emergence of the Taiwan issue and the need to cope with a U.S. military intervention (which the Chinese believed was inevitable if they decided to take military action) provided this impetus. What followed was a decade of intensive preparations marked by the procurement of new weapons and the adoption of new doctrines. These preparations completely transformed China's regional capabilities, which began to arouse serious concerns among American policymakers and defense officials regarding China's military capabilities and regional intentions, which seemed increasingly ominous. Numerous statements reflecting these concerns were forthcoming, best exemplified by the remarks of Secretary of Defense Donald Rumsfeld at a famous 2005 news conference in Singapore. He first observed that China was "improving its ability to project power" in the Asia-Pacific region. Then he added: "Since no nation threatens China, one must wonder: Why this growing investment? Why these continuing and expanding arms purchases? Why these continuing robust deployments?"[4]

Why indeed? What the Chinese viewed as defensive moves designed to counter a strongly presumed U.S. intervention over Taiwan, which the Chinese still consider an internal civil war-related issue, the George W. Bush administration has interpreted as an aggressive buildup that not only challenges American interests with respect to Taiwan, but also poses a long-term threat to the U.S. presence in the western Pacific. The Report on China's Military Power submitted to Congress by the Department of Defense in 2005 warned that China's "attempt to hold at risk U.S. naval forces...approaching the Taiwan Strait" potentially poses "a credible threat to modern militaries operating in the region." The Defense Department's 2006 *Quadrennial Defense Review* said that China had "the greatest potential to compete militarily" with the United States, and that its buildup "already puts regional militaries at risk." A top defense official stated that "China's military acquisitions...go beyond a Taiwan scenario and are intended to address other potential regional contingencies, such as conflict over resources or territory."[5] And a former official wrote in the *Washington Post* that "China has already changed Asia's balance of power. It is past time for America to get serious about deterring the potentially worst sorts of Chinese behavior and to provide allies in the region with reason for renewed confidence in the U.S. security umbrella."[6]

It is not clear as to what specific evidence such assessments are based on. It is particularly questionable whether China has "already changed Asia's balance of power," since without a U.S. presence, the balance remains in China's favor, while with a U.S. presence China remains plainly inferior militarily. Whatever its wider regional objectives, the Chinese buildup has so far been oriented toward capturing Taiwan and interdicting U.S. naval intervention. They have pursued a denial strategy for the maritime areas close to Taiwan and their borders, but they have not demonstrated an intention of maintaining a dominant presence in the western Pacific. The capabilities they are acquiring may have a marginal "dual use" purpose—such as sending signals to Japan, an American strategic ally, by penetrating Japanese waters with Chinese submarines or Japanese airspace with spy planes. But it is difficult to see how these capabilities can be used to advance broad Chinese interests in the Asia-Pacific region. In fact, if they induce Japan to discard the restraints on its present defense-only armed forces, the Chinese signals to Japan would be positively counterproductive.

China's military development in both quality and quantity—submarines and not aircraft carriers, diesel rather than nuclear submarines, for example—has not been directed toward mounting a challenge to the U.S. presence in the western Pacific. Its capabilities are far from adequate for that purpose. And, it should be noted, even China's limited interdiction capabilities have never been tested in battle—they are enveloped in a fog of no-war. As much as the Chinese would presumably like to evict the United States from the region, they know this is an unattainable goal. They will have to settle for less—a defensive strategy designed to protect the maritime approaches to China.

Such a strategy is dictated not only by security calculations, but also by China's political aspirations. If China cannot defend its own neighborhood, the same one in which the traditional Chinese empire held sway, it can hardly expect recognition as a paramount power in the region, to say nothing of its great power aspirations. China also needs to strengthen its maritime forces in order to secure its position on issues and areas in dispute with Japan, Vietnam, the Philippines, Malaysia, South Korea, and Brunei. Given their growing dependence on imported oil, natural gas, and other resources, the Chinese likewise need to protect their sea lines of communication, especially those from the Middle East, but the Chinese Navy at present is unable to accomplish this mission.[7] And looking further ahead, the Chinese are probably thinking of strategic challenges that might arise from the reappearance of Japanese militarism and the emergence of Japan as a regional military power or from the growing military power of India.

All these are reasons for sustaining long-term military modernization, but they do not provide a motive for an intense, rapid military buildup since they do not pose a strategic threat to China in the short term. Such a threat can come only from the United States, as it does over Taiwan. However, even though the possibility of war over Taiwan has receded, China's perception of a U.S. threat is not likely to recede significantly as well.

This is because U.S. apprehensions about China's buildup have already prompted it to adopt a "hedging" strategy against possible aggressive Chinese actions in the future by strengthening American forces in the western Pacific. These measures include adding at least one aircraft carrier and at least five nuclear submarines to the Pacific fleet over the next decade, which would place half the U.S. Navy's carriers and 60 percent of its submarines in the Pacific. Other measures include upgrading the U.S. missile defense system, transferring long-range bombers and attack submarines to Guam, stationing stealth bombers in South Korea, redeploying troops to Japan, and establishing new combat headquarters in Honolulu.[8] They also include efforts to strengthen ties and alliances with nations such as Japan, India, and Australia. And to make sure the Chinese get the message, in June 2006 the United States carried out a massive exercise near Guam in which three aircraft carriers, more than 40 surface vessels, and 160 aircraft participated, watched by an official delegation from the PLA.

The Chinese undoubtedly got the message—the United States is engaged in a major long-range military buildup aimed at China. As the Chinese government's 2004 *White Paper on Defense* put it, "Complicated security factors in the Asia-Pacific region are on the increase. The United States is realigning and reinforcing its military presence in this region by buttressing military alliances and accelerating deployment of missile defense systems."[9] They see the United States as building up its forces and strengthening strategic alliances in East and Central Asia in order to block China's rise to great power status in the region and beyond. Since the Chinese view their rise as rightful, they are probably echoing Secretary Rumsfeld's own question: "Since no nation threatens the U.S., why these continuing robust deployments?" In a proverbial case of self-fulfilling prophecy, the Chinese will presumably continue to build up their own forces — especially air and naval — as a "hedging" strategy aimed at countering U.S. military might in the western Pacific. Although the speed and scope of China's buildup may change in accordance with internal needs, political factors, and economic considerations, its direction most probably will not.

ENDNOTES - CHAPTER 12

1. For one account of the new situation, see *Washington Post*, June 15, 2006.

2. For hints of these assessments, see the Central News Agency report citing the Hong Kong *Ta Kung Pao*, June 5, 2006; *Jiefangjun Bao*, WWW-Text in English, April 28, 2006; Hong Kong *Zhongguo Tongxun*, June 9, 2006.

3. *Ibid*.

4. *Los Angeles Times*, June 4, 2005.

5. Peter W. Rodman, Assistant Secretary of Defense for International Security Affairs, Remarks before the U.S.-China Economic and Security Review Commission, March 16, 2006.

6. *Washington Post*, May 25, 2006.

7. Bernard D. Cole, "Waterways and Strategy: China's Priorities," *China Brief*, February 15, 2005.

8. Taipei, Central News Agency, July 13, 2005; *Washingtonpost.com*, July 22, 2005; *Washington Post*, September 17, 2005; *The Asahi Shimbun*, August 11, 2005.

9. *China's National Defense in 2004*, Xinhuanet, December 27, 2004.

ABOUT THE CONTRIBUTORS

KENNETH W. ALLEN is a Senior Analyst at the Center for Naval Analyses Corporation, a nonprofit research and analysis organization. He was previously a Senior Associate at the Henry L. Stimson Center and the Executive Vice-President of the U.S.-Taiwan Business Council. During his 21-year career in the U.S. Air Force, he was an Assistant Air Force Attaché in China and also worked for the Defense Intelligence Agency. He has written extensively on China's air force and China's foreign military relations.

DENNIS J. BLASKO, Lieutenant Colonel, U.S. Army (Retired), served 23 years as a Military Intelligence Officer and Foreign Area Officer specializing in China. He was an Army attaché in Beijing and Hong Kong from 1992 to 1996. Lieutenant Colonel Blasko also served in infantry units in Germany, Italy, and Korea; and in Washington at the Defense Intelligence Agency, Headquarters Department of the Army (Office of Special Operations), and the National Defense University War Gaming and Simulation Center. He has written numerous articles and chapters on the Chinese military and is the author of the book, *The Chinese Army Today: Tradition and Transformation for the 21st Century*, published by Routledge in 2006. Lieutenant Colonel Blasko is a graduate of the United States Military Academy and the Naval Postgraduate School.

MICHAEL R. CHAMBERS is Associate Professor of Political Science at Indiana State University. He has taught at St. Olaf College and was a visiting scholar at the John K. Fairbank Center for East Asian Research at Harvard University from 2003 to 2004. Since December

2004, Dr. Chambers has served as an editor of the journal *Asian Security*. He has published on China's relations with its East Asian neighbors and on China's alliance behavior, including articles in *Current History*, the *Journal of East Asian Studies*, and the *Journal of Contemporary China*. He is also the editor of *South Asia in 2020: Future Strategic Balances and Alliances* (2002). He was a member of the "Contending Perspectives: Southeast Asian and American Views of a Rising China" project sponsored by NBR and Singapore's IDSS. Dr. Chambers received his Ph.D. from Columbia University.

BERNARD D. COLE is a faculty member at the National War College, in Washington, DC. He served 30 years in the Navy as a Surface Warfare Officer, commanding a frigate and a destroyer squadron. Before retiring as a captain, he also served with the Third Marine Division in Vietnam and as Plans Officer for the U.S. Pacific Fleet. Dr. Cole has authored numerous articles and chapters, as well as four books: *Gunboats and Marines: The U.S. Navy in China*; *The Great Wall at Sea: China's Navy Enters the 21st Century*; *Oil for the Lamps of China: Beijing's 21st Century Search for Energy*; and *Taiwan's Security: History and Prospects*. He is currently writing a book on energy security in Asia. Dr. Cole holds an A.B. degree in History from the University of North Carolina, M.P.A. in National Security Affairs from the University of Washington, and a Ph.D. in History from Auburn University.

CORTEZ A. COOPER III joined Hicks and Associates, Inc., in August 2005, and serves as the Director for East Asia Studies in the Strategic Assessment Center. Since joining Hicks, Mr. Cooper has focused on providing comprehensive analysis of the evolving Asia-Pacific geo-strategic environment. Prior to joining Hicks and Associates, Mr. Cooper served in the U.S. Navy Executive Service as the Senior Analyst for the Joint Intelligence Center Pacific, U.S. Pacific Command (PACOM). Before his Hawaii assignment, Mr. Cooper was a Senior Analyst with CENTRA Technology, Inc., specializing in Asia-Pacific political-military affairs. Mr. Cooper's 20 years of military service included assignments as both an Army Signal Corps Officer and a China Foreign Area Officer. He served as a policy advisor and Team Chief in the Defense Prisoner of War/Missing Personnel Office, where he represented the Department of Defense at several rounds of talks with officials from China, North and South Korea, Vietnam, Laos, and Cambodia. He also served as a Branch Chief for the China Division at the Defense Intelligence Agency, where he provided senior policymakers with comprehensive studies on Chinese military modernization and numerous papers in support of U.S. delegations to Beijing. Mr. Cooper has a B.A. in Psychology from Davidson College, and an M.A. in Asian Studies from the University of Hawaii. He attended the U.S. Army Command and General Staff College, the Armed Forces Staff College, the Defense Language Institute, and the UK Ministry of Defence Chinese Language School.

DAVID M. FINKELSTEIN, a retired U.S. Army officer, is the Director of *Project Asia* as well as the China Studies Center at the CNA Corporation in Alexandria, Virginia. While on active duty, he held various command and staff positions in tactical field units in the United States and Korea, and positions at the Pentagon involving Chinese security affairs, and served on the faculty at West Point where he taught Chinese history. He is coeditor of multiple volumes on China, including: *China's Leadership in the 21st Century: The Rise of the Fourth Generation* (M. E. Sharpe, 2003), *Chinese Warfighting: The PLA Experience Since 1949* (M. E. Sharpe, 2003), *Swimming in a New Sea: Civil-Military Issues in Today's China* (M. E. Sharpe, 2006); *Chaos Under Heaven: Continuity and Change in the PRC Media* (2007); and *China's Revolution in Doctrinal Affairs: Evolving Trends in the Operational Art in of the Chinese People's Liberation Army* (Washington, DC: December 2005). Numerous articles by Dr. Finkelstein have appeared in journals in the United States, Europe, and China. His historical study, *From Abandonment to Salvation: Washington's Taiwan Dilemma, 1949-50* (George Mason University Press, 1993), was widely noted. Dr. Finkelstein is a graduate of the U.S. Military Academy, the U.S. Army Command and General Staff College, the U.S. Army War College, and the U.S. Army Foreign Area Officer Course at the JFK Center for Military Assistance and Unconventional Warfare. He received his Ph.D. in Chinese history from Princeton University and studied Mandarin at Nankai University in Tianjin, China.

ELLIS JOFFE is Professor Emeritus of Chinese Studies and International Relations at the Hebrew University of Jerusalem and adjunct professor at Tel Aviv University. He has written two books and many articles and book chapters on the Chinese military.

ROY KAMPHAUSEN is Vice President for Political and Security Affairs and Director of National Bureau of Asia Research (Washington, DC Office). Prior to joining NBR, Mr. Kamphausen served as a U.S. Army officer where his assignments included Country Director for China-Taiwan-Mongolia Affairs in the Office of the Secretary of Defense (OSD), and intelligence analyst and China Branch Chief in the Directorate for Strategic Plans and Policy (J5), Joint Chiefs of Staff. He served two tours at the Defense Attaché Office of the U.S. Embassy in the People's Republic of China. His areas of concentration include China's People's Liberation Army (PLA), U.S.-China defense relations, U.S. defense and security policy toward Asia, and East Asian security issues. His recent research has embraced PLA modernization, Taiwan defense and security issues, changing U.S. defense policy and posture in Asia, and the implications of China as a "responsible stakeholder" in East Asian security. Mr. Kamphausen received a B.A. in Political Science from Wheaton College and holds a Master's in International Affairs from Columbia University. He studied Chinese at the Defense Language Institute and Beijing's Capital Normal University.

KEVIN M. LANZIT is a program manager with the Institute for Physical Sciences. Prior to retirement from the military, he served in a variety of operational flying and staff assignments with the U.S. Air Force. As a Foreign Area Officer, his military service included two duty tours with the Defense Attaché Office at the U.S. Embassy in the People's Republic of China. From 2005 to 2006, he served as a senior military analyst to the U.S.-China Economic and Security Review Commission. Mr. Lanzit holds a Master's degree in Management from the University of Southern California.

MICHAEL MCDEVITT, Rear Admiral, U.S. Navy (Retired), is a Vice President and Director of the Center for Strategic Studies, a division of the Center for Naval Analyses (CNA), a not-for- profit federally funded research center in Washington, DC. The Center for Strategic Studies conducts research and analyses that focus on strategy, political-military issues, and regional security studies. During his Navy career, Rear Admiral McDevitt held four at-sea commands, including an aircraft carrier battlegroup. He was the Director of the East Asia Policy office for the Secretary of Defense during the first Bush Administration. He also served for 2 years as the Director for Strategy, War Plans, and Policy (J-5) for USCINCPAC. Rear Admiral McDevitt concluded his 34-year active duty career as the Commandant of the National War College in Washington, DC. In addition to his management and leadership responsibilities as the founder of the Center for Strategic Studies, he has been an active participant in conferences and workshops regarding security issues in East Asia, and has had a number of papers published in edited volumes on this subject. Rear Admiral McDevitt received a B.A. in U.S. History from the University of Southern California and a Master's Degree in American Diplomatic History from Georgetown University. He is also a graduate of the National War College.

EVAN S. MEDEIROS is currently a Senior Political Scientist at the RAND Corporation in the Washington, DC, office. He specializes in research on China's foreign and national security policy, U.S.-China relations, and Chinese defense industrial issues. Prior to joining RAND, Dr. Medeiros was a Senior Research Associate for East Asia at the Center for Nonproliferation Studies at the Monterey Institute of International Studies. During 2000, he was a visiting fellow at the Institute of American Studies at the China Academy of Social Sciences (CASS) in Beijing, and an adjunct lecturer at China's Foreign Affairs College. He has also conducted research on Asian security issues at the Carnegie Endowment for International Peace from 1993 to 1995 and has consulted for the National Intelligence Council. Dr. Medeiros' most recent publications are a *Washington Quarterly* article on U.S. and Chinese "hedging strategies" in Asia and three RAND studies titled *A New Direction for China's Defense Industry* (MG-334-AF); *Chasing the Dragon: Assessing China's System of Export Controls on WMD-related Goods and Technologies* (MG-353), and *Modernizing China's Military: Opportunities and Constraints* (MG-260-AF). In November 2003, he published (with M. Taylor Fravel) an article in *Foreign Affairs* tited "China's New Diplomacy" which has received broad notice. His other publications have appeared in *The China Quarterly, Current History, Foreign Affairs, Issues and Studies: A Journal of Chinese Studies and International Affairs, The Nonproliferation Review, The International Herald Tribune, the Washington Post, The Washington Quarterly,* IISS's *Strategic Comments, Jane's Intelligence Review,* the *Los Angeles Times, YaleGlobal,* the *Christian Science Monitor, The Boston Globe, Defense News, Asia Times,* and the *San Diego Union-Tribune.* Dr. Medeiros holds a B.A. in Analytic Philosophy from Bates College in Lewiston, ME; an M.A. in China Studies from the University of London's School of Oriental and African Studies (SOAS), London, England; an M.Phil in International Relations from the University of Cambridge (where he was a Fulbright Scholar), Cambridge, England; and a Ph.D. in International Relations from the School of Economics and Political Science, London, England.

ERIK QUAM worked as a research intern for Dr. Phil Saunders at the Institute for National Security Studies at National Defense University, Washington, DC, in 2006. He completed the first year of his Master's degree at the Monterey Institute of International Studies, where he is working on nonproliferation and security in East Asia in 2007. He received his B.A. from the University of Minnesota, majoring in International Relations, History, and Chinese. After graduating from the University of Minnesota, he spent 1 year studying Chinese in Taipei, Taiwan, and the following year at the Hopkins-Nanjing Center in Nanjing, China.

PHILLIP C. SAUNDERS has been a Senior Research Professor at the National Defense University's Institute for National Strategic Studies since January 2004. He previously worked at the Monterey Institute of International Studies, where he served as Director of the East Asia Nonproliferation Program at the Center for Nonproliferation Studies from 1999 to 2004, and taught courses on Chinese politics, Chinese foreign policy, and East Asian security. Dr. Saunders has conducted research and consulted on East Asian security issues for Princeton University, the Council on Foreign Relations, RAND, and the National Committee on U.S.-China Relations. His current research projects include Chinese influence in East Asia; deterrence and U.S. alliances in Asia; and China's global activities. Dr. Saunders has published numerous articles on China and Asian security in journals including *International Security, China Quarterly, The China Journal, Survival, Asian Survey, Pacific Review*, and *Orbis*. His most recent publications are *China's Global Activism: Strategy, Drivers, and Tools* (National Defense University Press) and "Visions of Order: Japan and China in U.S. Strategy." Dr. Saunders attended Harvard College and received his MPA and Ph.D. in International Relations from the Woodrow Wilson School at Princeton University.

ANDREW SCOBELL is Associate Research Professor at the Strategic Studies Institute, U.S. Army War College, and Adjunct Professor of Political Science at Dickinson College. He joined the Strategic Studies Institute in 1999 and is the institute's specialist on Asia-Pacific security. Prior to his current position, Dr. Scobell taught at the University of Louisville, Kentucky, and Rutgers University, New Jersey. He is the author of *China's Use of Military Force: Beyond the Great Wall and the Long March* (Cambridge University Press, 2003), and numerous other publications. Dr. Scobell holds a Ph.D. in Political Science from Columbia University.

JOHN M. SHALIKASHVILI has served on the National Bureau of Asian Research's Board of Directors since 1998. A retired U.S. Army General, he served as the 13th Chairman of the Joint Chiefs of Staff, Department of Defense, from October 1993 until September 1997. In this capacity, he was the senior officer in the U.S. military, serving as the principal military advisor to the President, the Secretary of Defense, and the National Security Council, and chairing the senior military body tasked with providing strategic direction to the U.S. military. Prior to an appointment as the Chairman of the Joint Chiefs of Staff, General Shalikashvili served as the Commander-in-Chief of all U.S. forces in Europe and as NATO's 10th Supreme Allied Commander, Europe (SACEUR). In 1958, he was drafted as a private into the U.S. Army and, upon graduation from Officer Candidate School in 1959, was commissioned a second lieutenant in the Artillery. For the next 38 years, he served in a variety of command and staff positions in the continental United States, Alaska, Belgium, Germany, Italy, Korea, Turkey, and Vietnam. General Shalikashvili graduated from Bradley University.

LARRY M. WORTZEL has spent 36 years specializing mainly in U.S.-China relations, U.S. policy in East Asia, security issues in Asia, and the Chinese armed forces. He is a commissioner on the congressionally appointed U.S.-China Economic and Security Review Commission. Dr. Wortzel's 32-year military career comprised 7 years as an infantryman and 25 years in military intelligence. He served overseas in China, South Korea, Singapore, Thailand, and Morocco. In the United States, he was a strategist in the Pentagon, and a counterintelligence officer in the Office of the Secretary of Defense, and was director of the Strategic Studies Institute of the U.S. Army War College. Dr. Wortzel retired from the Army as a colonel in 1999, after which he was Director of Asian Studies and Vice President for Foreign Policy and Defense Studies of The Heritage Foundation, a Washington, DC, think tank. He is the author of two books on China and has edited and contributed chapters to six other books on China. He has shared his expert knowledge with Fox News, CNN, the BBC, MSNBC, the *Los Angeles Times, Chosin Ilbo, Dong A Ilbo, Cai Jing,* and *Sankei Shimbun*. Dr. Wortzel is a graduate of the U.S. Army War College, and earned his Ph.D. in political science at the University of Hawaii.